Foundations of the Classical Theory
of Partial Differential Equations

Springer
Berlin
Heidelberg
New York
Barcelona
Budapest
Hong Kong
London
Milan
Paris
Santa Clara
Singapore
Tokyo

Yu. V. Egorov   M. A. Shubin

Foundations of the Classical Theory of Partial Differential Equations

Springer

Consulting Editors of the Series:
A. A. Agrachev, A. A. Gonchar, E. F. Mishchenko,
N. M. Ostianu, V. P. Sakharova, A. B. Zhishchenko

Title of the Russian edition:
Itogi nauki i tekhniki, Sovremennye problemy matematiki,
Fundamental'nye napravleniya, Vol. 30,
Differentsial'nye uravneniya s chastnymi proizvodnymi 1
Publisher VINITI, Moscow 1988

Second Printing 1998 of the First Edition 1992, which was originally
published as Partial Differential Equations I,
Volume 30 of the Encyclopaedia of Mathematical Sciences.

Die Deutsche Bibliothek - CIP-Einheitsaufnahme

**Foundations of the classical theory of partial differential
equations** / ed.: Yu. V. Egorov ; M. A. Shubin. - 1. ed., 2. printing. -
Berlin ; Heidelberg ; New York ; Barcelona ; Budapest ; Hongkong ;
London ; Mailand ; Paris ; Santa Clara ; Singapur ; Tokio : Springer,
1998
    (Encyclopaedia of mathematical sciences ; Vol. 30)
    ISBN 3-540-63825-3

Mathematics Subject Classification (1991): 35-02

ISBN 3-540-63825-3 Springer-Verlag Berlin Heidelberg New York

SPIN: 10654770
46/3143-5 4 3 2 1 0 – Printed on acid-free paper.

List of Editors, Authors and Translators

Editor-in-Chief

R.V. Gamkrelidze, Russian Academy of Sciences, Steklov Mathematical Institute, ul. Gubkina 8, 117966 Moscow, Institute for Scientific Information (VINITI), ul. Usievicha 20 a, 125219 Moscow, Russia; e-mail: gam@ipsun.ras.ru

Consulting Editors
Authors

Yu. V. Egorov, U.F.R. M.I.G., Université Paul Sabatier, 118, route de Narbonne, 31062 Toulouse Cedex, France; e-mail: egorov@mip.ups-tlse.fr
M. A. Shubin, Department of Mathematics, Northeastern University, Boston, MA 02115, USA; e-mail: shubin@neu.edu

Translator

R. Cooke, Department of Mathematics, University of Vermont, Burlington, Vermont 05405, USA

Linear Partial Differential Equations.
Foundations of the Classical Theory

Yu. V. Egorov, M. A. Shubin

Translated from the Russian
by R. Cooke

Contents

Preface

This volume contains a general introduction to the classical theory of linear partial differential equations for nonspecialist mathematicians and physicists.

Examples of partial differential equations are found as early as the papers of Newton and Leibniz, but the systematic study of them was begun by Euler. From the time of Euler on the theory of partial differential equations has occupied a central place in analysis, mainly because of its direct connections with physics and other natural sciences, as well as with geometry. In this connection the theory of linear equations has undergone a very profound and diverse development.

The present volume is introductory to a series of volumes devoted to the theory of linear partial differential equations. We could not encompass all aspects of the classical theory, and we did not try to do so. In writing this volume we did not hesitate to repeat ourselves in those situations where it seemed to us that repetition would facilitate the reading. However we have attempted to give a sketch of all the ideas that seemed fundamental to us, making no claim to completeness, of course. The reader who wishes to form a deeper acquaintance with some aspect of the theory discussed here may turn to the following, more specialized volumes in this series. In particular, many of the ideas of the modern theory are described in the authors' article published in the next volume.

The bibliography of this volume also makes no claim to completeness. We have attempted to cite as far as possible only textbooks, monographs, and survey articles.

The authors thank B. R. Vajnberg, who wrote Sect. 2.7, and M. S. Agranovich, who read this volume in manuscript and made many valuable remarks that enabled us to improve the exposition.

Chapter 1. Basic Concepts

§1. Basic Definitions and Examples

1.1. The Definition of a Linear Partial Differential Equation. The general *linear partial differential equation* is an equation of the form

$$Au = f, \qquad (1.1)$$

where f is a known function (possibly vector-valued) in a region $\Omega \subset \mathbb{R}^n$ and A is a *linear differential operator* defined in Ω, i.e., an operator of the form

$$A = \sum_{|\alpha| \le m} a_\alpha(x) D^\alpha, \qquad (1.2)$$

where α is a multi-index, i.e., $\alpha = (\alpha_1, \ldots, \alpha_n)$, $\alpha_j \ge 0$ are integers, $D^\alpha = D_1^{\alpha_1} D_2^{\alpha_2} \cdots D_n^{\alpha_n}$, $D_j = i^{-1} \partial/\partial x_j$, $i = \sqrt{-1}$, $|\alpha| = \alpha_1 + \cdots + \alpha_n$, a_α are functions on Ω (possibly matrix-valued), and $u = u(x)$ is an unknown function on Ω. The smallest possible number m is called the *order* of (1.1) and of the operator (1.2). Sometimes a more general form of (1.1) is useful:

$$\sum_{|\alpha| + |\beta| \le m} D^\alpha \Big(a_{\alpha\beta}(x) D^\beta u \Big) = f. \qquad (1.3)$$

Equation (1.3) is equivalent to (1.1) in the case of sufficiently smooth coefficients $a_{\alpha\beta}$.

The most commonly occurring equations, and those which play the greatest role in mathematical physics, are second-order equations (i.e., equations of the form (1.1) or (1.3) with $m = 2$).

1.2. The Role of Partial Differential Equations in the Mathematical Modeling of Physical Processes. Partial differential equations are a fundamental tool of investigation in modern mathematical physics. This fact is explained by the extensive possibilities for using them to describe the dependence of phenomena under investigation on a large number of parameters of various kinds. At the same time such equations occupy a central place in mathematical analysis.

In studying a physical phenomenon the first thing to do is to isolate the quantities that characterize it. Such quantities may be density, velocity, temperature, and the like. The next task is to choose and state mathematically the physical laws that can be applied as the foundation of a theory of the given phenomenon and which are usually the result of generalization from experiments and observations. These laws must be as simple and as free from contradiction as possible. As a rule these laws can be written in the form of relations between the fundamental characteristics of the phenomenon and their derivatives at a given point of space and at a given instant of time. The possibility of such an expression is essentially a consequence of the localness of all known interactions, although in deriving the equations it is often convenient at first to use some integral conservation laws (for example, conservation of mass, momentum, energy, electric charge, and the like) and only later to pass to the local equations, assuming some smoothness of the quantities being studied. Let us give some examples of such a derivation of the equations describing physical processes.

1.3. Derivation of the Equation for the Longitudinal Elastic Vibrations of a Rod (cf. Tikhonov and Samarskij 1977). Consider a homogeneous elastic rod with cross-sectional area S made of material of density ρ. We direct the x-axis along this rod (Fig. 1a), and we shall assume that each section is displaced only in the direction of the x-axis. We denote by $u(t, x)$ the longitudinal displacement at the instant t of the section of the rod whose points have coordinate x when in equilibrium, so that at the instant t they will have coordinate $x + u(t, x)$. We shall try to trace the motion of the section lying over the interval $[x, x + \Delta x]$ of the x-axis when in equilibrium, neglecting all external forces acting on it except elastic forces arising in the sections joining this segment to the remainder of the rod. Let us find these elastic forces. We remark that at the instant t the segment in question has length $l = u(t, x + \Delta x) - u(t, x) + \Delta x$, and its lengthening in comparison with its equilibrium position is $\Delta l = u(t, x + \Delta x) - u(t, x)$, so that the relative lengthening has the form

$$\frac{\Delta l}{l} = \frac{u(t, x + \Delta x) - u(t, x)}{\Delta x}.$$

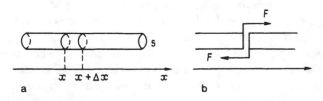

Fig. 1.1

In the limit as $\Delta x \to 0$ we obtain the result that the relative lengthening of the infinitesimal segment situated over the point having coordinate x when in equilibrium is $u_x(t,x) = \frac{\partial u}{\partial x}(t,x)$ (in the theory of elasticity this quantity is called the *deformation gradient*; for general three-dimensional deformations this role is played by the deformation tensor). By Hooke's Law, which is a linear approximation in the one-dimensional elasticity problem under consideration, the elastic force F acting on the left-hand piece of the rod in the section under consideration (cf. Fig. 1b) is $ESu_x(t,x)$, where the coefficient E, which characterizes the elastic properties of the material of which the rod is made, is called the *Young's modulus*. Thus the forces acting on the segment under consideration are $ESu_x(t,x+\Delta x)$ (from the right) and $-ESu_x(t,x)$ (from the left), so that the total external force is $ES[u_x(t,x+\Delta x)-u_x(t,x)]$. Since the total momentum of this segment is obviously $\int_x^{x+\Delta x} \rho Su_t(t,\xi)\,d\xi$, we have by Newton's Second Law

$$\frac{d}{dt}\int_x^{x+\Delta x} \rho Su_t(t,\xi)\,d\xi = ES[u_x(t,x+\Delta x) - u_x(t,x)].$$

Assuming that u has continuous derivatives up to second order, we can differentiate under the integral sign, then divide both sides by Δx and let Δx tend to 0, from which we obtain the *one-dimensional wave equation*

$$u_{tt} = c^2 u_{xx}, \tag{1.4}$$

where the constant $c = \sqrt{\frac{E}{\rho}}$ has an interpretation as the speed of propagation of elastic waves (sound) in the rod.

1.4. Derivation of the Equation of Heat Conduction (cf. Vladimirov 1967, Tikhonov and Samarskij 1977). Consider a homogeneous medium consisting of a substance of density ρ in three-dimensional space. Let $u(t,x)$ be the temperature of this medium at the point $x \in \mathbb{R}^3$ at the instant t. We shall assume that u is a sufficiently smooth function of t and x. The derivation of the equation for u is based on Fourier's law of heat transmission: If a small surface of area ΔS is given, then in a small interval of time Δt a quantity of heat

$$\Delta Q \cong -k\frac{\partial u}{\partial n}\Delta S\Delta t, \tag{1.5}$$

passes through the surface in the direction of the normal n. Here k is a coefficient depending on the substance in question and is called its *coefficient of thermal conductivity*. Now let Ω be some distinguished volume of the medium (a bounded region with a piecewise smooth boundary in $\mathbb{R}^3$). The law of conservation of energy in Ω during the time interval $[t, t+\Delta t]$ has the form

$$\int_\Omega c[u(t+\Delta t, x) - u(t,x)]\rho\, dx = \int_t^{t+\Delta t} \int_{\partial\Omega} k\frac{\partial u}{\partial n}\, dS\, dt,$$

where c is the specific heat capacity of the substance, $\partial\Omega$ is the boundary of the region Ω, n is the exterior normal to $\partial\Omega$, and dx is the usual element of volume in $\mathbb{R}^3$. By the divergence theorem the right-hand side can be transformed into a volume integral

$$\int_t^{t+\Delta t} \int_\Omega k\Delta u\, dx\, dt,$$

where $\Delta = \partial^2/\partial x_1^2 + \partial^2/\partial x_2^2 + \partial^2/\partial x_3^2$ is the Laplacian operator in $\mathbb{R}^3$. Dividing both sides now by Δt and by the volume of the region Ω and passing to the limit as $\Delta t \to 0$ and the region Ω shrinks to the point x, we obtain the *heat equation*

$$u_t = a^2 \Delta u, \tag{1.6}$$

where $a^2 = k/c\rho$. This equation also describes a diffusion process in liquids and gases (with a suitable interpretation of the function u and the coefficient a).

1.5. The Limits of Applicability of Mathematical Models. Naturally the mathematical description of real phenomena using differential equations, like all mathematical models, is an idealization. For example, Hooke's Law in 1.3 holds only approximately, and the Fourier law of heat transmission (1.5) can be made more precise by taking account of the molecular structure of the substance. Therefore the deductions obtained using the study, and even the exact solution, of the differential equations obtained, are also approximate. For example, the heat equation (1.6) predicts an infinite velocity of propagation for heat (even from a point source), which of course is absurd. At the same time, this effect has very little influence on computations of heat transmission in engineering, where the mathematical theory of the heat equation is used quite successfully. The situation is the same with many other mathematical models, in particular with models based on partial differential equations.

It may happen that deductions obtained by considering a mathematical model differ significantly from the results of experiment. Such a disagreement is an indication that the mathematical model is incomplete and is grounds for replacing it with a model based on the application of other laws that take more precise account of the characteristics of the object under study.

Additional assumptions about the smoothness of the functions describing the behavior of the fundamental parameters are usually introduced in the derivation of differential equations in order to simplify the mathematical expression of the laws of nature. These assumptions, however, are not always justified or suitable. In the modern theory of differential equations this difficulty has led to the creation of the concept of a generalized solution, reflecting a transition from the differential equation to the integro-differential equation, which often arises earlier in the process of constructing the mathematical model under study.

1.6. Initial and Boundary Conditions. As a rule a mathematical model is created in order to reflect properties of physical processes taking place in some bounded portion of space. In such a situation the connection with processes taking place outside the distinguished portion of space cannot be entirely ignored and must be reflected in the construction of the mathematical model. Relations that hold between the values of the parameters being studied and their derivatives on the boundary of the region are called *boundary conditions*. Thus if, say, a rod of length l is being considered whose endpoints have coordinates 0 and l when in equilibrium, then the boundary conditions at the left-hand endpoint $x = 0$ may, for example, have the following form (cf. Fig. 2):

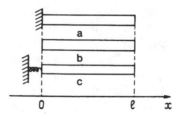

Fig. 1.2

a) $u|_{x=0} = 0$ (fixed endpoint);

b) $u_x|_{x=0} = 0$ (free endpoint);

c) $(ESu_x - ku)|_{x=0} = 0$ (elastically fixed endpoint, i.e., a spring of elasticity k that is in equilibrium when its right-hand endpoint is at the point $x = 0$ is attached to the left-hand endpoint of the rod. Here the left-hand endpoint of the spring is rigidly fixed). Similar boundary conditions can be written for the right-hand endpoint $x = l$.

For the heat equation describing a medium occupying the region $\Omega \subset \mathbf{R}^3$ one may take as boundary conditions one of the following relations:

a) $u|_{\partial\Omega} = \varphi$ (boundary maintained at a given temperature φ);

b) $\frac{\partial u}{\partial n}|_{\partial\Omega} = \varphi$ (prescribed heat flux through the boundary);

c) $[\frac{\partial u}{\partial n} - \gamma(u_0 - u)]|_{\partial\Omega} = 0$ (heat exchange with an environment at temperature u_0 takes place at the boundary).

In studying processes taking place over time, the course of the process is studied beginning at a certain instant. Here the prehistory of the process,

which is partly reflected in the form of the relations between the values of
the parameters in question and their derivatives at the initial instant of time,
is essential. These relations are called *initial conditions*.

For example, the natural initial conditions for the one-dimensional wave
equation (1.4) are obtained by prescribing the initial position and velocity of
all points of the rod:

$$u|_{t=0} = \varphi(x), \quad u_t|_{t=0} = \psi(x) \tag{1.7}$$

(if the rod has length l and is situated as indicated above, then it is necessary
to assume here that $x \in [0, l]$). For the heat equation (1.6) it is natural to
prescribe the initial temperature distribution:

$$u|_{t=0} = \varphi(x), \tag{1.8}$$

where $x \in \Omega$ in the case when the medium being studied occupies the region
Ω.

We note finally that since physical laws usually lead to nonlinear relations between parameters, it becomes necessary to study nonlinear differential
equations and nonlinear boundary conditions. When this is done, however, as
a rule serious mathematical difficulties arise. Therefore it is frequently necessary to sacrifice precision in constructing a mathematical model and neglect
small nonlinear increments or pass to a linearization in the neighborhood of
some given solution, reducing the problem to a linear one. Linearization is
also important in studying stability questions for the solutions of nonlinear
equations. This accounts for the important role of linear partial differential
equations in mathematical physics.

1.7. Examples of Linear Partial Differential Equations. We shall now give
some important examples of linear partial differential equations that arise as
the equations of mathematical physics.

Example 1.1. (The multidimensional wave equation). This equation has the
form

$$u_{tt} = c^2 \Delta u, \tag{1.9}$$

where $u = u(t, x)$, $t \in \mathbb{R}$, $x \in \mathbb{R}^n$, $\Delta = \partial^2/\partial x_1^2 + \partial^2/\partial x_2^2 + \cdots + \partial^2/\partial x_n^2$
is the Laplacian on $\mathbb{R}^n$, and $c > 0$ is some constant (of course the solution
may be defined not for all t and x, but only in some region of variation of
the coordinates t and x). For $n = 3$ the equation (1.9) describes a great
variety of processes of wave propagation in the situation when the space
is homogeneous and isotropic for the waves in question. In this situation c
is the speed of wave propagation. For example, all the components of the
electric field intensity and the magnetic field in a vacuum satisfy (1.9) (in
this case c is the speed of light), as do the pressure and density of a gas
under small (acoustic) vibrations of the gas, and the like. For $n = 2$ this

equation describes, for example, the small vibrations of an elastic membrane (here $u(t, x)$ denotes the transverse displacement of a point of the membrane). For $n = 1$ the equation, as we have already seen, describes the longitudinal vibrations of a rod; it also describes the small transverse vibrations of a string (Vladimirov 1967, Tikhonov and Samarskij 1977).

To understand why the solutions of (1.9) have the character of propagating waves and ascertain the meaning of the coefficient c, it is necessary to write the so-called *dispersion law* for this equation: the relations between the frequency ω and the wave vector k under which a sinusoidal plane wave $u(t, x) = e^{i(\omega t - k \cdot x)}$ is a solution of the equation (here $k \cdot x = k_1 x_1 + k_2 x_2 + \cdots + k_n x_n$). Substitution in the equation obviously gives $\omega^2 = c^2 |k|^2$, i.e., $u(t, x) = \exp[i|k|(ct - \frac{k}{|k|} \cdot x)]$. The surfaces of constant phase are the planes $ct - \frac{k}{|k|} \cdot x = $ const, and for fixed t each such surface gives a plane in $\mathbf{R}^n$ moving in the direction of the vector k with speed c. By superposition (taking the sum) of sinusoidal waves of the form described it is possible to obtain other solutions of (1.9) (and even, in a certain sense, all solutions of it). Therefore solutions of sinusoidal wave type play an important role (this is always the case in the study of equations with constant coefficients).

For $n = 1$ the general solution of (1.9) has the form

$$u(t, x) = f(x - ct) + g(x + ct),$$

where f and g are arbitrary functions of one variable. More precisely, this holds for a solution u defined in a (plane) region (t, x) that intersects each line of the form $x - ct = $ const or $x + ct = $ const in a (connected) interval (possibly empty). The smoothness of the functions f and g corresponds to the smoothness of the solution u (for example, if $u \in C^2$, then f and g are also in C^2). To prove this it suffices to introduce the new variables $\xi = x - ct$ and $\eta = x + ct$, in terms of which (1.9) assumes the form $u_{\xi\eta} = 0$.

The natural initial conditions for (1.9) are the initial conditions (1.7).

A generalization of (1.9) is the equation

$$\rho(x)u_{tt} = \operatorname{div}(A(x)\operatorname{grad} u) - q(x)u, \qquad (1.10)$$

where $\rho(x) > 0$, $A(x)$ is a matrix-valued function (having values in the set of positive definite symmetric $n \times n$ matrices), and $q(x) \geq 0$. For $n = 3$ this equation describes the propagation of waves in an inhomogeneous and anisotropic medium with dissipation of energy characterized by the coefficient q. For $n = 1$ and $n = 2$ the equation describes the vibration of an inhomogeneous string and an inhomogeneous anisotropic membrane respectively. If we take account of external forces (for example, the force of gravity), (1.10) assumes the somewhat more general form of the inhomogeneous linear equation

$$\rho u_{tt} = \operatorname{div}(A \operatorname{grad} u) - qu + f, \qquad (1.10')$$

where $f = f(t, x)$ (the coefficients ρ, A, and q in (1.10) and (1.10')) can in principle depend on t).

We remark that (1.10) does not, generally speaking, have solutions in the form of sinusoidal plane waves; but if we study the medium under a microscope with $q \equiv 0$ – more precisely, if we limit ourselves to a piece on which the coefficients ρ and A can be considered constant – such solutions will exist and possess properties similar to those of the corresponding solutions of (1.6). The theory of hyperbolic equations makes it possible to make this heuristic reasoning rigorous: in that theory it is proved that there exist solutions to (1.10) of wave type, although these waves no longer propagate in straight lines.

We note further that in analogy with (1.10) and (1.10') one may write the inhomogeneous heat equation in an inhomogeneous and anisotropic medium

$$\rho u_t = \operatorname{div}\,(A\operatorname{grad} u) - qu + f, \tag{1.11}$$

where f has an interpretation as the density of the external heat sources and ρ, A, and q are local characteristics of the medium.

Example 1.2. (The Laplace and Poisson Equations). The *Laplace equation* has the form

$$\Delta u = 0, \tag{1.12}$$

where $u = u(x)$, $x \in \mathbb{R}^n$, and Δ is the Laplacian on $\mathbb{R}^n$ introduced earlier. The corresponding inhomogeneous equation

$$\Delta u = \rho \tag{1.12'}$$

(ρ is a known function) is called the *Poisson equation*. The Laplace and Poisson equations arise in a variety of problems. For example the steady-state temperature distribution (i.e., one that does not change with time) in a homogeneous medium and the permanent shape of a stretched membrane obviously satisfy the Laplace equation, while the analogous temperature distribution in the presence of heat sources (with unchanging density) and the shape of a membrane in the presence of stationary external forces satisfy the Poisson equation. The potential of an electrostatic field satisfies the Poisson equation (1.12') with a function ρ proportional to the charge density (consequently in a region without charges it satisfies Laplace's equation).

Thus the Laplace and Poisson equations describe steady states of various objects. There is consequently no need to prescribe initial conditions for them, and the natural boundary conditions are posed as in the corresponding nonsteady-state problem. Therefore the natural boundary conditions for the Laplace and Poisson equations in a bounded region $\Omega \subset \mathbb{R}^n$ are the *Dirichlet condition*

$$u|_{\partial\Omega} = \varphi, \tag{1.13}$$

the *Neumann condition*

$$\frac{\partial u}{\partial n}\Big|_{\partial\Omega} = \varphi, \tag{1 14}$$

and the *third boundary condition*

$$\left(\frac{\partial u}{\partial n} - \gamma u\right)\Big|_{\partial\Omega} = \varphi, \tag{1.15}$$

where γ is a function on $\partial\Omega$.

We shall also give an important generalization of the Laplacian in $\mathbb{R}^n$: the *Laplace-Beltrami operator* on an n-dimensional Riemannian manifold $\mathcal{M}$, also denoted by Δ and defined by the formula

$$\Delta u = \frac{1}{\sqrt{g}} \sum_{i,j=1}^{n} \frac{\partial}{\partial x_i}\left(\sqrt{g}\, g^{ij} \frac{\partial u}{\partial x_j}\right), \tag{1.16}$$

where $(x_1, x_2, \ldots, x_n)$ are arbitrary local coordinates on $\mathcal{M}$ and $\|g^{ij}\|$ is the matrix inverse to the matrix $\|g_{ij}\|$ consisting of the components of the metric tensor, $g = \det \|g_{ij}\|$. The Laplace and Poisson equations have a meaning on any Riemannian manifold, and on a Riemannian manifold with boundary it makes sense to talk about the boundary conditions (1.13)–(1.15).

It is possible to introduce a Laplacian on the space $\Lambda^p(\mathcal{M})$ of smooth exterior p-forms on $\mathcal{M}$:

$$\Delta = d\delta + \delta d, \tag{1.17}$$

where $d : \Lambda^p(\mathcal{M}) \to \Lambda^{p+1}(\mathcal{M})$ is the exterior differential, and δ the operator formally adjoint to it. The Laplace operator and the corresponding Laplace and Poisson equations play an important role in geometry and topology (cf., for example, Warner 1983, Chap. 6).

Example 1.3. (The Helmholtz equation). This name is given to the equation

$$(\Delta + k^2)u = 0, \tag{1.18}$$

where $u = u(x)$, $x \in \mathbb{R}^n$, Δ is the Laplacian on $\mathbb{R}^n$, and $k > 0$. This equation arises in the study of the solutions of the wave equation (1.9) having the special form $e^{i\omega t}u(x)$, where $\omega = k/c$. The same equation is important in the study of various spectral problems, for example the eigenvalue problem for the Laplace operator. The simplest such problem is the eigenvalue problem in a bounded region $\Omega \subset \mathbb{R}^n$ with the Dirichlet condition on $\partial\Omega$:

$$\begin{cases} -\Delta u = \lambda u, \\ u|_{\partial\Omega} = 0. \end{cases} \tag{1.19}$$

It is easy to prove that this equation may have nonzero solutions only for $\lambda > 0$, from which it is clear that u satisfies Helmholtz' equation.

Example 1.4. (The Maxwell equations and the telegraph equations). The *Maxwell equations* are a system of equations for the vectors $\mathbf{E} = (E_1, E_2, E_3)$

and $\mathbf{H} = (H_1, H_2, H_3)$ giving the electric and magnetic field intensities in some medium. In the Gaussian CGS system of units the system has the form (Landau and Lifshits 1973)

$$\operatorname{div} \mathbf{D} = 4\pi\rho,$$
$$\operatorname{div} \mathbf{B} = 0,$$
$$\operatorname{curl} \mathbf{E} = -\frac{1}{c}\frac{\partial \mathbf{B}}{\partial t}, \tag{1.20}$$
$$\operatorname{curl} \mathbf{H} = \frac{4\pi}{c}\mathbf{j} + \frac{1}{c}\frac{\partial \mathbf{D}}{\partial t},$$

where ρ is the electric charge density, c is the speed of light in a vacuum, and the case of a field in a vacuum $\mathbf{D} = \mathbf{E}$, $\mathbf{B} = \mathbf{H}$, $\mathbf{j} = 0$, while for any isotropic medium

$$\mathbf{D} = \varepsilon\mathbf{E}, \quad \mathbf{B} = \mu\mathbf{H}, \quad \mathbf{j} = \sigma\mathbf{E} + \mathbf{j}_{\text{ext}},$$

where ε is the dielectric permittivity of the medium, μ is the magnetic permeability of the medium, σ is the specific conductivity (ε, μ, and σ may be functions of t and x), and $\mathbf{j}_{\text{ext}}$ is the external current density, i.e., currents maintained by any forces other than those of the electric field (for example, by a magnetic field or by diffusion).

The Maxwell system is the foundation of the theory of electromagnetic waves and serves as the basis for radiotechnic calculations, for example for the theory of wave conductors. Boundary and initial conditions for it are usually written based on physical considerations.

In particular the *telegraph equations*, which are important in electrical engineering and describe the variation in current strength and intensity in a conductor (Landau and Lifshits 1982, Sect. 91), are deduced from the Maxwell equations:

$$\begin{cases} \dfrac{\partial i}{\partial x} + C\dfrac{\partial v}{\partial t} + Gv & = 0, \\[2mm] \dfrac{\partial v}{\partial x} + L\dfrac{\partial i}{\partial t} + Ri & = 0, \end{cases}$$

where x is a coordinate along the conductor, v is the potential at the given point of the conductor (measured from an arbitrary initial level), i is the current strength, R is the resistance per unit length, L is the inductance per unit length, C is the capacitance per unit length, and G is the conductance per unit length.

Example 1.5. (The Schrödinger Equation). The *Schrödinger equation* is the fundamental equation of nonrelativistic quantum mechanics. In the simplest case for a particle without spin in an external field it has the form

$$i\hbar\frac{\partial \psi}{\partial t} = -\frac{\hbar^2}{2m}\Delta\psi + V(x)\psi, \tag{1.21}$$

where $x \in \mathbf{R}^3$, $\psi = \psi(t, x)$ is the wave function (or, as it is sometimes called, the "psi-function") of a quantum particle, giving the complex amplitude characterizing the presence of the particle at each point x (in particular $|\psi(t, x)|^2$ is interpreted as the probability density for the particle to be at the point x at the instant t), m is the mass of the particle, $\hbar$ is Planck's constant, and $V(x)$ is the external field potential (a real-valued function). For (1.21) the natural initial condition is

$$\psi|_{t=0} = \psi_0(x),$$

and its solution is formally written in terms of ψ_0 in the form

$$\psi(t, \cdot) = e^{-\frac{i}{\hbar} tH} \psi_0, \tag{1.22}$$

where the operator

$$H = -\frac{\hbar^2}{2m} \Delta + V(x) \tag{1.23}$$

is called the *Schrödinger operator* and has an interpretation as the energy operator of the particle under consideration. In the same way as the wave equation leads to Helmholtz' equation, the time-dependent Schrödinger equation (1.21) for solutions of the form $e^{-\frac{i}{\hbar} Et} \psi(x)$ (here E is a constant) gives the equation

$$\left[-\frac{\hbar^2}{2m} \Delta + V(x)\right] \psi(x) = E\psi(x), \tag{1.24}$$

called the *steady-state Schrödinger equation* (it describes states with fixed energy E).

Instead of boundary conditions for the equations (1.21) and (1.24) it is customary to use certain natural conditions limiting the rate of increase of ψ at infinity and depending on the character of the potential. For (1.21) it is customary to require the inclusion $\psi(t, \cdot) \in L_2(\mathbf{R}^3)$ for each fixed t. Equation (1.24) is usually solved in this class (this is done, for example, in the case when the potential increases at infinity: the relation $V(x) \to +\infty$ as $|x| \to \infty$ guarantees that the operator H has a *discrete spectrum*, i.e., there exists a complete orthogonal system of eigenfunctions $\psi_j \in L_2(\mathbf{R}^3)$, $j = 1, 2, \ldots$ with eigenvalues $E_j \to +\infty$ as $j \to +\infty$) or in the class of bounded functions having a definite asymptotic behavior at infinity (this approach is important in scattering theory related to the case of a potential decaying at infinity (cf. Sect. 2.7 below)).

The behavior of the potential V at infinity is determined by the character of the quantum mechanical problem under consideration. The oscillations of a particle in a potential well are described by an increasing potential, a typical example of which is the potential of a harmonic oscillator $V(x) = |x|^2$, or more generally, $V(x) = \sum_{j=1}^{3} \beta_j x_j^2$; the corresponding equations (1.21) and (1.24) in the class of functions decreasing on x can be solved explicitly (the eigenfunctions of the corresponding operator H are expressed in terms of Hermite functions). A decaying potential V corresponds to a scattering problem for a particle in a field formed by one or more other particles. Equations of the form (1.21) and (1.24) can describe not only a single particle but also a system of several such particles In the case of N particles one must take $x = (x^{(1)}, \ldots, x^{(N)})$, $x^{(j)} \in \mathbb{R}^3$, so that $x \in \mathbb{R}^{3N}$ and instead of (1.21) one must write

$$i\hbar \frac{\partial \psi}{\partial t} = -\sum_{j=1}^{N} \frac{\hbar^2}{2m_j} \Delta_j \psi + V(x)\psi,$$

where Δ_j is the Laplacian on $x^{(j)}$. In this situation we customarily have

$$V(x) = \sum_{i<j} V_{ij}(x^{(i)} - x^{(j)}).$$

The steady-state equation (1.24) is rewritten similarly. In studying the system of N particles one must take account of their characteristics: if the particles are identical and are bosons, then the function $\psi = \psi(x^{(1)}, \ldots, x^{(N)})$ must be symmetric, i.e., must not change when the arguments $x^{(1)}, \ldots, x^{(N)}$ are permuted. But if the particles are fermions, one must consider ψ antisymmetric, i.e., one must remember that it reverses sign when any two of the points $x^{(j)}$ are interchanged.

Spin is taken into account by studying equations of the form (1.21) and (1.24) in certain spaces of vector-valued functions.

Example 1.6. (The Klein-Gordon-Fock and Dirac equations (Berestetskij 1980)). Equation (1.21) with a time variable t actually present is not relativistically invariant (i.e., invariant under the Poincaré group of transformations acting on $\mathbb{R}^4$ – the Lorentz group together with the translations on $\mathbb{R}^4$) for any potential V. The simplest relativistically invariant equation is the wave equation (1.9) in the case $n = 3$ and with constant c equal to the speed of light. However, it works only for describing massless particles – photons. Its generalization to the case of particles of finite mass $m > 0$ is the *Klein-Gordon-Fock equation*

$$(\hbar^2 \Box + m^2 c^4)\psi = 0. \tag{1.25}$$

where $\Box$ is the *wave operator* or *d'Alembert operator* or *d'Alembertian*

$$\Box = \frac{\partial^2}{\partial t^2} - c^2 \Delta$$

(Δ being the Laplacian on $\mathbb{R}^3$). A solution ψ of this equation should not be interpreted as a wave function; it is more accurate to regard (1.25) as a field equation (similar to the wave equation or the Maxwell equations) and subject it to quantization, i.e., regard ψ as an operator function.

The dispersion relation for the Klein-Gordon-Fock equation has the form $E^2 = p^2 c^2 + m^2 c^4$ (this is the condition for the exponential $\exp(\frac{i}{\hbar}(Et + p \cdot x))$, where $p = (p_1, p_2, p_3)$, to be a solution), whence $E = \pm mc^2$ for a resting particle. In quantum field theory the state of a particle with negative energy is interpreted as the state of an antiparticle possessing positive energy, but opposite electric charge. We note also that the energy band $-mc^2 < E < mc^2$ is "forbidden."

It is a defect of (1.25) that it is of second order in t and its solution ψ is not determined uniquely by giving its value for $t = 0$, as is the case for the Schrödinger equation. A system free of this defect is the Dirac system, which is constructed using matrices γ^μ, $\mu = 0, 1, 2, 3$, such that the operator

$$\sum_{\mu=0}^{3} \gamma^\mu \frac{\partial}{\partial x_\mu}$$

yields the d'Alembertian $\square$ when squared (here $x_0 = ct$), i.e., matrices γ^μ satisfying the anticommutativity relations

$$\gamma^\mu \gamma^\nu + \gamma^\nu \gamma^\mu = 2g^{\mu\nu},$$

where $g^{\mu\nu} = 0$ for $\mu \neq \nu$, $g^{00} = -g^{11} = -g^{22} = -g^{33} = 1$ (i.e., $g^{\mu\nu}$ is the matrix of the quadratic form obtained when the differentiations $\partial/\partial x^\mu$ in the d'Alembertian are replaced by variables ξ_μ). The simplest matrices γ^μ, and the most commonly used, are the 4×4 matrices of the form

$$\gamma^0 = \begin{pmatrix} I & 0 \\ 0 & I \end{pmatrix}, \quad \gamma^j = \begin{pmatrix} 0 & \sigma_j \\ -\sigma_j & 0 \end{pmatrix}, \quad j = 1, 2, 3,$$

where I is the 2×2 identity matrix and σ_j are the Pauli matrices

$$\sigma_1 = \begin{pmatrix} 0 & 1 \\ 1 & 0 \end{pmatrix}, \quad \sigma_2 = \begin{pmatrix} 0 & -i \\ i & 0 \end{pmatrix}, \quad \sigma_3 = \begin{pmatrix} 1 & 0 \\ 0 & -1 \end{pmatrix}.$$

The *Dirac equation* is the equation for a bispinor (a 4-component vector-valued function ψ) having the form

$$\left(i\hbar \sum_{\mu=0}^{3} \gamma^\mu \frac{\partial}{\partial x_\mu} - mc \right) \psi = 0. \tag{1.26}$$

Equation (1.26) describes a free particle (with spin 1/2 or $-1/2$) and simultaneously its antiparticle. In taking account of the mutual interaction an additional term must be added to it, whose form in quantum electrodynamics, the theory of electroweak interaction, and quantum chromodynamics is determined by the requirement of gauge invariance, which essentially consists of the requirement that $\partial/\partial x_\mu$ be replaced by a covariant derivative with respect to some connection. For example in quantum electrodynamics $\partial/\partial x_\mu$ is replaced by $\partial/\partial x_\mu + ieA_\mu/\hbar c$, where A_μ is the four-dimensional electromagnetic field potential (a vector-valued function on $\mathbb{R}^4$).

Example 1.7. (The Cauchy-Riemann equations and the $\bar{\partial}$-equation). In the complex plane $\mathbb{C}$ of the variable $z = x + iy$ consider a complex-valued function $u \in C^1$ (i.e., u has continuous partial derivatives on x and y). Its differential can be written in the form

$$du = \frac{\partial u}{\partial z}\, dz + \frac{\partial u}{\partial \bar{z}}\, d\bar{z}, \tag{1.27}$$

where $dz = dx + i\, dy$, and $d\bar{z} = dx - i\, dy$, and the functions $\frac{\partial u}{\partial z}$ and $\frac{\partial u}{\partial \bar{z}}$ are uniquely determined by this notation for du; in particular

$$\frac{\partial}{\partial \bar{z}} = \frac{1}{2}\left(\frac{\partial}{\partial x} + i\frac{\partial}{\partial y} \right). \tag{1.28}$$

The function u is holomorphic (i.e., expandable in a Taylor series of powers of the variable z in a neighborhood of each point of its domain) if and only if it satisfies the *Cauchy-Riemann equation*

$$\frac{\partial u}{\partial \bar{z}} = 0 \tag{1.29}$$

(for the real and imaginary parts of the function u this equation gives a system of two real equations, also called the *Cauchy-Riemann system*). In the theory of functions the corresponding inhomogeneous equation is also used:

$$\frac{\partial u}{\partial \bar{z}} = f. \tag{1.29'}$$

We shall exhibit a multidimensional generalization of this equation. In $\mathbb{C}^n$ (or on an n-dimensional complex manifold) exterior forms of type p, q are defined, i.e., exterior forms having (in local coordinates) the form (cf., for example, Hörmander 1973, Sect. 2.1)

$$u = \sum_{i_1,\ldots,i_p, j_1,\ldots,j_q} a_{i_1,\ldots,i_p, j_1,\ldots,j_q}(z, \bar{z})\, dz_{i_1} \wedge \cdots \wedge dz_{i_p} \wedge d\bar{z}_{j_1} \wedge \cdots \wedge d\bar{z}_{j_q},$$

where the indices i_k and j_k range over values from 1 to n and $a_{i_1,\ldots,i_p, j_1,\ldots,j_q}$ are smooth functions (in any local coordinates). We denote by $\Lambda^{p,q}$ the set of all such forms. Then the exterior derivative d defines a mapping $d: \Lambda^{p,q} \to \Lambda^{p+1,q} \oplus \Lambda^{p,q+1}$; and expanding du for $u \in \Lambda^{p,q}$ into a sum $du = \partial u + \bar{\partial} u$, where $\partial u \in \Lambda^{p+1,q}$ and $\bar{\partial} u \in \Lambda^{p,q+1}$, we obtain two operators

$$\partial: \Lambda^{p,q} \to \Lambda^{p+1,q}, \quad \bar{\partial}: \Lambda^{p,q} \to \Lambda^{p,q+1}.$$

For $n = 1$ the operator $\bar{\partial}$ (on $\Lambda^{0,0}$) becomes the operator $\partial/\partial \bar{z}$, so that the analogue of (1.29') has the form

$$\bar{\partial} u = f, \quad f \in \Lambda^{p,q+1}, \tag{1.30}$$

where the unknown is the form $u \in \Lambda^{p,q}$. Equation (1.30) is called the $\bar{\partial}$-*equation* and the problem of solving it is called the $\bar{\partial}$-*problem*. This problem plays an important role in multidimensional complex analysis and mathematical physics.

An important quality of partial differential equations is their *universality* – a single equation can describe many physical phenomena of completely different natures. For example Laplace's equation occurs in hydrodynamics, the theory of heat conduction, the theory of analytic functions, geometry, probability, etc.

1.8. The Concept of Well-Posedness of a Boundary-value Problem. The Cauchy Problem. We shall now discuss the important concept of well-posedness for a boundary-value problem. This concept was first introduced by Hadamard. Just from the examples given above it is clear that the number of boundary and initial conditions can be different for different equations and depends essentially on the order of the equation. In this situation if the number of conditions is insufficient, they may be satisfied by functions having no relation to the physical problem being studied. But if the number is excessive, it may happen that the problem has no solutions. The mathematical model can be considered satisfactory only in the case when for some class of data of the problem, i.e., functions occurring in the initial and boundary conditions, the boundary-value problem has a solution and the solution is unique. However, even this is not sufficient. In each boundary-value problem connected with a real physical phenomenon the data of the problem are found using measurements, which cannot be perfect and always have some error. The problem can be considered well-posed only in the case when a small change in the data of the problem leads to a small change in the solution. As it happens this is not the case for every boundary-value problem even when there exists a unique solution.[1]

Example 1.8. (Hadamard's Example). In the plane $\mathbb{R}^2$ of the variables (t, x) we consider Laplace's equation in the region $t > 0$

$$\frac{\partial^2 u}{\partial t^2} + \frac{\partial^2 u}{\partial x^2} = 0$$

with the condition

$$u(0, x) = 0, \quad \frac{\partial u}{\partial t}(0, x) = \varphi(x). \tag{1.31}$$

It can be shown that the solution u of this problem (for example, of class C^2 for $t \geq 0$) is unique. The sequence of functions

$$u_n(t, x) = e^{-\sqrt{n}} \sin nx \, e^{+nt}$$

---

[1] We note that problems that are not well-posed in the indicated sense are encountered in applications, but they are beyond the scope of the present work (Tikhonov and Arsenin 1979).

satisfies Laplace's equation and the initial condition (1.31) with

$$\varphi = \varphi_n(x) = e^{-\sqrt{n}} n \sin nx.$$

It is clear that for each $\varepsilon > 0$ there exists a number N_ε such that

$$\sup_x |\varphi_n(x)| \le \varepsilon$$

for $n \ge N_\varepsilon$. However for any $t_0 > 0$, no matter how small,

$$\sup_x |u_n(t_0, x)| = e^{nt_0 - \sqrt{n}} \to \infty$$

as $n \to \infty$. The difficulty cannot be rectified even by requiring the derivatives of the function $\varphi_n(x)$ up to order m to be small: for any number $\varepsilon > 0$ and $m \in \mathbb{N}$ the inequality

$$\sup_x \sum_{j \le m} |\varphi_n^{(j)}(x)| < \varepsilon$$

holds for $n \ge N = N_{\varepsilon,m}$.

This example shows that it is important to take account of the structure of the equation when posing a boundary-value problem. Moreover in defining well-posedness an important role is played by the proper choice of function spaces for the solution of the problem under consideration.

In the most general form the commonest definition of well-posedness has the following form. Let U, V, and F be topological vector spaces with $U \subset V$. We denote by u a (vector-valued) function that satisfies the boundary-value problem and by f the vector data of the problem, i.e., the vector that includes the right-hand side of the differential equation and also the boundary and initial conditions.

The boundary-value problem is said to be *well-posed* if:

(1) for each element $f \in F$ there exists a solution $u \in U$ of the boundary-value problem being studied;

(2) the solution is unique;

(3) the solution u as an element of the space V depends continuously on $f \in F$.

The choice of spaces U, V, and F depends on the particular problem being studied. We now give several examples.

Example 1.9. (The Cauchy problem for the one-dimensional wave equation). For the one-dimensional wave equation consider the problem

$$\begin{aligned} u_{tt} = c^2 u_{xx}, \quad x \in \mathbb{R}, \quad 0 \le t \le T, \\ u|_{t=0} = \varphi(x), \quad u_t|_{t=0} = \psi(x), \quad x \in \mathbb{R}, \end{aligned} \tag{1.32}$$

which is called the *Cauchy problem*. A solution of it of class C^2 (i.e., a solution $u \in C^2([0,T] \times \mathbb{R})$ exists and is unique for any $\varphi \in C^2(\mathbb{R})$ and $\psi \in C^1(\mathbb{R})$ and can be given in explicit form by *d'Alembert's formula*

$$u(t, x) = \frac{1}{2}\Big[\varphi(x - ct) + \varphi(x + ct)\Big] + \frac{1}{2c} \int_{x-ct}^{x+ct} \psi(\xi)\, d\xi, \qquad (1.33)$$

which is easily deduced from the representation of a solution u given in 1.7. It is clear from this formula that the problem (1.32) is well-posed: the solution u depends continuously on φ and ψ in suitable norms. More precisely for $k \in \mathbb{Z}_+$ we introduce the Banach spaces $C_b^k = C_b^k(\Omega)$, defined for any region $\Omega \subset \mathbb{R}^n$ as the spaces of functions for which all derivatives of order $\le k$ exist and are continuous and bounded in Ω; the norm in $C_b^k(\Omega)$ is defined by the formula

$$\|v\|_{C_b^k(\Omega)} = \sum_{|\alpha| \le k} \sup_{x \in \Omega} |D^\alpha v(x)|. \qquad (1.34)$$

Then if $\varphi \in C_b^k(\mathbb{R})$ and $\psi \in C_b^{k-1}(\mathbb{R})$ in the problem (1.32), where $k \ge 2$, it follows that $u \in C_b^k([0, T] \times \mathbb{R})$ and that

$$\|u\|_{C_b^k([0,T] \times \mathbf{R})} \le C\Big(\|\varphi\|_{C_b^k(\mathbf{R})} + \|\psi\|_{C_b^{k-1}(\mathbf{R})}\Big). \qquad (1.35)$$

This means that the Cauchy problem (1.32) is well-posed in the sense described above: We can take $U = V = C_b^k([0, T]) \times \mathbb{R})$, $F = C_b^k(\mathbb{R}) \times C_b^{k-1}(\mathbb{R})$ (here the role of f is played by the pair $\{\varphi, \psi\}$). For V we can take an even larger space in which $C_b^k([0, T] \times \mathbb{R})$ is continuously embedded, for example the space $C_b^l([0, T] \times \mathbb{R})$ with $l \le k$, the space $C^k([0, T] \times \mathbb{R})$, or the space $L_{\text{loc}}^p([0, T] \times \mathbb{R})$, $1 \le p < +\infty$. Here $C^k([0, T] \times \mathbb{R})$ is a *Fréchet space*, i.e., a complete countably-normed space (Rudin 1973, Chap. 1), consisting of the functions of class C^k on $[0, T] \times \mathbb{R}$ with topology defined by the seminorms

$$\|u\|_{C^k(K)} = \sum_{|\alpha| \le k} \sup_{x \in K} |D^\alpha u(x)|, \qquad (1.36)$$

where K is an arbitrary compact subset of $[0, T] \times \mathbb{R}$ (the space $C^k(\Omega)$, where Ω is any region in $\mathbb{R}^n$ is defined similarly); $L_{\text{loc}}^p([0, T] \times \mathbb{R})$ is the Fréchet space consisting of the functions that belong to $L_p(K)$ on any compact set $K \subset [0, T] \times \mathbb{R}$; its topology is defined by the seminorms

$$\|u\|_{L_p(K)} = \left(\int_K |u(x)|^p\, dx\right)^{1/p} \qquad (1.37)$$

(the space $L_{\text{loc}}^p(\Omega)$, where Ω is any region in $\mathbb{R}^n$, is defined similarly).

We can also take $F = C^\infty(\mathbb{R}) \times C^\infty(\mathbb{R})$ and $U = V = C^\infty([0, T] \times \mathbb{R})$ with the standard Fréchet topology of uniform convergence of each derivative on compact sets; in general in $C^\infty(\Omega)$, where $\Omega \subset \mathbb{R}^n$, the topology is defined by the seminorms (1.36), where the number $k \in \mathbb{Z}_+$ and the compact set $K \subset \Omega$ are taken arbitrarily.

We see that the spaces U, V, and F occurring in the definition of well-posedness can be chosen in very many ways (although, of course, not every space will work). This situation is typical. Spaces of type C^k are often chosen as the spaces U, V, and F; but in many cases it is convenient (and sometimes even necessary) to use other spaces (for example the Hölder or Sobolev spaces, cf. Sects. 2.2 and 2.3). At the same time, in the definition of well-posedness it is reasonable to take spaces whose description does not depend too strongly on the properties of solutions of the problem under consideration. Hadamard's example shows that the Cauchy problem for Laplace's equation (with initial data (1.31)) cannot be well-posed if spaces of C^k type are taken as U, V, and F; in fact almost none of the natural spaces work in this situation, and for that reason the problem is considered ill-posed. Nevertheless it is possible to choose the spaces U, V, and F so that the problem becomes well-posed. For example, as the space F one can take the space Z of functions on $\mathbb{R}$ that are Fourier transforms of functions in $C_0^\infty(\mathbb{R})$ and as U and V the spaces of functions of class $C^2([0, T], Z)$, i.e. functions that are twice continuously differentiable on $[0, T]$ with values in Z. An explicit description of the space Z is given by the Paley-Wiener-Schwartz Theorem (cf. Hörmander 1983–1985, Theorem 7.3.1). The topology in Z should be transferred by the Fourier transform from $C_0^\infty(\mathbb{R})$; for information on the topology of $C_0^\infty(\mathbb{R})$ cf. Sect. 2.1. An application of the Fourier transform on x shows that under such a choice of spaces U, V, and F the Cauchy problem (1.31) for the Laplace equation becomes well-posed in the sense described above. The defect of the space Z is that it is not trivial to describe it in the language of the original functions (i.e., without the Fourier transform): it is the space of entire functions of exponential type and first order that decay rapidly along the real axis. Such a description does not allow us to regard the problem as well-posed, since the smallness of a change in the initial data in the topology of the space Z, indeed even the property of belonging to the space Z, is very difficult to control.

Example 1.10. (The Dirichlet problem for Laplace's equation). Let Ω be a bounded region in $\mathbb{R}^n$ with smooth or piecewise smooth boundary Γ. Then the *Dirichlet problem for Laplace's equation* (1.12)–(1.13) in the region Ω is well-posed if we take $U = V = C^2(\Omega) \cap C(\bar{\Omega})$ and $F = C(\Gamma)$, since it has a unique solution and the *maximum-modulus principle* holds:

$$\max_{x \in \Omega} |u(x)| = \max_{x \in \Gamma} |u(x)|. \tag{1.38}$$

In fact the solution is even infinitely-differentiable and analytic in Ω; hence with the same $F = C(\Gamma)$ we can take, for example, $U = V = C^\infty(\Omega) \cap C(\bar{\Omega})$. (The topology on the intersection of two spaces is always defined using the union of the families of seminorms of the spaces.) We note that matters are somewhat more complicated in the Dirichlet problem for the Poisson equation (1.12'), (1.13): prescribing a pair $\{\rho, \varphi\} \in C(\bar{\Omega}) \times C(\Gamma)$, we do not in general obtain solutions u of class $C^2(\bar{\Omega})$ (or even of class $C^2(\Omega) \cap C(\bar{\Omega})$: here one must use other spaces (for example, Hölder or Sobolev spaces; for more details see Chap. 2).

Example 1.11. (The Cauchy problem for the heat equation). For the heat equation consider the problem

$$\begin{cases} u_t & = a^2 \Delta u, \quad x \in \mathbb{R}^n, \quad t \in [0, T], \\ u|_{t=0} & = \varphi(x), \quad x \in \mathbb{R}^n \end{cases} \tag{1.39}$$

which is called the *Cauchy problem* for this equation. In this case, in contrast to Example 1.9, the solution cannot be sought in local spaces; for example, the solution is not unique, even in the space $C^\infty([0, T] \times \mathbb{R}^n)$. However if we impose certain conditions on the behavior of u as $|x| \to \infty$, we can make the problem a well-posed one. For example, if $\varphi \in F = C_b(\mathbb{R}^n)$, then a solution $u \in U = V = C^\infty((0, T] \times \mathbb{R}^n) \cap C_b([0, T] \times \mathbb{R}^n)$ exists and is unique; it is given by *Poisson's formula*

$$u(t, x) = (2a\sqrt{\pi t})^{-n} \int\limits_{\mathbb{R}^n} e^{-|x-y|^2/4a^2 t} \varphi(y) \, dy. \tag{1.40}$$

For an initial function $\varphi \in C_0^\infty(\mathbb{R}^n)$ formula (1.40) is easily obtained using the Fourier transform on x as the formula for one of the solutions; for an arbitrary function $\varphi \in C_b(\mathbb{R}^n)$ one can verify directly that formula (1.40) gives a solution; the uniqueness of the solution can be established using Holmgren's principle (cf. Sect. 1.2). For the solutions u of this class one can prove the *maximum principle*

$$\inf_{x \in \mathbb{R}^n} \varphi(x) \le u(t, x) \le \sup_{x \in \mathbb{R}^n} \varphi(x) \tag{1.41}$$

(for example, it can be deduced from formula (1.40) although, being a much more general fact, it can be obtained from general considerations). The physical meaning of the second inequality in the maximum principle is that the maximal temperature under heat exchange without sources of heat cannot exceed the maximum temperature at the initial instant, and the left-hand inequality has a similar meaning. In studying physical problems it is natural to confine oneself to solutions for which the relation (1.41), which is natural from the physical point of view, holds. Thus the solutions that are unbounded on x, which account for the nonuniqueness of the solution of the problem (1.39), should be considered as having no physical meaning.

Now let us consider the general Cauchy problem for an equation of order m with constant coefficients in $\mathbb{R}^{n+1} = \mathbb{R}_t^1 \times \mathbb{R}_x^n$:

$$p(D_t, D_x)u = f, \tag{1.42}$$

where $p = p(\tau, \xi)$ is a polynomial of degree m. For such an equation the question of the well-posedness of the Cauchy problem can be studied more or less completely. We shall exhibit only the simplest information about the available results here (for more details see the article of L. R. Volevich and S. G. Gindinkin, *The Cauchy Problem*, in one of the following volumes of this series).

We first assume that (1.42) contains a term $D_t^m u$ (i.e., $p(1, 0) \ne 0$ or, in other words, the coefficient of τ^m in the polynomial $p(\tau, \xi)$ is not zero;

dividing by this coefficient, we may assume it equal to 1). In this case we say that the plane $t = 0$ is *noncharacteristic* for the operator $p(D_t, D_x)$. Assume that (1.42) has a solution $u \in C^m(\mathbf{R}^{n+1})$ equal to 0 when $t \leq 0$ for any function $f \in C_0^\infty(\mathbf{R}^{n+1})$ (i.e., any function $f \in C^\infty(\mathbf{R}^{n+1})$ equal to 0 outside some compact set) that vanishes for $t \leq 0$. Then if $\lambda_1(\xi), \ldots, \lambda_m(\xi)$ are all the roots of the equation $p(\tau, \xi) = 0$ with respect to τ, there exists C such that

$$\text{Im} \lambda_j(\xi) > -C \quad \text{for } \xi \in \mathbf{R}^n, \quad j = 1, \ldots, m. \tag{1.43}$$

We denote by p_m the leading homogeneous part of the polynomial p, also called the *principal symbol of the operator* $p(D_t, D_x)$. To be specific if

$$p(\eta) = \sum_{|\alpha| \leq m} p_\alpha \eta^\alpha$$

(here α is an $(n+1)$-dimensional multi-index and $\eta = (\tau, \xi)$), then

$$p_m(\eta) = \sum_{|\alpha| = m} p_\alpha \eta^\alpha.$$

From the condition (1.43) for p a similar condition follows for p_m, which, as is easily seen, is equivalent to the condition that all the roots of the equation $p_m(\tau, \xi) = 0$ with respect to τ for $\xi \in \mathbf{R}^n$ are real. In this case the polynomial p_m and the operator $p_m(D_t, D_x)$ are called *hyperbolic* (or *nonstrictly hyperbolic*).

Thus the condition of (nonstrict) hyperbolicity of the principal part is necessary for the existence of a solution of the problem just described, which is in essence equivalent to the Cauchy problem in its usual formulation.

$$\begin{cases} p(D_t, D_x)u = f, \quad t > 0, \\ u|_{t=0} = \varphi_0(x), \quad \dfrac{\partial u}{\partial t}\Big|_{t=0} = \varphi_1(x), \ldots, \dfrac{\partial^{m-1}u}{\partial t^{m-1}}\Big|_{t=0} = \varphi_{m-1}(x). \end{cases} \tag{1.44}$$

(If $u \in C^m(\mathbf{R}^{n+1})$ satisfies (1.42) and is equal to 0 for $t < 0$, then in an obvious way it is a solution of the problem (1.44) with $\varphi_0 = \varphi_1 = \ldots = \varphi_{m-1} = 0$; conversely, if $u \in C^m$ for $t \geq 0$ and u is a solution of the problem (1.44), where $\varphi_0, \varphi_1, \ldots, \varphi_{m-1} \in C^\infty(\mathbf{R}^{n-1})$, then the function equal to $u(t, x) - \sum_{j=0}^{m-1} \varphi_j(x) t^j / j!$ for $t \geq 0$ and 0 for $t \leq 0$ belongs to $C^m(\mathbf{R}^{n+1})$ and is a solution of an equation of the form (1.42), but with a different function f.)

When condition (1.43), which is stronger than the condition of nonstrict hyperbolicity, holds, the polynomial $p(\tau, \xi)$ and the operator $p(D_t, D_x)$ are often called *hyperbolic*. This condition is necessary and sufficient for the existence of a unique smooth solution (of class C^j, $j \geq m$, for $t \geq 0$) of the problem (1.44) with sufficiently smooth data (for example, for $f \in C^{j+r}$, for $t \geq 0$, $\varphi_k \in C^{m-k+j+r}$, $k = 0, 1, \ldots, m-1$, where $r = [(n+1)/2] + 1$). This solution will depend continuously on the right-hand side f and the initial data φ_k if the latter vary continuously in the topology of the corresponding spaces C^{j+r} and $C^{m-k+j+r}$ respectively (in fact in this case there is a finite region of dependence: for any compact set $K \subset \{(t,x) : t \geq 0\}$ there exist compact sets $K_1 \subset \{(t,x) : t \geq 0\}$ and $K_2 \subset \mathbb{R}^n$ such that $u|_K$ depends only on the restrictions $f|_{K_1}$ and $\varphi_k|_{K_2}$, $k = 0, 1, \ldots, m - 1$). Thus in this case the Cauchy problem (1.44) is well-posed under a suitable choice of spaces of type C^l for the spaces U, V, and F in the definition of well-posedness.

The hyperbolicity condition (1.43) necessarily holds if the principal part p_m is *strictly hyperbolic*, i.e., if the roots $\mu_1(\xi), \ldots, \mu_m(\xi)$ of the equation $p_m(\tau, \xi) = 0$ with respect to τ for $\xi \neq 0$ are real and distinct. In this case the operator $p(D_t, D_x)$ itself is also called *strictly hyperbolic* (the hyperbolicity condition (1.43) in this case holds not only for p_m but also for any polynomial obtained by adding any terms of degree at most $m - 1$ to the polynomial p_m). The well-posedness of the Cauchy problem and the existence of a finite region of dependence hold also in the case of strictly hyperbolic equations with variable coefficients.

The Cauchy problem for the heat equation (Problem (1.39)) cannot be studied from the same angle, since the initial plane $t = 0$ is characteristic (the equation does not contain D_t^2). Let us consider a more general situation: assume that $p_m(1, 0) = 0$; then, as in the case of the heat equation, the equation $p(D_t, D_x)u = 0$ has a non-trivial solution $u \in C^\infty(\mathbb{R}^{n+1})$ equal to 0 for $t \leq 0$, i.e., uniqueness fails for the Cauchy problem. Let us assume that the term of highest degree in τ in the polynomial $p(\tau, \xi)$ has the form τ^k, i.e., (1.42) has been solved with respect to the highest derivative on t. Then it is natural to consider the Cauchy problem

$$\begin{cases} p(D_t, D_x)u = f, \\ u|_{t=0} = \varphi_0(x), \quad \dfrac{\partial u}{\partial t}\Big|_{t=0} = \varphi(x), \ldots, \dfrac{\partial^{k-1} u}{\partial t^{k-1}}\Big|_{t=0} = \varphi_{k-1}(x). \end{cases}$$
$$(1.45)$$

Consider the roots $\lambda_1(\xi), \ldots, \lambda_k(\xi)$ of the equation $p(\tau, \xi) = 0$ on τ. The condition

$$\operatorname{Im} \lambda_j(\xi) > -C \quad \text{for } \xi \in \mathbb{R}^n, \quad j = 1, \ldots, k, \tag{1.46}$$

which generalizes (1.43), is called the condition for *Petrovskij well-posedness*. This condition guarantees the existence of a solution of the problem (1.45) for sufficiently smooth data functions $f, \varphi_0, \varphi_1, \ldots, \varphi_{k-1}$ equal to 0 outside some compact set. The solution is unique when suitable restrictions are imposed on the growth of $u(t, x)$ as $|x| \to +\infty$. For example it suffices that this growth be at most polynomial, i.e., that in each strip $[0, T] \times \mathbb{R}^n$ some estimate of the form

$$|u(t, x)| \leq C(1 + |x|)^N \tag{1.47}$$

hold. Thus the condition for Petrovskij well-posedness guarantees that the Cauchy problem (1.45) is well-posed in suitable spaces whose description takes account of the growth of the functions as $|x| \to +\infty$. In the case of variable coefficients the condition of Petrovskij well-posedness must be replaced by a stronger condition – the condition of parabolicity (cf. Sect. 2.5).

§2. The Cauchy-Kovalevskaya Theorem and Its Generalizations

2.1. The Cauchy-Kovalevskaya Theorem. The first proof of the existence and uniqueness of a solution to the Cauchy problem for an ordinary differential equation of the general form

$$\frac{du}{dt} = f(t, u), \quad u(t_0) = u_0, \tag{2.1}$$

was found by Cauchy under the assumption that the function f is holomorphic in a neighborhood of the point (t_0, u_0). He proved that under this assumption there exists one and only one solution $u(t)$ holomorphic in a neighborhood of the point t_0.

The idea of this proof is very simple. If $u(t) = \sum_{j=0}^{\infty} a_j (t - t_0)^j$ is a solution of the problem (2.1), then $a_0 = u_0$, $a_1 = f(t_0, u_0)$, and all the subsequent coefficients $a_2, a_3, \ldots$ can be found by differentiating both sides of the differential equation and setting $t = t_0$. For example

$$a_2 = \frac{1}{2} \left(\frac{\partial f(t, u)}{\partial t} + \frac{\partial f(t, u)}{\partial u} f(t, u) \right) \Big|_{\substack{t = t_0 \\ u = u_0}}.$$

Thus the coefficients a_j are determined uniquely from the equation and the initial data, i.e., the solution is unique.

To prove the existence it suffices to show that the power series with coefficients found in this way converges in some neighborhood of the point t_0. Here one can use the *method of majorants*. The Cauchy problem

$$\frac{dv}{dt} = \frac{M}{\left(1 - \dfrac{t - t_0}{r}\right)\left(1 - \dfrac{v - u_0}{r}\right)}, \quad v(t_0) = u_0, \tag{2.1'}$$

is solved using separation of variables. At the same time it can be shown that for sufficiently large M and sufficiently small r the series $\sum_{j=0}^{\infty} b_j (t - t_0)^j$, whose sum is the function $v(t)$ *majorizes* the series for $u(t)$, i.e., $|a_j| \leq b_j$ for all j. It follows from this that the series for $u(t)$ converges, and so there exists a solution of the problem (2.1).

Cauchy's theorem was generalized to partial differential equations by S. V. Kovalevskaya.

This theorem applies to a class of solutions that are now called *equations of Kovalevskaya type* and have the following form:

$$\frac{\partial^{n_i} u_i}{\partial t^{n_i}} = f_i(t, x, u, \frac{\partial u}{\partial t}, \frac{\partial u}{\partial x}, \ldots), \quad i = 1, \ldots, m, \qquad (2.2)$$

where $x = (x_1, \ldots, x_{n-1})$, $u = (u_1, \ldots, u_m)$, and for each $i = 1, \ldots, m$ the function f_i depends on the derivatives of the functions u_j only up to order n_j, is independent of $\partial^{n_j} u_j / \partial t^{n_j}$, and is an analytic function of all its arguments.

The *Cauchy problem* is to construct a solution of the system (2.2) that assumes prescribed initial values for $t = 0$:

$$\frac{\partial^k u_i}{\partial t^k}(0, x) = \varphi_{i,k}(x), \quad k = 0, 1, \ldots, n_i - 1, \quad i = 1, \ldots, m. \qquad (2.2')$$

It is assumed that the functions $\varphi_{i,k}(x)$ are analytic in a neighborhood of the point $x = 0$. It is clear that the initial conditions (2.2′) make it possible to compute the values of all arguments of the functions f_i in a neighborhood of the point $x = 0$ when $t = 0$. Fixing the values of these derivatives for $t = 0$ and $x = 0$, we shall assume that the functions f_i are analytic in a neighborhood of these fixed values.

Theorem 1.1 (Cauchy-Kovalevskaya, (cf. Petrovskij 1961)). *Under the assumptions made above the Cauchy problem (2.2), (2.2′), has one and only one solution $u(t, x)$ that is analytic in a neighborhood of the point $t = 0$, $x = 0$.*

The content of this theorem is quite simple. If an analytic function $u(t, x)$ satisfies (2.1) and conditions (2.2′), then its derivatives of all orders are determined uniquely at the point $t = 0$, $x = 0$. In addition if the order of differentiation of the function u_i on the variable t does not exceed $n_i - 1$, then these derivatives are found from conditions (2.2′). The remaining derivatives are determined using the differential equations (2.2). Thus we can find all the coefficients of the Taylor series of the unknown solution at the point $(0, 0)$, from which the uniqueness of an analytic solution follows. To establish the existence of a solution it is necessary to prove that the power series constructed for the functions u_i, whose coefficients are determined in the indicated way, converge. This technically complicated proof is based on the method of constructing majorants.

The Cauchy-Kovalevskaya Theorem applies to equations of a very general form and is widely used in the modern general theory of partial differential equations. If the derivatives of a solution are taken as new unknowns, the problem (2.2)–(2.2′) can be reduced to the following Cauchy problem for a system of quasilinear first-order differential equations:

$$\frac{\partial v_i}{\partial t} = \sum_{j=1}^{N} a_{ij}(t, x, v_1, \ldots, v_N) \frac{\partial v_i}{\partial x_j} + f_i(t, x, v_1, \ldots, v_N), \qquad (2.3)$$

$$i = 1, \ldots, N;$$

$$v_i(0, x) = \varphi_i(x), \quad i = 1, \ldots, N. \qquad (2.4)$$

In the important special case when (2.2) are linear in the functions u_j and their derivatives the system (2.3) will also be linear. In this case we can estimate the radius of convergence of the power series that give the solution. To be specific, let us consider, for example, (1.1), where the operator A of the form (1.2) has analytic coefficients a_α that can be analytically continued to the polydisk

$$\Omega_{R,\delta} = \{z \in \mathbf{C}^n : |z_j| < R \text{ for } j < n, |z_n| < \delta R\},$$

and if $\alpha_0 = (0,0,\ldots,0,m)$, then $a_{\alpha_0} \equiv 1$. Further suppose

$$2(2^n e)^m \sum_{\alpha \neq \alpha_0} R^{m-|\alpha|} \delta^{m-\alpha_n} |a_\alpha(z)| \leq 1, \quad z \in \Omega_{R,\delta} \tag{2.5}$$

and the right-hand side f is also holomorphic and bounded in $\Omega_{R,\delta}$. Then (1.1) with the zero Cauchy conditions on the hypersurface $x_n = 0$

$$D_n^j u \Big|_{x_n=0} = 0, \quad j = 0, 1, \ldots, m-1, \tag{2.6}$$

has a unique analytic solution holomorphic in the polydisk $\Omega_{R/2,\delta}$ and satisfying there the estimate

$$\sup_{\Omega_{R/2,\delta}} |u(z)| \leq 2(R\delta)^m \sup_{\Omega_{R,\delta}} |f(z)| \tag{2.7}$$

(Hörmander 1963). In particular if a_α and f are entire functions, we can take R as large as desired and take δ so small that condition (2.5) holds. In this situation the δ found and, consequently, the radius of convergence of the power series giving the solution u will be independent of the right-hand side f (provided it is an entire function).

Passing to the inhomogeneous Cauchy problem for the same equation (1.1) with the conditions

$$D_n^j u \Big|_{x_n=0} = \varphi_j(x), \quad j = 0, 1, \ldots, m-1, \tag{2.8}$$

we can reduce it to the problem with the Cauchy conditions (2.6) if we change the right-hand side f by the procedure described in 1.8. For the Cauchy problem with conditions (2.8) the same assertions are true as in the case of conditions (2.6); here it is necessary to assume that the data φ_j are holomorphic and bounded in the polydisk

$$\Omega_R = \{z \in \mathbf{C}^{n-1} : |z_j| < R, \quad j = 1, 2, \ldots, n-1\},$$

and the estimate (2.7) will assume the form

$$\sup_{\Omega_{R/2,\delta}} |u(z)| \leq 2(R\delta)^m \sup_{\Omega_{R,\delta}} |f(z)| + 2 \sum_{j=0}^{m-1} \frac{(R\delta)^{m+j}}{j!} \sup_{\Omega_R} |\varphi_j(z)|. \tag{2.9}$$

If f and φ_j are entire functions we can again choose any $R > 0$ and then $\delta = \delta(R)$ independent of f and φ_j (but depending on the operator A) such that the radius of convergence of the Taylor series for the solution is independent of the choice of entire functions f and φ_j.

2.2. An Example of Nonexistence of an Analytic Solution. The reasoning of 2.1, which was used for the proof of the uniqueness of an analytic solution of the Cauchy problem, is applicable to systems of the form (2.1) even in the case when the right-hand sides may contain derivatives $D_t^k D_x^\alpha u_j$ with $k + |\alpha| > n_j$, $k < n_j$ (we previously required $k + |\alpha| \le n_j$, $k < n_j$). However, in this case an analytic solution may fail to exist for some data of the problem.

Example 1.12. Consider the Cauchy problem for the heat equation

$$u_t = u_{xx}, \quad x \in (-1, 1), \quad t > 0; \quad u|_{t=0} = \frac{1}{1-x}, \quad x \in (-1, 1). \quad (2.10)$$

If this problem had a solution analytic in some neighborhood of the origin, its Taylor series would be

$$u(t, x) = \exp\left(t\frac{d^2}{dx^2}\right)(1 + x + x^2 + \cdots),$$

from which it follows in particular that the Taylor series of the function $u(t, 0)$ at the point $t = 0$ would have the form of the series

$$\sum_{k=0}^{\infty} (2k)! t^k,$$

which diverges for any $t \ne 0$. Thus the Cauchy problem (2.10) under consideration has no analytic solution in any neighborhood of the origin. The reason is that the right-hand side of the equation contains a derivative of order 2, higher than the maximal order of the derivative on t, which is 1 (so that the heat equation is not an equation of Kovalevskaya type).

2.3. Some Generalizations of the Cauchy-Kovalevskaya Theorem. Characteristics. Consider the case when the system of equations does not have the special form (2.1) and the initial conditions are given on a smooth analytic hypersurface Γ. In a region $\Omega \subset \mathbb{R}^n$ consider the system

$$F_i(x, u, \frac{\partial u}{\partial x}, \ldots, \partial^\beta u, \ldots) = 0, \quad i = 1, \ldots, m, \quad (2.11)$$

where $u = (u_1, \ldots, u_m)$ and assume that the functions F_i are analytic and depend on the derivatives of the functions u_j of orders up to n_j. Assume that the "initial" conditions have the form

$$\frac{\partial^k u_i}{\partial \nu^k}(x) = \varphi_{i,k}(x), \quad x \in \Gamma, \quad k = 0, 1, \ldots, n_i - 1; \quad i = 1, \ldots, m, \quad (2.12)$$

where ν is the direction of the normal to Γ.

To solve this problem we reduce it to a problem of the form (2.1), (2.2). It is not always possible to do this; it is only possible under additional hypotheses which will be of interest to us. In a neighborhood of some point $P_0 \in \Gamma$ we introduce a local coordinate system $(t, y_1, \ldots, y_{n-1})$ in such a way that these coordinates are expressed as analytic functions of x and so that the variable t varies in the direction of the normal to Γ, so that the surface Γ is defined by the equation $t = 0$ and the coordinates $y_1, \ldots, y_{n-1}$ are coordinates on Γ for $t = 0$. Using conditions (2.12) it is easy to compute all the derivatives of the function u_j at $t = 0$ up to order $n_j - 1$, so that in the new coordinates the initial conditions assume the form (2.2). Moreover differentiating the initial conditions makes it possible to find all the derivatives $D^\alpha_{t,y} u_j$ having order at most $n_j - 1$ on t (any order on y) at $t = 0$. Thus at $t = 0$ it is possible to find the value of all arguments of the function F_i from the initial conditions except the derivatives in the direction of ν of maximal order n_j for u_j, so that at $t = 0$ these equations can be written in the form

$$\Phi_i \left(y, \frac{\partial^{n_1} u_1}{\partial \nu^{n_1}}, \ldots, \frac{\partial^{n_m} u_m}{\partial \nu^{n_m}} \right) = 0, \quad i = 1, \ldots, m. \quad (2.13)$$

The derivatives $\frac{\partial^{n_j} u_j}{\partial \nu^{n_j}}$ can be chosen at $t = 0$ from the set of analytic functions satisfying (2.13). Afterwards, to reduce the system of equations (2.11) to the form (2.1) it suffices to require that the Jacobian

$$\frac{\partial(\Phi_1, \ldots, \Phi_m)}{\partial \left(\dfrac{\partial^{n_1} u_1}{\partial \nu^{n_1}}, \ldots, \dfrac{\partial^{n_m} u_m}{\partial \nu^{n_m}} \right)}$$

be nonzero at $t = 0$. If the surface Γ is given by the equation $S(x) = 0$ and $\partial S/\partial x \neq 0$, where $\frac{\partial S}{\partial x} = (\frac{\partial S}{\partial x_1}, \ldots, \frac{\partial S}{\partial x_m})$, then this condition can be written in the form

$$\det \left\| \sum_{|\beta|=n_j} \frac{\partial F_i}{\partial(\partial^\beta u_j)} \left(\frac{\partial S}{\partial x} \right)^\beta \right\|_{i,j=1,\ldots,m} \neq 0 \quad \text{for } t = 0. \quad (2.14)$$

Here the arguments of the function $\partial F_i/\partial(\partial^\beta u_j)$ are computed according to the following rule: the derivatives $\partial^\gamma u_j$ for $|\gamma| \leq n_j$ are computed from the initial conditions (2.12) and equation (2.12) taking account of the choice made above. It is clear that under the condition (2.14) the problem (2.11), (2.12) reduces to the equivalent problem (2.2), (2.2$'$), and the Cauchy-Kovalevskaya Theorem holds.

Condition (2.14) has a particularly simple meaning in the case when the functions F_i depend linearly on the derivatives $\partial^\beta u_j$ for $|\beta| = n_j$, i.e., have the form

$$F_i(x, \partial^\beta u) = \sum_{j=1}^{m} \sum_{|\beta|=n_j} a_{\beta,ij}(x) \partial^\beta u_j + f_i(x, \partial^\gamma u_j),$$

where $|\gamma| < n_j$. It is violated at those points where

$$\det \left\| \sum_{|\beta|=n_j} a_{\beta,ij}(x) \left(\frac{\partial S}{\partial x}\right)^\beta \right\| = 0. \qquad (2.15)$$

In this situation we say that the normal to Γ has a *characteristic direction* at such points. If condition (2.15) holds at each point on Γ, then the surface Γ is called *characteristic* or *a characteristic*. From what was said above it is clear that in this case (2.11) generates at $t = 0$ a nontrivial relation of the form

$$R(\varphi_{1,0}, \dots, \varphi_{m,0}, \dots, \partial^\gamma \varphi_{i,j}, \dots) = 0,$$

i.e., the functions $\varphi_{i,j}$ cannot be prescribed arbitrarily, but must satisfy certain relations. The Cauchy-Kovalevskaya theory is inapplicable in such a case. For a linear equation (1.1) the condition for being characteristic assumes the form

$$\sum_{|\alpha|=m} a_\alpha(x) \left(\frac{\partial S}{\partial x}\right)^\alpha = 0 \quad \text{for } S(x) = 0. \qquad (2.16)$$

The following generalization is possible, however. Suppose the equation has the form

$$D^\beta u = \sum_{|\alpha| \le |\beta|} a_\alpha D^\alpha u + f, \qquad (2.17)$$

where a_α and f are analytic functions in some neighborhood of the origin. We introduce the conditions

$$D_j^k u = \varphi_{jk} \quad \text{for } x_j = 0, \quad \text{if } 0 \le k < \beta_j, \quad j = 1, \dots, n, \qquad (2.18)$$

where φ_{jk} are analytic functions. We remark that the number of these conditions is $|\beta|$.

Theorem 1.2 (Hörmander 1973). *The problem (2.17), (2.18) has a unique solution that is analytic in a neighborhood of the origin if at least one of the following conditions holds:*

1^0. *β does not belong to the convex hull of the indices α for which $a_\alpha \not\equiv 0$.*

2^0. *The sum $\sum_{|\alpha|=|\beta|} |a_\alpha(0)|$ is less than some positive number depending only on $|\beta|$.*

2.4. Ovsyannikov's Theorem. The hypotheses of the Cauchy-Kovalevskaya Theorem can be weakened by replacing the requirement of analyticity in all the variables on the right-hand side of the equation with the condition of continuity in t and analyticity in the other arguments. In this case the solution is an analytic function of x that is continuously differentiable with respect to t. Other possible generalizations of this kind can be obtained from a theorem of L. V. Ovsyannikov, which we give in a linear variant (Egorov 1985).

Let E_s be Banach spaces for $0 \le s \le 1$, $\|\cdot\|_s$ the norm in E_s with $E_s \subset E_{s'}$ for $s' < s$ and $\|u\|_{s'} \le \|u\|_s$. Let $A = A(t)$ be a continuous function of t on $[-T, T]$ with values in the Banach space of bounded linear operators from E_s into $E_{s'}$ and $\|A(t) : E_s \to E_{s'}\| \le C(s - s')^{-1}$ for any s and s' with $0 \le s' < s \le 1$.

Theorem 1.3. *Let $u_0 \in E_1$, and let f be a continuous function on $[-T, T]$ with values in E_1. There exists a function $u = u(t)$ defined for $|t| < \min(T, (Ce)^{-1})$ with values in E_0 that is continuously differentiable with values in E_s for $|t| \le (Ce)^{-1}(1 - s)$ for all s, $0 \le s < 1$, and*

$$\frac{du}{dt} = A(t)u + f(t), \quad |t| < \min\left(T, (Ce)^{-1}\right), \quad u(0) = u_0.$$

If in addition $A(t)$ and $f(t)$ are analytic functions of t for $|t| < T$, then $u(t)$ also depends analytically on t. If $u(t) \in C^1$ on the interval $(-T', T')$ with values in E_s for some s with $0 < s \le 1$, $\frac{du}{dt} = A(t)u$ for $|t| < T'$, and $u(0) = 0$, then $u(t) = 0$ for $|t| < T'$.

Let us show how Theorem 1.3 can be applied to the study of the Cauchy problem for the system of first-order differential equations

$$\frac{\partial u}{\partial t} = \sum_{j=1}^{n} A_j(t, x) \frac{\partial u}{\partial x_j} + B(t, x)u + f(t, x) \tag{2.19}$$

with initial condition

$$u(0, x) = \varphi(x). \tag{2.20}$$

Here u, φ, and f are vector-valued functions with values in $\mathbb{C}^N$, and A_j and B are square matrices of order N. As indicated in 2.1, the Cauchy problem reduces to this form in the case of a general linear system of equations of Kovalevskaya type of arbitrary order.

Assume that for $|t| < T$ the functions $A_j(t, x)$, $B(t, x)$, and $f(t, x)$ admit an analytic continuation on x to functions that are holomorphic in a bounded region $\Omega \subset \mathbb{C}^N$, continuous in $\bar{\Omega}$, and depend continuously on (t, x) under the supremum norm on $\bar{\Omega}$. Further suppose that the function φ possesses the same properties of being holomorphic in Ω and continuous in $\bar{\Omega}$. Let $\Omega_0 \Subset \Omega$ (i.e., $\bar{\Omega}_0$ is a compact subset of Ω) and

$$\Omega_s = \left\{ z : z \in \mathbb{C}^n, \operatorname{dist}(z, \Omega_0) \le s \right\},$$

and moreover $\Omega_1 = \bar{\Omega}$. Let E_s be the Banach space of functions g that are continuous in Ω_s and holomorphic in the interior of Ω_s and let $\|g\|_s = \sup_{z \in \Omega} |g(z)|$. Then $E_s \subset E_{s'}$ for $s > s'$ and $\|u\|_{s'} \le \|u\|_s$. It is easy to verify using Cauchy's theorem that

$$\left\| \frac{\partial g(z)}{\partial z_j} \right\|_{s'} \le \frac{1}{(s-s')d} \|g\|_s, \quad (j = 1, \ldots, n),$$

so that

$$\|A(t)g\|_{s'} \le \frac{C}{(s-s')d} \|g\|_s \quad \text{for } 0 \le s' < s.$$

Applying Theorem 1.3, we obtain a theorem on the solvability of the Cauchy problem (2.19), (2.20) in the class of analytic functions of x that depend continuously on t. If we assume that A_j, B, and f are analytic in both t and x, the solution obtained is also an analytic function of t and x.

2.5. Holmgren's Theorem (cf. Petrovskij 1961, Sect. 1.4; Bers and Schechter 1964, Sect. 3.4). The Cauchy-Kovalevskaya Theorem applies to the rather narrow class of systems of the form (2.2) with analytic right-hand sides, and the initial values are also required to be analytic. Such hypotheses are often not attainable and may conflict with the nature of the physical phenomena being studied. However, as Holmgren was the first to remark, the Cauchy-Kovalevskaya Theorem makes it possible to prove the uniqueness of the solution of the Cauchy problem for a linear system of the form (2.19) without any assumption on the analyticity of the initial values and the solution. In particular the solution corresponding to analytic initial data is unique, so that any solution in this case is analytic.

We shall use the example (2.19), (2.20) of the Cauchy problem to illustrate Holmgren's scheme. Let $u = u(t, x)$ be a solution of this problem of class C^1 defined for $t \ge 0$ in some neighborhood Ω of the origin, and let $f \equiv 0$ and $\varphi \equiv 0$. We extend u by setting $u(t, x) = 0$ for $(t, x) \in \Omega$ and $t < 0$. It is clear then that, as before, $u \in C^1(\Omega)$. We change from the variables (t, x) to new variables (s, y) connected with (t, x) by an analytic diffeomorphism, setting $y = x$, $s = t + |x|^2$ (Fig. 1.3a).

The function $v(s, y) = u(t, x)$ satisfies the system (2.19) transformed into the new coordinates for $-T \le s \le T$, $y \in \Omega'$, where Ω' is a neighborhood of zero in $\mathbf{R}_y^{n-1}$ and $T > 0$ is sufficiently small; moreover $v(s, y) \equiv 0$ for $0 \le s \le |y|^2$ (cf. Fig. 1.3b). The transformed system (2.19) has the same form, but with different matrices A_j and B. Therefore we may assume from the outset that $u(t, x) = 0$ for $t \le |x|^2$ and it is not necessary to introduce the new variables s and y. Consider the "conjugate" system of equations

$$\frac{\partial \psi}{\partial t} = \sum_{j=1}^{n} \frac{\partial}{\partial x_j} \left(A_j^*(t, x)\psi \right) - B^*(t, x)\psi, \tag{2.21}$$

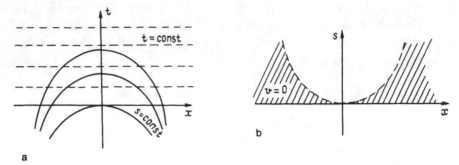

Fig. 1.3

where A_j^* and B^* are the matrices that are Hermitian-conjugate to the matrices A_j and B respectively. If $\psi = \psi(t, x)$ is any C^1 solution of this system defined in a neighborhood of the origin, then for small t the integral

$$\big(u(t, \cdot), \psi(t, \cdot)\big) = \int \langle u(t, x), \psi(t, x) \rangle \, dx$$

is defined. (Here the brackets $\langle \cdot, \cdot \rangle$ on the right-hand side denote the Hermitian scalar product in $\mathbb{C}^n$, and the integration extends over a compact set since $u(t, x) = 0$ for $|x|^2 \geq t$.) Now integration by parts gives

$$\frac{d}{dt}\big(u(t, \cdot), \psi(t, \cdot)\big) = \Big(\frac{\partial u(t, \cdot)}{\partial t}, \psi(t, \cdot)\Big) + \Big(u(t, \cdot), \frac{\partial \psi(t, \cdot)}{\partial t}\Big)$$

$$= \int \Big\langle \sum_{j=1}^n A_j \frac{\partial u}{\partial x_j} + Bu, \psi \Big\rangle \, dx$$

$$+ \int \Big\langle u, \sum_{j=1}^n \frac{\partial}{\partial x_j}(A_j^* \psi) - B^* \psi \Big\rangle dx = 0.$$

Thus $(u(t, \cdot), \psi(t, \cdot)) = \text{const}$ and consequently $(u(t, \cdot), \psi(t, \cdot)) = 0$ for all small t. We now find an analytic solution ψ of (2.21) for which $\psi(\delta, x) = P(x)$, where P is an arbitrary polynomial. If δ is sufficiently small, such a solution is defined in a fixed neighborhood of the origin in (t, x)-space by the Cauchy-Kovalevskaya Theorem (with the improved radius of convergence mentioned above).

As a result we obtain $(u(\delta, \cdot), P(\cdot)) = 0$. Since this holds for all polynomials P, we have $u(\delta, x) \equiv 0$, i.e., $u = 0$ in a neighborhood of the origin, which proves a uniqueness theorem for solutions of class C^1 of the problem (2.19), (2.20) with analytic coefficients A_j and B.

§3. Classification of Linear Differential Equations. Reduction to Canonical Form and Characteristics

3.1. Classification of Second-Order Equations and Their Reduction to Canonical Form at a Point. In the study of linear partial differential equations in mathematical physics three basic types of equations are distinguished: elliptic, parabolic, and hyperbolic. The simplest examples of these types are respectively

$$\text{Laplace's equation:} \quad \Delta u = 0, \quad \text{where } \Delta = \sum_{j=1}^{n} \partial^2/\partial x_j^2; \quad (3.1)$$

$$\text{the heat equation:} \quad u_t - \Delta u = 0; \quad (3.2)$$

$$\text{the wave equation:} \quad u_{tt} - \Delta u = 0. \quad (3.3)$$

(Equation (3.1) is considered in $\mathbb{R}^n$ and (3.2) and (3.3) in $\mathbb{R}^{n+1}$.) Consider the general linear second-order equation in $\mathbb{R}^n$

$$\sum_{i,j=1}^{n} a_{ij}(x)\frac{\partial^2 u}{\partial x_i \partial x_j} + \cdots = 0, \quad (3.4)$$

where the coefficients $a_{ij}(x) \equiv a_{ji}(x)$ are real and the dots indicate terms of lower order (terms containing only u and $\partial u/\partial x_j$ but no second derivatives of u). We introduce the quadratic form associated with (3.4)

$$\sum_{i,j=1}^{n} a_{ij}(x)\xi_i\xi_j. \quad (3.5)$$

By direct computation it can be verified that this quadratic form is invariant under a change of variables $y = f(x)$ if the vector $\xi = (\xi_1, \ldots, \xi_n)$ is transformed using the matrix T'^{-1}, which is the transposed inverse of the Jacobian matrix $T = f'(x)$ of the change of variables under consideration at the point x. In other words the quadratic form (3.5) is well-defined if we regard ξ as a cotangent vector (or a covariant vector) at the point x. In particular the invariants of linear transformations of the quadratic form (rank, number of positive coefficients, and number of negative coefficients of square terms in its canonical form) are invariant under a change of variables in the equation. If we also allow the equation (3.4) to be multiplied by a nonzero real number (or a nowhere-vanishing real-valued function), the positive and negative coefficients of the canonical form of (3.5) may yet change places. This gives meaning to the following definition.

Definition 1.4. a) Equation (3.4) is called *elliptic* at the point x if the canonical form of the quadratic form (3.5) contains n positive or n negative coefficients, i.e., the form is either positive-definite or negative-definite.

b) Equation (3.4) is called *hyperbolic* at the point x if the quadratic form (3.5) has rank n and its canonical form contains (possibly after a change of sign) $n-1$ positive coefficients and 1 negative coefficient.

c) Equation (3.4) is called *parabolic* at the point x if the quadratic form (3.5) has rank $n-1$ and becomes nonnegative-definite after a possible change of sign, i.e., its canonical form contains $n-1$ positive or $n-1$ negative coefficients.

If one of the conditions a), b), c) holds for all $x \in \Omega$, where Ω is a region in $\mathbb{R}^n$, we speak of *ellipticity, hyperbolicity,* or *parabolicity* respectively in the region Ω.

We note that the terms of first order play an important role in the study of parabolic equations. Therefore in the more detailed study of parabolic equations in Chap. 2 we shall use stronger parabolicity conditions than the condition in c) (cf. also 1.3.3).

The canonical form of the quadratic form (3.5) is determined by the eigenvalues of the symmetric matrix $\|a_{ij}(x)\|_{i,j=1}^n$. To be specific (3.4) is elliptic at a point x if and only if all the eigenvalues are of the same sign. It is hyperbolic if and only if $n-1$ of the eigenvalues are of the same sign and one is of the opposite sign. Finally, it is parabolic if one of its eigenvalues is zero and the other $n-1$ are of the same sign.

Sometimes *ultrahyperbolic equations* are used in theoretical questions (cf., for example, John 1955, Chap. V). These are equations for which the rank of the quadratic form (3.5) is n and the numbers p and q of positive and negative coefficients respectively in the canonical form are such that $p \geq 2$ and $q = n - p \geq 2$.

Equations of mixed type are also encountered in mathematical physics, i.e., equations having different type at different points of the region Ω under consideration. For example *Tricomi's Equation*

$$yu_{xx} + u_{yy} = 0, \tag{3.6}$$

considered in $\mathbb{R}^2$ is elliptic for $y > 0$, hyperbolic for $y < 0$, and parabolic on the line $y = 0$. This equation arises in describing the motion of a body in a gas with speed approximately the speed of sound: the region of ellipticity $y > 0$ corresponds to subsonic motion, and the region of hyperbolicity $y < 0$ corresponds to supersonic motion.

Fixing the point x, we can arrange for the quadratic form (3.5) to assume canonical form by a linear change of variable in (3.4). This means that the equation itself will assume the following canonical form at the point x:

$$\sum_{j=1}^{r} \pm \frac{\partial^2 u}{\partial x_j^2} + \cdots = 0, \tag{3.7}$$

where r is the rank of the quadratic form (3.5). In particular, if the initial equation was elliptic, all the signs in (3.7) will be the same, so that, changing sign if necessary, we arrive at an equation whose principal part at the point

x is the same as in Laplace's equation (3.1). For a hyperbolic equation the principal part at the point x in the canonical form will be as in the wave equation in $\mathbb{R}^n$, and for a parabolic equation the principal part will become the Laplacian on $n-1$ variables in $\mathbb{R}^n$.

In general it is not possible to reduce an equation to the form (3.7) in a whole region, as opposed to a single point, by the transformation just described, even if the equation is of constant type. For example if (3.4) is elliptic, introducing a Riemannian metric with components $g^{ij} = a_{ij}$, we see that the Laplacian of this metric has the same principal part as the operator given by the left-hand side of (3.4). Under changes of variables all the invariants of the Riemannian metric (for example the sectional curvature) are preserved. In particular a local reduction to the form (3.7) is possible if and only if the metric is locally Euclidean; and this, in turn, is equivalent to the identical vanishing of the curvature tensor. Permitting also a multiplication of the equation by a nonvanishing function, we can carry out a reduction to the form (3.7) if and only if the metric is conformally Euclidean. This also is by no means always the case when $n \geq 3$. This is heuristically clear from the fact that the principal part of (3.4) contains $n(n+1)/2$ arbitrary functions a_{ij} ($i \leq j$) and in the reduction the change of variables and multiplication by a function give only $n+1$ arbitrary functions. Thus if $n(n+1)/2 > n+1$, i.e., $n \geq 3$, one would not expect the reduction of the general equation (3.4) to the form (3.7) to be possible. For $n = 2$ no such contradiction arises, and, as we shall see below, a local reduction is possible under natural restrictions. For any n an obvious reduction to the form (3.7) in a region is possible for equations with constant coefficients in the principal part.

3.2. Characteristics of Second-Order Equations and Reduction to Canonical Form of Second-Order Equations with Two Independent Variables. In 2.3 we gave the general definition of characteristics. For a second-order linear equation of the form (3.4) a characteristic is a hypersurface Γ (a submanifold of codimension 1) in $\mathbb{R}^n$ whose normal vector $\xi = (\xi_1, \ldots, \xi_n)$ at any point $x \in \Gamma$ satisfies the condition

$$\sum_{i,j=1}^{n} a_{ij} \xi_i \xi_j = 0, \tag{3.8}$$

i.e., causes the quadratic form (3.5) associated with the equation to vanish. If the surface Γ is defined by the equation $S = 0$, where S is a real-valued function such that $\mathrm{grad}\, S \neq 0$ on Γ, then $\xi_j = \partial S / \partial x_j$ and it is clear that under a change of variable the components of the vector ξ transform like the components of a covariant vector. Therefore the concept of a characteristic is independent of the choice of curvilinear coordinates. In particular, let $n = 2$, so that the equation has the form

$$a u_{xx} + 2b u_{xy} + c u_{yy} + \cdots = 0, \tag{3.9}$$

where a, b, c, etc., are functions of x and y defined in some region. Equation (3.9) is elliptic if and only if $b^2 - ac < 0$, hyperbolic if and only if $b^2 - ac > 0$, and parabolic if and only if $b^2 - ac = 0$. The characteristics of (3.9) are the curves along which

$$a \, dy^2 - 2b \, dx \, dy + c \, dx^2 = 0. \tag{3.10}$$

(This relation is obtained by substituting the vector $(dy, -dx)$ normal to the characteristic into the associated quadratic form.) It follows from this that a hyperbolic equation (3.9) has two families of real characteristics, which can be written locally in the form $\varphi_1(x,y) = C_1$, $\varphi_2(x,y) = C_2$, where C_1 and C_2 are arbitrary constants and $d\varphi_1$ and $d\varphi_2$ are linearly independent at each point (cf. Fig. 1.4), i.e., through each point there passes precisely one characteristic of each of the two families, and the two characteristics intersect each other.

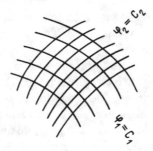

Fig. 1.4

If, for example, $a \neq 0$, then these two families of characteristics are obtained as the solutions of two differential equations

$$y' = \frac{b \pm \sqrt{b^2 - ac}}{a}; \tag{3.11}$$

then φ_1 and φ_2 are the first integrals of these equations. If we introduce new independent variables $z = \varphi_1(x,y)$ and $w = \varphi_2(x,y)$, (3.9) will contain neither u_{zz} nor u_{ww} in the new coordinates, since the lines $z = C_1$ and $w = C_2$ are characteristics. Therefore, after being divided by the coefficient of u_{zw}, it will assume the following form.

$$u_{zw} + \cdots = 0. \tag{3.12}$$

Introducing instead the coordinates $p = z + w$ and $q = z - w$, we arrive at another canonical form

$$u_{pp} - u_{qq} + \cdots = 0, \tag{3.12'}$$

which is a particular case of (3.7), but here the reduction can be done in the entire region.

If (3.9) is parabolic everywhere in some region, it has a single family of real characteristics $\varphi(x,y) = C$, where $d\varphi \neq 0$. Then in the new coordinates

$z = \varphi(x, y)$, $w = \psi(x, y)$, where ψ is any function with the property that $d\varphi$ and $d\psi$ are linearly independent, (3.9) will not contain u_{zz}. But then it will also not contain u_{zw}, since the rank of the quadratic form must remain equal to 1. Therefore it assumes the following canonical form:

$$u_{ww} + \cdots = 0. \tag{3.13}$$

If (3.9) is elliptic, it has no real characteristics, but there may be complex characteristics. For simplicity let the coefficients a, b, and c be analytic; then there exists a first integral $\varphi_1(x, y) + i\varphi_2(x, y) = C$ of one of the equations (3.11) such that φ_1 and φ_2 are real, and $d\varphi_1 \neq 0$ (then $d\varphi_2 \neq 0$ automatically and the differentials $d\varphi_1$ and $d\varphi_2$ are linearly independent). If $z = \varphi_1 + i\varphi_2$, then arguments formally the same as in the hyperbolic case show that in the variables z and $\bar{z}$ (3.9) assumes the form

$$u_{z\bar{z}} + \cdots = 0, \tag{3.14}$$

or in the variables $p = \varphi_1(x, y)$ and $q = \varphi_2(x, y)$

$$u_{pp} + u_{qq} + \cdots = 0. \tag{3.14'}$$

By applying more delicate methods (Courant and Hilbert 1962), we can reduce an elliptic equation to the form (3.14) or (3.14'), provided that the coefficients of (3.9) satisfy a Lipschitz condition.

If the lower-order terms in the canonical form (3.12') or (3.13) are absent, then the corresponding equation is solvable: for a hyperbolic equation the general solution will locally have the form

$$u = f(z) + g(w), \tag{3.15}$$

and for a parabolic equation it will have the form

$$u = f(z) + g(z)w, \tag{3.16}$$

where f and g are arbitrary functions of one variable. One can often take account of the lower-order terms in the hyperbolic case by perturbation theory – solving a suitable boundary-value problem for an exact equation by the method of successive approximations. This serves as the basis of Riemann's method (cf., for example, Smirnov 1981). For an elliptic equation we obtain Laplace's equation when the lower-order terms are absent; its solution is locally the sum of an analytic and a conjugate-analytic function.

$$u = f(z) + g(\bar{z}), \tag{3.17}$$

which are convergent power series in z and $\bar{z}$ respectively.

3.3. Ellipticity, Hyperbolicity, and Parabolicity for General Linear Differential Equations and Systems. Consider the general linear differential operator

$$A = \sum_{|\alpha| \le m} a_\alpha(x) D^\alpha \tag{3.18}$$

in a region $\Omega \subset \mathbb{R}^n$ (cf. the notation in 1.1) and the corresponding equation

$$Au = f. \tag{3.19}$$

We introduce the *principal symbol* of the operator A, defined by the formula

$$a_m(x, \xi) = \sum_{|\alpha| = m} a_\alpha(x) \xi^\alpha. \tag{3.20}$$

For $m = 2$ it differs only in sign from the quadratic form (3.5) used above. The principal symbol is a well-defined function on the cotangent bundle $T^*\Omega$; this means that if the components of the vector ξ transform like the components of a cotangent vector at the point x in a change of variable, the value of the principal symbol will not change.

Definition 1.5. The operator (3.18) and the equation (3.19) are called *elliptic at the point x* if $a_m(x, \xi) \ne 0$ for all $\xi \in \mathbb{R}^n \setminus \{0\}$. If this holds for all $x \in \Omega$, the operator A and (3.19) are called *elliptic in the region Ω* or simply *elliptic*.

Instead of the scalar functions a_α one can consider $N \times N$-matrix-valued functions. Then (3.19) turns into a system of N equations with N unknown functions (in this case u and f are N-component vector-valued functions on Ω), and the principal symbol becomes a matrix function. In this case the matrix operator and system (3.19) are called *elliptic* (or *Petrovskij-elliptic*) *at the point x* if

$$\det a_m(x, \xi) \ne 0, \quad \xi \in \mathbb{R}^n \setminus \{0\}, \tag{3.21}$$

and *elliptic in the region Ω* (or simply *elliptic*) if this holds for all $x \in \Omega$. I. G. Petrovskij has studied still more general systems, which he also called *elliptic*. These systems are defined by condition (3.21), but the matrix a_m is constructed differently. It contains polynomials in ξ of maximal order m_j in the jth column of the matrix, $j = 1, \dots, N$. If a_α is a rectangular matrix of dimension $N \times N_1$, then (3.19) becomes a system of N equations in N_1 unknown functions. The case of *overdetermined systems*, when $N > N_1$, occurs frequently. Such a system is called *elliptic* (at the point $x \in \Omega$) if for $\xi \in \mathbb{R}^n \setminus \{0\}$ the matrix $a_m(x, \xi)$ has maximal rank N_1 (or, what is the same, does not annihilate any nonzero vector in $\mathbb{R}^{N_1}$). Examples of overdetermined elliptic systems are the Cauchy-Riemann system $\bar\partial u = f$, where u is a scalar function (cf. Sect. 1.1) and the system of electrostatic equations

$$\operatorname{div}(\varepsilon \mathbf{E}) = 4\pi\rho, \quad \operatorname{curl} \mathbf{E} = 0, \quad \mathbf{E} = (E_1, E_2, E_3), \tag{3.22}$$

which is obtained from the Maxwell equations (1.20) by assuming that all the fields are independent of t.

In addition we encounter systems of equations in which it is natural to ascribe different orders to different components of the vector u and the vector f on the right-hand side. The corresponding concept of ellipticity is called *Douglis-Nirenberg ellipticity* and consists of the following. One must assume that $A = (A_{ij})_{i,j=1}^{N}$, where A_{ij} is a differential operator of order $s_j - t_i$, $(s_1, \ldots, s_N)$ and $(t_1, \ldots, t_N)$ being certain collections of integers. We form the matrix $\sigma(x,\xi) = (a_{s_j-t_i}^{(ij)}(x,\xi))_{i,j=1}^{N}$ consisting of the principal symbols $a_{s_j-t_i}^{(ij)}$ (homogeneous of degree $s_j - t_i$ in the variable ξ) of the operators A_{ij}. Then Douglis-Nirenberg ellipticity of the operator A and the system (3.19) (at the point x) means that $\det \sigma(x,\xi) \neq 0$ for $\xi \in \mathbb{R}^n \setminus \{0\}$.

Hyperbolicity of an equation or a system is usually defined in the presence of a distinguished variable (which is usually the time variable) or at least a distinguished direction (when a distinguished variable t is present, this direction is taken as the direction of the t-axis).

Definition 1.6. An operator A of the form (3.18) and the equation (3.19) are called *hyperbolic in the direction of the vector* ν (at the point x) if $a_m(x,\nu) \neq 0$ (i.e., the direction ν is noncharacteristic) and for any vector $\xi \in \mathbb{R}^n$ not proportional to ν all the roots λ of the equation

$$a_m(x,\xi + \lambda\nu) = 0 \qquad (3.23)$$

are real. The operator A and the equation (3.19) are called *strictly hyperbolic in the direction of the vector* ν (at the point x) if all the roots of (3.23) (there are m of them, by virtue of the characteristic condition) are real and distinct.

If the condition of hyperbolicity (or strict hyperbolicity) in the direction ν holds for all $x \in \Omega$, reference to the point x is dropped and we speak of hyperbolicity (or strict hyperbolicity) in Ω. We note that the requirement of hyperbolicity (or strict hyperbolicity) acquires an invariant meaning (independent of the choice of coordinates) if instead of a single vector we consider a covector field $\nu = \nu(x)$.

If a second-order equation (of the form (3.4)) is hyperbolic in the sense of Definition 1.4, then it is strictly hyperbolic in the sense of Definition 1.6 in any direction ν that is timelike, i.e., such that the corresponding quadratic form (whose canonical form contains 1 positive and $n - 1$ negative squares) is positive at this vector.

Hyperbolicity for a matrix-valued operator A of the form (3.18) (of dimension $N \times N$) and the corresponding system (3.19) is defined similarly: the condition for being noncharacteristic has the form $\det a_m(x,\nu) \neq 0$, and instead of (3.23) in this case one must consider the equation

$$\det a_m(x,\xi + \lambda\nu) = 0. \qquad (3.23')$$

First-order systems with a distinguished variable t are frequently encountered. They have the form

$$\frac{\partial u}{\partial t} + \sum_{j=1}^{n} A_j \frac{\partial u}{\partial x_j} + Bu = f, \tag{3.24}$$

where u is an N-component vector-valued function, A_j and B are $N \times N$ matrices (depending on t and x), and f is a known vector-valued function of t and x. The condition of hyperbolicity (resp. strict hyperbolicity) of such a system (with respect to the direction of the t-axis) means that for any real $\xi_1, \ldots, \xi_N$ all the eigenvalues of the matrix $\sum_{j=1}^{N} \xi_j A_j$ are real (resp. real and distinct). In particular if all the matrices A_j are symmetric, the system (3.24) is hyperbolic (such systems are called *symmetric hyperbolic systems*).

We now give the most frequently used definition of a parabolic equation or system. In describing parabolicity, as a rule, one assumes the presence of a distinguished variable t. The equation

$$\frac{\partial u}{\partial t} = \sum_{|\alpha| \le 2b} a_\alpha(t,x) D_x^\alpha u + f(t,x) \tag{3.25}$$

is called *parabolic* or *2b-parabolic* (*Petrovskij 2b-parabolic*) if

$$\mathrm{Re} \sum_{|\alpha|=2b} a_\alpha(t,x)\xi^\alpha < 0, \quad \xi \in \mathbb{R}^n \setminus \{0\}. \tag{3.26}$$

The more general equation

$$\frac{\partial^p u}{\partial t^p} = \sum_{\substack{|\alpha|+2b\alpha_0 \le 2bp \\ \alpha_0 < p}} a_{\alpha,\alpha_0}(t,x) D_x^\alpha D_t^{\alpha_0} u + f(t,x) \tag{3.27}$$

is called *parabolic* or *Petrovskij 2b-parabolic* if all the roots $\lambda_j = \lambda_j(t,x,\xi)$ of the equation

$$\lambda^p - \sum_{|\alpha|+2b\alpha_0=2bp} a_{\alpha,\alpha_0}(t,x)\xi^\alpha \lambda^{\alpha_0} = 0 \tag{3.28}$$

satisfy the condition

$$\mathrm{Re}\, \lambda_j(t,x,\xi) < 0, \quad \xi \in \mathbb{R}^n \setminus \{0\}. \tag{3.29}$$

Finally for the system of equations

$$\frac{\partial^{p_j} u_j}{\partial t^{p_j}} = \sum_{k=1}^{N} \sum_{\substack{|\alpha|+2b\alpha_0 \le 2bp_k \\ \alpha_0 < p_k}} a_{\alpha,\alpha_0}^{j,k}(t,x) D_x^\alpha D_t^{\alpha_0} u_k + f_j(t,x), \quad j=1,\ldots,N,$$

$$\tag{3.30}$$

the condition of Petrovskij $2b$-parabolicity means that inequality (3.29) holds for all the roots λ_j of the equation

$$\det \left\| \sum_{|\alpha|+2b\alpha_0=2bp_k} a^{j,k}_{\alpha,\alpha_0}(t,x)\xi^\alpha\lambda^{\alpha_0} - \lambda^{n_j}\delta_{jk} \right\|^N_{j,k=1} = 0, \quad \xi \in \mathbb{R}^n\setminus\{0\}. \quad (3.31)$$

3.4. Characteristics as Solutions of the Hamilton-Jacobi Equation. In studying equations with two independent variables we have seen that an important role is played by characteristics and how useful it is to know the families of characteristics of the form $S(x) = C$, where C is an arbitrary constant. For the general equation (3.19) of high order such a family of level lines of the function S consists of characteristics if and only if the function S satisfies the *Hamilton-Jacobi equation*

$$a_m\left(x, \frac{\partial S}{\partial x}\right) = 0, \qquad (3.32)$$

where a_m is the principal symbol of the operator A defining the equation under consideration and $\frac{\partial S}{\partial x} = (\frac{\partial S}{\partial x_1}, \ldots, \frac{\partial S}{\partial x_n})$. The characteristics of general hyperbolic equations also play an important role: they define the wave fronts that serve to describe the connection between wave optics and geometrical optics. The solutions of (3.32) are needed to find the short-wave asymptotics of the solutions of hyperbolic equations. We shall now briefly describe the method of integrating the general *Hamilton-Jacobi equation* (for details cf. Hörmander 1983–1985).

$$H\left(x, \frac{\partial S}{\partial x}\right) = 0. \qquad (3.33)$$

(3.33) is closely connected with the Hamiltonian system of $2n$ ordinary differential equations

$$\begin{cases} \dot{x} = \dfrac{\partial H}{\partial \xi}, \\ \dot{\xi} = -\dfrac{\partial H}{\partial x}. \end{cases} \qquad (3.34)$$

To be specific, if S is a solution of (3.33), then the graph of the gradient (or differential) of the function S

$$\Gamma = \left\{ (x, S_x(x)) \right\}$$

(where $S_x = \partial S/\partial x$) is invariant with respect to the flow defined by the system (3.34), since if $(x(t), \xi(t))$ is a solution of (3.34), and $\xi(0) = S_x(x(0))$, then

$$\frac{d}{dt}\Big(\xi(t) - S_x\big(x(t)\big)\Big)\Big|_{t=0} = \dot{\xi}(0) - S_{xx}\big(x(0)\big)\dot{x}(0)$$

$$= -H_x\big(x(0)\big) - S_{xx}\big(x(0)\big)H_\xi\big(x(0), S_x(x(0))\big)$$

$$= -\frac{d}{dx}H\big(x, S_x(x)\big)\Big|_{x=x(0)} = 0.$$

Now let $\mathcal{M}$ be a smooth $(n-1)$-dimensional manifold in $\mathbb{R}^n$ and let the values of the function S be defined on $\mathcal{M}$. The derivatives of the function S in the directions tangent to $\mathcal{M}$ can be computed directly, while the derivative along the normal to $\mathcal{M}$ must be found from (3.33). Suppose such values of the derivatives of the function S on $\mathcal{M}$ are defined and are smooth functions. Then an $(n-1)$-dimensional submanifold $\Gamma_0 = \{(x, \partial S/\partial x), x \in \mathcal{M}\}$ over $\mathcal{M}$ arises in $\mathbb{R}^{2n}$. Let the integral curves of the Hamiltonian system (3.34) passing through points of Γ_0 be such that their projections on the space of the variables x are transversal to $\mathcal{M}$. Then we can recover Γ over some neighborhood in $\mathcal{M}$ and thereby also recover the function S itself over some neighborhood of the manifold $\mathcal{M}$ by the formula

$$S(x) = S(x_0) + \int_{x_0}^{x} \xi\, dx = S(x_0) + \int_{x_0}^{x} (\xi_1\, dx_1 + \cdots + \xi_n\, dx_n), \qquad (3.35)$$

where the integral extends along the trajectory of the Hamiltonian system (3.34) starting at the point $(x_0, \xi_0) \in \Gamma_0$ and ending at the point (x, ξ) (the point (x_0, ξ_0) is chosen using this condition on the trajectory).

We note that in the case of most importance for us, when the function $H = H(x, \xi)$ is positive-homogeneous in ξ (of any degree m), the solution $S = S(x)$ of the Hamilton-Jacobi equation (3.33) is constant along the projections $x(t)$ of the trajectories $(x(t), \xi(t))$ of the Hamiltonian system (3.34), lying on the graph of the gradient of the function S, since by Euler's identity

$$\frac{d}{dt}S\big(x(t)\big) = S_x\big(x(t)\big) \cdot \dot{x}(t) = \xi(t) \cdot H_\xi\big(x(t)\big) = mH\big(x(t), \xi(t)\big) = 0.$$

If the function H is the principal symbol of the operator A, so that the Hamilton-Jacobi equation (3.33) is the equation of the characteristics, then the trajectories of the Hamiltonian system (3.34) are called *bicharacteristics* (of the function H, the operator A, or the equation $Au = f$).[2]

---

[2] The name *bicharacteristics* is often limited to the trajectories along which $H = 0$ (in earlier works the latter were called *null-bicharacteristics*).

Chapter 2. The Classical Theory

§1. Distributions and Equations with Constant Coefficients

1.1. The Concept of a Distribution (cf. Schwartz 1950–1951). In analysis
and mathematical physics one frequently encounters difficulties connected
with the nondifferentiability of various functions. The theory of distributions
makes it possible to get rid of these difficulties (at least in studying linear
differential equations with sufficiently smooth coefficients). Many concepts
and theorems assume greater simplicity in the theory of distributions and
are freed of inessential restrictions.

The origin of the concept of a distribution can be explained, for example, as follows. Suppose there is a physical quantity $f(x)$ that is a function
of the point x in the space $\mathbb{R}^n$ (for example, temperature, pressure, or the
like). If we wish to measure this quantity at the point x_0 using some device
(a thermometer, a manometer, etc.), then we are actually measuring some
average value of $f(x)$ taken over a neighborhood of the point x_0 – an integral
$\int f(x)\varphi(x)\,dx$, where $\varphi(x)$ is a function characterizing the measuring device
and "smeared" over a neighborhood of the point x_0. The idea arises of dis-
pensing entirely with the function $f(x)$ and considering instead the linear
functional that assigns to each test function φ the number [1]

$$\langle f, \varphi \rangle = \int f(x)\varphi(x)\,dx. \tag{1.1}$$

Considering now arbitrary linear functionals (not necessarily of this form),
we arrive at the concept of a *distribution*.

The simplest and at the same time the most important example of a distribution not defined by a formula of the form (1.1) with an ordinary function
$f(x)$ is the *Dirac δ-function*. Dirac himself described it as a function $\delta(x)$
such that $\delta(x) = 0$ for $x \neq 0$, $\delta(0) = \infty$, and $\int \delta(x)\,dx = 1$. The fundamental
property of the δ-function is the equality

$$\int \delta(x)\varphi(x)\,dx = \varphi(0),$$

from which it can be seen that the δ-function can be understood as the
functional assigning to each test function $\varphi(x)$ the number $\varphi(0)$.

---

[1] Here and in what follows dx is standard Lebesgue measure and in all cases where the
region of integration is not indicated, the integration extends over all x belonging to the
natural region of definition of the integrand.

1.2. The Spaces of Test Functions and Distributions (cf. Schwartz 1950–1951; Gel'fand and Shilov 1958–1959; Edwards 1965; Shilov 1965; Rudin 1973; Vladimirov 1979; Hörmander 1983–1985). The choice of the spaces of *test functions* and distributions is determined by the problem under consideration. We indicate here the simplest methods of making such a choice, leaving other possibilities for a more specialized article.

Let Ω be an open subset of $\mathbb{R}^n$. We introduce the following notation:

$\mathcal{E}(\Omega) = C^\infty(\Omega)$ is the space of infinitely differentiable functions in Ω;

$\mathcal{D}(\Omega) = C_0^\infty(\Omega)$ is the space of infinitely differentiable functions with compact support contained in Ω, i.e., functions $\varphi \in C^\infty(\Omega)$ for which there exists a compact set $K \subset \Omega$ such that $\varphi|_{\Omega \setminus K} = 0$.

In general the *support* of a continuous function $\varphi : \Omega \to \mathbb{C}$ is defined as the closure (in Ω) of the set of $x \in \Omega$ such that $\varphi(x) \neq 0$. The support of a function φ is denoted $\operatorname{supp} \varphi$. Thus $\operatorname{supp} \varphi$ is the smallest closed set $F \subset \Omega$ for which $\varphi|_{\Omega \setminus F} = 0$ or, what is the same, the complement of the largest open set $G \subset \Omega$ for which $\varphi|_G = 0$. The space $\mathcal{D}(\Omega)$ thus consists of precisely those $\varphi \in C^\infty(\Omega)$ for which $\operatorname{supp} \varphi$ is a compact subset of Ω.

If K is a compact set in $\mathbb{R}^n$, we introduce the further notation $\mathcal{D}(K) = C_0^\infty(K)$ for the space of functions $\varphi \in C^\infty(\mathbb{R}^n)$ such that $\operatorname{supp} \varphi \subset K$.

It is clear that $\mathcal{D}(\Omega)$ is the union of the spaces $\mathcal{D}(K)$ over all compact subsets $K \subset \Omega$.

Finally let $\mathcal{S}(\mathbb{R}^n)$ be the *Schwartz space* consisting of functions that, together with all their derivatives, decay faster than any power of $|x|$ as $|x| \to \infty$, i.e., functions $\varphi \in C^\infty(\mathbb{R}^n)$ such that $\sup\limits_{x \in \mathbb{R}^n} |x^\alpha D^\beta \varphi(x)| < +\infty$ for any multi-indices α and β.

The spaces $\mathcal{D}(\Omega)$, $\mathcal{E}(\Omega)$, and $\mathcal{S}(\mathbb{R}^n)$ will be used as the spaces of test functions, and distributions will be linear functionals on these spaces. However, we do not need all the linear functionals, only those that are continuous with respect to the natural topologies of these spaces. We shall now describe these topologies.

The space $\mathcal{E}(\Omega)$ is a Fréchet space (a complete countably-normed space) whose topology can be given using the seminorms

$$p_{m,K}(\varphi) = \sum_{|\alpha| \leq m} \sup_{x \in K} |D^\alpha \varphi(x)|,$$

where K is a compact set in Ω and $m \in \mathbb{Z}_+$. We recall that this means that a fundamental system of neighborhoods of zero consists of all sets

$$\mathcal{U}_{m,K,\varepsilon} = \{\varphi : \varphi \in \mathcal{E}(\Omega), \quad p_{m,K}(\varphi) < \varepsilon\},$$

where $\varepsilon > 0$. Of course we can restrict ourselves to the collection $\varepsilon_n = 1/n$, $n = 1, 2, 3, \ldots$, and also to a countable system of compact sets $K_1, K_2, K_3, \ldots$, if it is chosen so that $K_1 \subset K_2 \subset K_3 \subset \cdots$ and for any point $x_0 \in \Omega$ there exists j such that x_0 is an interior point of the compact set K_j. Like any topology that is defined using a countable number of seminorms, the topology in $\mathcal{E}(\Omega)$ just described can be defined using the metric

$$\rho(\varphi, \psi) = \sum_{l=1}^{\infty} \frac{1}{2^l} \frac{p_l(\varphi - \psi)}{1 + p_l(\varphi - \psi)},$$

where $\{p_l \mid l = 1, 2, \ldots\}$ is a system of seminorms defining the topology (in the present case we can take $p_l = p_{l, K_l}$). Convergence of a sequence is described in terms of the seminorms as follows: $\varphi_k \to \varphi$ as $k \to \infty$ if $\lim\limits_{k \to \infty} p_l(\varphi_k - \varphi) = 0$ for each fixed $l = 1, 2, \ldots$ In the case of the spaces $\mathcal{E}(\Omega)$ this means that for any multi-index α and any compact set K the sequence $\{D^\alpha \varphi_k \mid k = 1, 2, \ldots\}$ converges to $D^\alpha \varphi$ uniformly on K. Therefore the topology described in $\mathcal{E}(\Omega)$ is called the *topology of uniform convergence together with derivatives on compact sets*.

Now consider the *set* $\mathcal{E}'(\Omega)$ of continuous linear functionals on the space $\mathcal{E}(\Omega)$ (or, as it is called for short, the *dual* or *conjugate* space to the space $\mathcal{E}(\Omega)^2$). The value of the functional f at the element φ will be denoted here and below by $\langle f, \varphi \rangle$. The continuity of the functional $f \in \mathcal{E}'(\Omega)$ is described in the usual manner in terms of the topology or in terms of convergence. (For example, in terms of convergence continuity means that if $\varphi_k \to \varphi$ in $\mathcal{E}(\Omega)$, then $\langle f, \varphi_k \rangle \to \langle f, \varphi \rangle$; of course because of linearity it suffices to verify continuity at the point $\varphi = 0$.) In addition continuity can be described as follows in terms of the seminorms: there exist l and C such that

$$|\langle f, \varphi \rangle| \leq C p_l(\varphi),$$

where $\varphi \in \mathcal{E}(\Omega)$ is arbitrary and l and C are independent of φ.

If $f \in L^1_{\mathrm{comp}}(\Omega)$, i.e, $f \in L^1(\Omega)$ and there exists a compact set $K \subset \Omega$ such that $f|_{\Omega \setminus K} = 0$ (in this case we shall say that f has compact support), then f defines a functional $f \in \mathcal{E}'(\Omega)$ by formula (1.1). Thus we obtain an imbedding $L^1_{\mathrm{comp}}(\Omega) \subset \mathcal{E}'(\Omega)$.

The space $\mathcal{S}(\mathbb{R}^n)$ also has a natural Fréchet space topology defined by the system of seminorms

$$p_l(\varphi) = \sum_{|\alpha|, |\beta| \leq l} \sup_{x \in \mathbb{R}^n} |x^\alpha D^\beta \varphi(x)|, \quad l = 1, 2, 3, \ldots$$

The space dual to it is denoted by $\mathcal{S}'(\mathbb{R}^n)$. Its elements are called *tempered distributions*. Among them, in particular, are all the ordinary measurable functions $f(x)$ that satisfy the inequality

$$|f(x)| \leq C(1 + |x|)^N \tag{1.2}$$

---

[2] In general for any topological vector space E we shall denote by E' the dual space to it, which consists of the continuous linear functionals on E.

with some constants C and N (they define functionals belonging to $\mathcal{S}'(\mathbb{R}^n)$ according to the formula (1.1)), as well as the functions belonging to $L^p(\mathbb{R}^n)$ for some $p \in [1, +\infty]$.

On the space $\mathcal{D}(K)$, where K is a compact subset of $\mathbb{R}^n$, we introduce the Fréchet topology induced by the topology of $\mathcal{E}(\mathbb{R}^n)$, i.e., the topology defined by the seminorms $p_{m,K}$, $m = 1, 2, 3, \ldots$, while on the space $\mathcal{D}(\Omega)$ we introduce the *inductive limit topology* of the spaces $\mathcal{D}(K_l)$, $l = 1, 2, \ldots$, where $K_1 \subset K_2 \subset K_3 \subset \cdots$ is the same sequence of compact sets as described above. To be specific, in this case $\mathcal{D}(\Omega) = \bigcup_{l=1}^{\infty} \mathcal{D}(K_l)$, and a convex *balanced*[3] set $\mathcal{U} \subset \mathcal{D}(\Omega)$ is considered to be a neighborhood of zero if and only if its intersection with each of the spaces $\mathcal{D}(K_l)$ is a neighborhood of zero in $\mathcal{D}(K_l)$ (in the Fréchet topology of the latter). It can be shown that convergence of the sequence $\varphi_k \to \varphi$ in $\mathcal{D}(\Omega)$ is equivalent to the existence of a compact set $K \subset \Omega$ such that $\varphi_k \in \mathcal{D}(K)$ for all k while $\varphi_k \to \varphi$ in $\mathcal{D}(K)$. It can also be established that a linear functional f on $\mathcal{D}(\Omega)$ is continuous if and only if its restriction to any subspace $\mathcal{D}(K)$ is continuous (in the Fréchet topology of the latter space). Therefore the inclusion $f \in \mathcal{D}'(\Omega)$ means that f is a linear functional on $\mathcal{D}(\Omega)$ and the condition $\lim_{k \to \infty} \varphi_k = 0$ in $\mathcal{D}(\Omega)$ implies that $\lim_{k \to \infty} \langle f, \varphi_k \rangle = 0$.

A function $f \in L^1_{\mathrm{loc}}(\Omega)$ (i.e., a function f that is integrable over every compact set $K \subset \Omega$) defines a functional $f \in \mathcal{D}'(\Omega)$ by formula (1.1). We thus obtain an imbedding $L^1_{\mathrm{loc}}(\Omega) \subset \mathcal{D}'(\Omega)$.

By abuse of notation we often write a distribution as an ordinary function $f(x)$, and instead of $\langle f, \varphi \rangle$ we write $\int f(x)\varphi(x)\,dx$. It is clear that no confusion can arise, since for ordinary functions $\langle f, \varphi \rangle$ is defined by just such an integral. When this is done the formulas relating to properties of the Dirac δ-function, for example, acquire a precise meaning. The Dirac function (cf. Sect. 1.1) can be regarded as an element of $\mathcal{D}'(\Omega)$ or $\mathcal{E}'(\Omega)$ in the case when $0 \in \Omega$, and also as an element of $\mathcal{S}'(\mathbb{R}^n)$.

There are many other useful spaces besides these spaces of test functions and distributions. For example, in the definition of the spaces $\mathcal{E}(\Omega)$ and $\mathcal{D}(\Omega)$ we can take the functions whose derivatives satisfy definite estimates (with constants depending in a definite way on the indices of the derivatives) instead of all smooth functions, for example functions of the so-called Gevrey classes (or ultradifferentiable functions). The corresponding continuous linear functionals are usually called *ultradistributions*. As the space of test functions, one can also take some space of analytic functions. In particular, following this route, one could obtain the *hyperfunctions* as the continuous linear functionals. For more information on the theory of ultradistributions and hyperfunctions we refer the reader to specialized articles and monographs (cf., for example, Komatsu 1977, Hörmander 1983–1985).

---

[3] This means that if $\varphi \in \mathcal{U}$, then $\lambda\varphi \in \mathcal{U}$ for all $\lambda \in \mathbb{C}$ with $|\lambda| \leq 1$.

1.3. The Topology in the Space of Distributions (cf. Schwartz 1950–1951; Gel'fand and Shilov 1958–1959; Edwards 1965; Shilov 1965; Rudin 1973; Hörmander 1983–1985). It is useful to introduce a topology in the spaces of distributions. This is done by the standard methods of the theory of topological vector spaces. There are several ways of introducing a topology. The most important for us is the *weak topology* defined by the seminorms

$$p_\varphi(f) = |\langle f, \varphi \rangle|, \quad \varphi \in E, \quad f \in E',$$

where $E = \mathcal{D}(\Omega)$, $\mathcal{E}(\Omega)$, or $\mathcal{S}(\mathbb{R}^n)$, and E' is the corresponding conjugate space of distributions. In the majority of cases one can use *weak convergence*, which we shall refer to as simply *convergence*, instead of this topology. Weak convergence is defined as follows: if $\{f_k | k = 1, 2, 3, \dots\}$ is a sequence of functionals of E', we shall write that $f_k \to f$ if $\langle f_k, \varphi \rangle \to \langle f, \varphi \rangle$ for any $\varphi \in E$. The functional f so defined is obviously linear. An important fact deducible from uniform boundedness principles (theorems of Banach-Steinhaus type) is that this implies the continuity of the functional f, i.e., $f \in E'$. This fact is called *weak completeness* of the space E' (or more precisely *sequential weak completeness* of this space).

We give an important example of convergence of distributions. Let $\varphi \in C_0^\infty(\mathbb{R}^n)$, $\varphi \geq 0$ everywhere, $\varphi(x) = 0$ for $|x| \geq 1$, and $\int \varphi(x)\, dx = 1$. Set $\varphi_\varepsilon(x) = \varepsilon^{-n}\varphi(x/\varepsilon)$, where $\varepsilon > 0$. Then $\varphi_\varepsilon \geq 0$, $\operatorname{supp}\varphi_\varepsilon \subset \{x | |x| \leq \varepsilon\}$, and $\int \varphi_\varepsilon(x)\, dx = 1$. It is easy to verify that $\varphi_\varepsilon(x) \to \delta(x)$ as $\varepsilon \to +0$ in $\mathcal{D}'(\Omega)$ and $\mathcal{E}'(\Omega)$ (we assume that $0 \in \Omega$) or in $\mathcal{S}'(\mathbb{R}^n)$.

We shall call a family $\{\varphi_\varepsilon\}$ of the structure just described a δ-shaped family of functions. It is often more convenient to consider instead of such a family a δ-shaped sequence $\psi_k(x) = \varphi_{1/k}(x)$, $k = 1, 2, \dots$, $\psi_k(x) \to \delta(x)$ as $k \to \infty$. In general a δ-shaped sequence is often defined as a sequence of smooth functions $\psi_k(x)$ such that $\psi_k(x) \to \delta(x)$ as $k \to \infty$. Such sequences often arise in analysis. For example, in the theory of Fourier series it is proved that the *Dirichlet kernel*

$$D_k(t) = \frac{1}{2\pi} \frac{\sin(k + \frac{1}{2})t}{\sin \frac{t}{2}},$$

defined by the condition that $\langle D_k, \varphi \rangle$ is the kth partial sum of the Fourier series of the function φ at $t = 0$, has the property of being δ-shaped, for example, in $\mathcal{D}'\big((-\pi, \pi)\big)$. Similarly the *Fejér kernel*

$$F_k(t) = \frac{1}{2\pi k} \frac{\sin^2 \frac{k}{2}t}{\sin^2 \frac{t}{2}},$$

defined by the condition that $\langle F_k, \varphi \rangle$ is the arithmetic mean of the first k partial sums of the Fourier series at $t = 0$, forms a δ-shaped sequence.

Using δ-shaped families or sequences we can approximate distributions by smooth distributions. To be specific, consider, for example, the *average* of the distribution $f \in \mathcal{E}'(\Omega)$

$$f_\varepsilon(x) = \int f(y)\varphi_\varepsilon(x - y)\, dy = \langle f, \varphi_\varepsilon(x - \cdot)\rangle.$$

It is easy to verify that $f_\varepsilon \in \mathcal{D}(\Omega)$. Here it can be shown that $f_\varepsilon \to f$ in $\mathcal{E}'(\Omega)$ as $\varepsilon \to +0$. Thus $\mathcal{D}(\Omega)$ is dense in $\mathcal{E}'(\Omega)$ in the weak topology of $\mathcal{E}'(\Omega)$. Combining the average with a cutoff by functions of increasing support, one can show easily that $\mathcal{D}(\Omega)$ is dense in $\mathcal{D}'(\Omega)$ and $\mathcal{D}(\mathbb{R}^n)$ is dense in $\mathcal{S}'(\mathbb{R}^n)$ (again in the corresponding weak topologies). These facts can be used, for example, to prove various properties of distributions "by continuity," i.e., starting from the corresponding properties of ordinary functions and passing to the limit.

Passage to the limit can also be used to define certain distributions. In this way, for example, one can define continuous linear functionals on $\mathbb{R}$:

$$\frac{1}{x + i0} = \lim_{\varepsilon \to +0} \frac{1}{x + i\varepsilon}, \qquad \frac{1}{x - i0} = \lim_{\varepsilon \to +0} \frac{1}{x - i\varepsilon}$$

(the limits exist, for example, in $\mathcal{S}'(\mathbb{R})$). The *Sokhotskij formulas* hold

$$\frac{1}{x + i0} = PV\left\{\frac{1}{x}\right\} - \pi i \delta(x), \qquad \frac{1}{x - i0} = PV\left\{\frac{1}{x}\right\} + \pi i \delta(x),$$

where $PV\left\{\dfrac{1}{x}\right\}$ is the continuous linear functional (in $\mathcal{S}'(\mathbb{R})$) defined by the formula

$$\left\langle PV\left\{\frac{1}{x}\right\}, \varphi\right\rangle = \lim_{\varepsilon \to +0} \int\limits_{|x| \geq \varepsilon} \frac{\varphi(x)}{x}\, dx.$$

In particular

$$\frac{1}{x + i0} - \frac{1}{x - i0} = -2\pi i \delta(x).$$

The distributions $\dfrac{1}{x \pm i0}$ and $PV\left\{\dfrac{1}{x}\right\}$ are different "regularizations" of the nonintegrable function $\dfrac{1}{x}$, i.e., they make it possible to give a meaning to the divergent integral $\displaystyle\int\limits_{-\infty}^{\infty} \frac{1}{x}\varphi(x)\, dx$. We see that this can be done in more than one way, so that the nonintegrable function $\dfrac{1}{x}$ can be associated with many distributions. The procedure of regularization is important if we wish to use $\dfrac{1}{x}$ as a distribution (for example, if we wish to differentiate it). Some such procedure is applicable to many other nonintegrable functions. Several regularization methods will be considered below.

We note also that the usual forms of convergence of locally integrable functions as a rule imply their convergence as distributions. For example, if $f_k \in L^1_{\text{loc}}(\Omega)$, $k = 1, 2, \ldots$ and $f_k \to f$ in the space $L^1(K)$ as $k \to \infty$, for any compact subset $K \subset \Omega$, then $f_k \to f$ in $\mathcal{D}'(\Omega)$. If the functions f_k satisfy the estimate (1.2) on $\mathbb{R}^n$ uniformly in k and $f_k(x) \to f(x)$ for almost all x (or $f_k \to f$ in the space $L^1(K)$ for any compact set $K \subset \mathbb{R}^n$), then $f_k \to f$ in $\mathcal{S}'(\mathbb{R}^n)$.

1.4. The Support of a Distribution. The General Form of Distributions (cf. Schwartz 1950–1951; Shilov 1965; Rudin 1973; Hörmander 1983–1985). Let Ω_1 and Ω_2 be two open subsets of $\mathbb{R}^n$ with $\Omega_1 \subset \Omega_2$. Then $\mathcal{D}(\Omega_1) \subset \mathcal{D}(\Omega_2)$ and if $f \in \mathcal{D}'(\Omega_2)$, we can restrict the functional f to $\mathcal{D}(\Omega_1)$ so as to obtain a distribution $f\big|_{\Omega_1} \in \mathcal{D}'(\Omega_1)$. The restriction operation so obtained possesses the following properties:

a) If $\Omega_1 = \Omega_2$, then $f\big|_{\Omega_1} = f$.

b) If $\Omega_1 \subset \Omega_2 \subset \Omega_3$ and $f \in \mathcal{D}'(\Omega_3)$, then

$$\left(f\big|_{\Omega_2}\right)\big|_{\Omega_1} = f\big|_{\Omega_1}.$$

c) Suppose a covering of the open set Ω by open sets Ω_j, $j \in \mathcal{J}$, is given, i.e., $\Omega = \bigcup_{j \in \mathcal{J}} \Omega_j$. Then $f \in \mathcal{D}'(\Omega)$ and $f\big|_{\Omega_j} = 0$, $j \in \mathcal{J}$, implies that $f = 0$.

d) Again let $\Omega = \bigcup_{j \in \mathcal{J}} \Omega_j$, and suppose a set of distributions $f_j \in \mathcal{D}'(\Omega_j)$ is given with $f_k\big|_{\Omega_k \cap \Omega_l} = f_l\big|_{\Omega_k \cap \Omega_l}$ for any $k, l \in \mathcal{J}$. Then there exists a distribution $f \in \mathcal{D}'(\Omega)$ such that $f\big|_{\Omega_j} = f_j$ for any $j \in \mathcal{J}$.

Properties a) and b) are obvious and properties c) and d) are easily proved using a partition of unity.

Properties a)–d) taken together mean that the family of spaces $\mathcal{D}'(\Omega')$ (where Ω' is an arbitrary open subset of Ω) form a *sheaf* on Ω. Property b) makes it possible to introduce unambiguously the largest open subset $\Omega' \subset \Omega$ for which $f\big|_{\Omega'} = 0$. Then the closed subset (of Ω) $\Gamma = \Omega \setminus \Omega'$ is called the *support* of the distribution f and is denoted supp f. It is easy to verify that if f is a continuous function on Ω, then supp f is the closure in Ω of the set $\{x | f(x) \neq 0\}$, so that the definition of support just introduced agrees with the standard definition for continuous functions. Moreover, since we have canonical imbeddings

$$\mathcal{E}'(\Omega) \subset \mathcal{D}'(\Omega), \quad \mathcal{S}'(\mathbb{R}^n) \subset \mathcal{D}'(\mathbb{R}^n),$$

induced by the imbeddings

$$\mathcal{D}(\Omega) \subset \mathcal{E}(\Omega), \quad \mathcal{D}(\mathbb{R}^n) \subset \mathcal{S}(\mathbb{R}^n),$$

it makes sense to talk about the support of distributions belonging to $\mathcal{E}'(\Omega)$ and $\mathcal{S}'(\mathbb{R}^n)$.

The support of the Dirac δ-function $\delta(x)$ is the point $\{0\}$. It can be proved that any distribution f with support at 0 is given by a formula

$$\langle f, \varphi \rangle = \sum_{|\alpha| \le N} c_\alpha (D^\alpha \varphi)(0), \tag{1.3}$$

where α is a multi-index, $N \ge 0$, and c_α are constants.

Further if $f \in \mathcal{D}'(\Omega)$, then the condition $f \in \mathcal{E}'(\Omega)$ is equivalent to the condition that supp f be a compact subset of Ω. Thus $\mathcal{E}'(\Omega)$ is the set of distributions with compact support in Ω.

Examples of distributions with compact support in Ω are the *Radon measures* on a compact set $K \subset \Omega$, which are most simply described as the continuous linear functionals μ on the Banach space $C(K)$ consisting of all continuous functions on K and having the usual sup-norm. By a well-known theorem of Riesz, such functionals can be written in the form of Lebesgue-Stieltjes integrals

$$\langle \mu, \varphi \rangle = \int_K \varphi \, d\mu, \quad \varphi \in C(K).$$

In particular the Dirac δ-function is such a measure (in this case $K = \{0\}$). If a compact set K is sufficiently regular (for example, if it is the closure of a region with a piecewise smooth boundary on some piecewise-smooth submanifold of Ω), then every distribution with support in K is defined by a formula

$$\langle f, \varphi \rangle = \sum_{|\alpha| \le N} \langle \mu_\alpha, D^\alpha \varphi \rangle, \tag{1.4}$$

where μ_α are Radon measures on K. This formula is an obvious generalization of the formula (1.3) written out above for distributions with support at the point 0. Any distribution f with compact support can be written in the same form, although in the general case the set K cannot be taken as supp f, but must be taken as the closure of some neighborhood of the compact set supp f. Moreover, one can even replace the measures μ_α by the functionals defined by continuous functions:

$$\langle f, \varphi \rangle = \sum_{|\alpha| \le N} \int_K f_\alpha(x) D^\alpha \varphi(x) \, dx, \tag{1.5}$$

where K is the closure of some neighborhood of the compact set supp f, $f_\alpha \in C(\Omega)$, and $f_\alpha(x) = 0$ for $x \in \Omega \setminus K$.

Any distribution $f \in \mathcal{D}'(\Omega)$ can be written in the form (1.4) or (1.5) in a neighborhood of any compact set (i.e., any restriction $f\big|_{\Omega'}$ can be written in such a form if $\overline{\Omega'}$ is a compact subset of Ω). Using a partition of unity we find that any distribution $f \in \mathcal{D}'(\Omega)$ can be written in the form of a locally finite sum $f = \sum\limits_{k=1}^{\infty} f_k$, where $f_k \in \mathcal{E}'(\Omega)$, and consequently f_k has the form (1.4)–(1.5).

An analogous assertion about the general form can be proved for tempered distributions. To be specific, every distribution $f \in \mathcal{S}'(\mathbb{R}^n)$ can be written in the form

$$\langle f, \varphi \rangle = \sum_{|\alpha| \leq N} \int_{\mathbb{R}^n} f_\alpha(x) D^\alpha \varphi(x) \, dx, \quad \varphi \in \mathcal{S}(\mathbb{R}^n), \tag{1.6}$$

where each of the functions f_α is continuous and satisfies an estimate of the form (1.2).

1.5. Differentiation of Distributions (cf. Schwartz 1950–1951; Gel'fand and Shilov 1958–1959; Shilov 1965; Hörmander 1983–1985). Operations with distributions are introduced so as to be the natural extension of operations with ordinary functions. As a rule such an extension turns out to be an extension by continuity; however, to construct it one must write down a formula defining the extension.

If $f \in C^1(\Omega)$ and $\varphi \in C_0^\infty(\Omega)$, then using integration by parts, we easily verify that

$$\int \frac{\partial f}{\partial x_j} \varphi \, dx = - \int f \frac{\partial \varphi}{\partial x_j} \, dx,$$

or

$$\langle \frac{\partial f}{\partial x_j}, \varphi \rangle = -\langle f, \frac{\partial \varphi}{\partial x_j} \rangle. \tag{1.7}$$

The last formula serves as the definition of the operator $\partial/\partial x_j$ on distributions f. To be specific, by definition, if $f \in \mathcal{D}'(\Omega)$, then the *derivative* $\dfrac{\partial f}{\partial x_j}$ is the functional whose value at the function $\varphi \in \mathcal{D}(\Omega)$ (the left-hand side of formula (1.7)) is determined using the right-hand side of (1.7). It is easy to see that the functional $\dfrac{\partial f}{\partial x_j}$ so defined is continuous (i.e., belongs to $\mathcal{D}'(\Omega)$). This follows from the obvious continuity of the operator $\dfrac{\partial}{\partial x_j} : \mathcal{D}(\Omega) \to \mathcal{D}(\Omega)$. Moreover it is clear from formula (1.7) that the operator $\dfrac{\partial}{\partial x_j} : \mathcal{D}'(\Omega) \to \mathcal{D}'(\Omega)$ is continuous in the weak topology (cf. Sect. 1.3). thus it can be regarded as an extension by continuity of the operator $\dfrac{\partial}{\partial x_j} : \mathcal{D}(\Omega) \to \mathcal{D}(\Omega)$.

In the same way the operator $\dfrac{\partial}{\partial x_j}$ is defined on $\mathcal{E}'(\Omega)$ and on $\mathcal{S}'(\mathbb{R}^n)$. On these spaces of distributions it can also be regarded as an extension by continuity (from $\mathcal{D}(\Omega)$ and $\mathcal{S}(\mathbb{R}^n)$ respectively). When this is done, we can take $\varphi \in \mathcal{E}(\Omega) = C^\infty(\Omega)$ in formula (1.7) for $f \in \mathcal{E}'(\Omega)$, and for $f \in \mathcal{S}'(\mathbf{R}^n)$ we can take $\varphi \in \mathcal{S}(\mathbf{R}^n)$, although it suffices to define the functional $\dfrac{\partial f}{\partial x_j}$ on functions of C_0^∞ (on Ω and $\mathbf{R}^n$ respectively).

Example 2.1. Let $H(x)$ be the *Heaviside function*, i.e., H is the following function on $\mathbf{R}$: $H(x) = 0$ for $x \le 0$ and $H(x) = 1$ for $x > 0$. In the usual way the function H defines a distribution of $\mathcal{D}'(\mathbf{R})$ and even a distribution in $\mathcal{S}'(\mathbf{R})$. Let us compute $H'(x)$ (i.e., $\dfrac{d}{dx}H$) in the distribution sense. For $\varphi \in \mathcal{D}(\mathbf{R})$ we have

$$\langle H', \varphi \rangle = -\langle H, \varphi' \rangle = - \int_0^\infty \varphi'(x)\,dx = \varphi(0) = \langle \delta, \varphi \rangle.$$

Thus $H'(x) = \delta(x)$.

Example 2.2. The function $\ln|x|$ is locally integrable on $\mathbf{R}$ and defines a distribution of $\mathcal{S}'(\mathbf{R})$. It is easy to verify that $(\ln|x|)' = PV\left\{\dfrac{1}{x}\right\}$. Thus one of the regularizations of the function $\dfrac{1}{x}$ is obtained in this way.

Applying the operators $\dfrac{\partial}{\partial x_j}$ successively, we obtain operators $\partial^\alpha = \dfrac{\partial^{|\alpha|}}{\partial x_1^{\alpha_1} \cdots \partial x_n^{\alpha_n}}$ on each of the spaces $\mathcal{D}'(\Omega)$, $\mathcal{E}'(\Omega)$, and $\mathcal{S}'(\mathbf{R}^n)$. We note that the order of differentiation can always be changed in distributions: $\dfrac{\partial}{\partial x_i}\left(\dfrac{\partial f}{\partial x_j}\right) = \dfrac{\partial}{\partial x_j}\left(\dfrac{\partial f}{\partial x_i}\right)$. This is obtained by continuity from the same fact for smooth functions (in $C_0^\infty(\Omega)$) or can be obtained using a direct computation from the same property of smooth functions using Eq. (1.7) as the definition.

It is not necessary to use successive differentiations to define the operator ∂^α on distributions; this definition can be carried out directly using the equality

$$\langle \partial^\alpha f, \varphi \rangle = (-1)^{|\alpha|} \langle f, \partial^\alpha \varphi \rangle. \tag{1.8}$$

This is of course equivalent to the method described above.

We note that since the operator $\dfrac{\partial}{\partial x_j}$ is continuous on the distributions (as are all the operators ∂^α), all convergent sequences and series of distributions can be differentiated termwise.

Example 2.3. It is known that if f and g belong to $C^2(\mathbf{R})$, then the function

$$u(t, x) = f(x - at) + g(x + at)$$

is a solution of the wave equation $\square u \equiv \dfrac{\partial^2 u}{\partial t^2} - a^2 \dfrac{\partial^2 u}{\partial x^2} = 0$. By passing to the limit we find that if it is known only that f and g belong to $L^1_{\text{loc}}(\mathbb{R})$, then the equation $\square u = 0$ holds as before in the sense of distributions (i.e., in $\mathcal{D}'(\mathbb{R}^2)$). This makes it possible to talk, for example, about discontinuous solutions of shock wave type and the like. This example shows that using distributions one can give a natural definition of *generalized* solutions of linear differential equations.

The theorems on the general form of distributions in Sect. 1.4 can be written more simply and naturally using the operation of differentiation. To be specific, every distribution with support at the point 0 has the form

$$f(x) = \sum_{|\alpha| \leq N} a_\alpha \partial^\alpha \delta(x),$$

where a_α are constants. Every distribution with compact support can be written in the form

$$f = \sum_{|\alpha| \leq N} \partial^\alpha f_\alpha,$$

where f_α are continuous functions of compact support. Every tempered distribution can be represented in the same form with continuous functions f_α satisfying an estimate of the form (1.2).

1.6. Multiplication of a Distribution by a Smooth Function. Linear Differential Operators in Spaces of Distributions (cf. Schwartz 1950–1951; Shilov 1965; Hörmander 1983–1985). In accordance with the general principle stated at the beginning of Sect. 1.5, multiplication of a distribution $f \in \mathcal{D}'(\Omega)$ by a smooth function $a \in C^\infty(\Omega)$ is defined by the formula

$$\langle af, \varphi \rangle = \langle f, a\varphi \rangle, \tag{1.9}$$

where $\varphi \in \mathcal{D}(\Omega)$. The question naturally arises: by which smooth functions a can any tempered distribution be multiplied without going outside the class of tempered distributions? It is easy to see that a necessary and sufficient condition for this is that the function a be a *multiplier* in $\mathcal{S}(\mathbb{R}^n)$, i.e., that multiplication by a be a continuous linear operator in $\mathcal{S}(\mathbb{R}^n)$. This in turn is equivalent to the estimates

$$\left| \partial^\alpha a(x) \right| \leq C_\alpha (1 + |x|)^{N_\alpha}, \tag{1.10}$$

where α is an arbitrary multi-index and C_α and N_α are constants depending on α. Multiplication by the function a in the class of distributions $\mathcal{S}'(\mathbb{R}^n)$ is also, of course, defined by formula (1.9), in which it is necessary to take $\varphi \in \mathcal{S}(\mathbb{R}^n)$. In particular one can multiply a distribution of $\mathcal{S}'(\mathbb{R}^n)$ by any polynomial in x.

Example 2.4. Let us compute $x\delta'(x)$, where $x \in \mathbb{R}$. We have:

$$\langle x\delta'(x), \varphi(x) \rangle = \langle \delta'(x), x\varphi(x) \rangle =$$
$$= -\langle \delta(x), \frac{d}{dx}(x\varphi(x)) \rangle = -\frac{d}{dx}(x\varphi(x))\big|_{x=0} = -\varphi(0),$$

so that $x\delta'(x) = -\delta(x)$.

A combination of multiplication and differentiation makes it possible to apply any differential operators

$$A = \sum_{|\alpha| \leq m} a_\alpha(x) D^\alpha$$

with coefficients $a_\alpha \in C^\infty(\Omega)$ to distributions in $\mathcal{D}'(\Omega)$ and operators of the same form with coefficients a_α that are multipliers in $S(\mathbb{R}^n)$ (i.e., satisfy estimates of the form (1.10)) to distributions in $S'(\mathbb{R}^n)$. One can describe the action of the operator A immediately using the *transposed operator* ^tA defined by the condition that

$$\langle Af, \varphi \rangle = \langle f, {}^tA\varphi \rangle \tag{1.11}$$

for $f, \varphi \in \mathcal{D}(\Omega)$. It is easy to see that the operator ^tA exists and is given by the formula

$$^tA\varphi = \sum_{|\alpha| \leq m} (-1)^{|\alpha|} D^\alpha(a_\alpha\varphi).$$

Assuming now that $f \in \mathcal{D}'(\Omega)$ and $\varphi \in \mathcal{D}(\Omega)$, we see that formula (1.11) gives an operator A on $\mathcal{D}'(\Omega)$. Similarly one can define an operator A of corresponding type in $S'(\mathbb{R}^n)$.

Multiplication of a distribution by a smooth function is connected with differentiation by the usual *Leibniz formula*

$$\frac{\partial}{\partial x_j}(af) = \frac{\partial a}{\partial x_j}f + a\frac{\partial f}{\partial x_j}.$$

Of course all the algebraic corollaries of this formula hold (for example, the formula that gives a higher-order derivative of the product af).

1.7. Change of Variables and Homogeneous Distributions (cf. Gel'fand and Shilov 1958–1959, Vol. 1, Chapter 4; Hörmander 1983–1985, Sect. 2.3.2). Suppose given a C^∞-diffeomorphism $\varkappa : \Omega \to \Omega_1$. It induces a mapping $\varkappa^* : C^\infty(\Omega_1) \to C^\infty(\Omega)$ taking the function f into $\varkappa^* f = f \circ \varkappa$. We extend this mapping to a continuous mapping $\varkappa^* : \mathcal{D}'(\Omega_1) \to \mathcal{D}'(\Omega)$. To do this we remark that if $f \in C^\infty(\Omega_1)$ and $\varphi \in \mathcal{D}(\Omega)$, then by a change of variables in the integral we obtain

$$\langle \varkappa^* f, \varphi \rangle = \int f(\varkappa(x)) \varphi(x)\, dx = \int f(z) \varphi(\varkappa^{-1}(z)) \left| \det \frac{\partial \varkappa^{-1}(z)}{\partial z} \right| dz,$$

where $\varkappa^{-1} : \Omega_1 \to \Omega$ is the mapping inverse to the mapping $\varkappa$ and $\dfrac{\partial \varkappa^{-1}(z)}{\partial z}$ is the Jacobian matrix of the mapping $\varkappa^{-1}$ at the point z. Thus if we introduce the operator $A = \varkappa^*$, then the transposed operator ${}^t A$ (in the sense of (1.11)) is given by the formula

$$({}^t A \varphi)(z) = ({}^t \varkappa^* \varphi)(z) = \left| \det \frac{\partial \varkappa^{-1}(z)}{\partial z} \right| \varphi(\varkappa^{-1}(z)), \qquad (1.12)$$

i.e., it is the composition of the change of variables and multiplication by a smooth function. Since ${}^t A$ defines a continuous mapping ${}^t A : \mathcal{D}(\Omega) \to \mathcal{D}(\Omega_1)$, the operator $A = \varkappa^*$ extends in the usual way to a continuous mapping

$$\varkappa^* : \mathcal{D}'(\Omega_1) \to \mathcal{D}'(\Omega).$$

It is easy to see that this mapping takes $\mathcal{E}'(\Omega_1)$ into $\mathcal{E}'(\Omega)$.

If $\Omega = \mathbf{R}^n$ and the operator ${}^t \varkappa^*$ defined by formula (1.12) maps $\mathcal{S}(\mathbf{R}^n)$ continuously into $\mathcal{S}(\mathbf{R}^n)$, we obtain a continuous mapping $\varkappa^* : \mathcal{S}'(\mathbf{R}^n) \to \mathcal{S}'(\mathbf{R}^n)$. This is the situation, for example, for a linear mapping $\varkappa$.

We now give some examples. Let $\varkappa$ be translation by the vector $(-x_0)$, i.e., $\varkappa(x) = x - x_0$. According to the general rule the shift operator $\varkappa^*$ in the spaces $\mathcal{D}'(\mathbf{R}^n)$ and $\mathcal{S}'(\mathbf{R}^n)$ is defined by the formula

$$\langle f(x - x_0), \varphi(x) \rangle = \langle f(x), \varphi(x + x_0) \rangle$$

(here $f(x - x_0)$ denotes the distribution $\varkappa^* f$). In particular, for example, the distribution $\delta(x - x_0)$ is given by

$$\int \delta(x - x_0) \varphi(x)\, dx = \varphi(x_0).$$

Another important example of a diffeomorphism of the space $\mathbf{R}^n$ into itself is the dilation $\varkappa_t : \mathbf{R}^n \to \mathbf{R}^n$ given by $\varkappa_t(x) = tx$ (here $t \in \mathbf{R} \setminus \{0\}$). Instead of $\varkappa_t^* f$ for the distribution $f = f(x)$, we shall write simply $f(tx)$. It follows from the general formula that

$$\langle f(tx), \varphi(x) \rangle = |t|^{-n} \langle f(x), \varphi(t^{-1}x) \rangle. \qquad (1.13)$$

The dilation makes it possible to introduce the concept of a *homogeneous distribution*. To be specific, a distribution $f \in \mathcal{S}'(\mathbf{R}^n)$ is called *homogeneous* (more precisely positive-homogeneous) of degree $s \in \mathbb{C}$ if

$$f(tx) = t^s f(x), \quad t > 0. \qquad (1.14)$$

Here $t^s = e^{s \ln t}$ is the standard branch of the function t^s. It is easy to see that the Dirac δ-function in $\mathbf{R}^n$ is homogeneous of degree $-n$. If $f \in L^1_{\mathrm{loc}}(\mathbf{R}^n)$, and f is homogeneous of degree s in the ordinary sense, i.e. (1.14) holds for almost all x, then the corresponding distribution is also homogeneous of degree s.

For example, the function $|x|^s$ in $\mathbb{R}^n$ with $\operatorname{Re} s > -n$ defines a homogeneous distribution of degree s.

Homogeneous distributions map to homogeneous distributions under differentiation. To be specific, if $f \in S'(\mathbb{R}^n)$ is homogeneous of degree s, then $\partial^\alpha f$ is homogeneous of degree $s - |\alpha|$. In particular, the derivative of the δ-function $\delta^{(\alpha)}(x) = \partial^\alpha \delta(x)$ is homogeneous of degree $-n - |\alpha|$.

It is an important question when an ordinary locally integrable function f on $\mathbb{R}^n \setminus \{0\}$ that is homogeneous of degree α can be extended to a homogeneous distribution $\hat{f}$ on $\mathbb{R}^n$. (The degree of homogeneity of the extension $\hat{f}$ must of necessity be α also.) For simplicity we shall assume that f is continuous on $\mathbb{R}^n \setminus \{0\}$. It turns out that if $\alpha \neq -n, -n-1, -n-2, \ldots$, then a *regularization* $\hat{f}$ exists and is unique: it can be obtained, for example, by analytic continuation of the integral

$$\langle f_\lambda, \varphi \rangle = \int f\left(\frac{x}{|x|}\right) |x|^\lambda \varphi(x)\, dx, \quad \varphi \in \mathcal{D}(\mathbb{R}^n),$$

on the parameter λ from the region $\operatorname{Re} \lambda > -n$, in which it converges absolutely, to the whole complex λ-plane (and, in particular, to the point α), then setting $\hat{f} = f_\alpha$. If $\alpha = -n$, a homogeneous regularization $\hat{f}$ exists if and only if

$$\int\limits_{|\omega|=1} f(\omega)\, dS_\omega = 0, \tag{1.15}$$

where dS_ω denotes the standard volume element on the unit sphere in $\mathbb{R}^n$. The desired regularization can be defined, for example, in the form of an integral taken in the sense of principal value

$$\langle \hat{f}, \varphi \rangle = \operatorname{PV}\left\{ \int f(x)\varphi(x)\, dx \right\} = \lim_{\varepsilon \to +0} \int\limits_{|x| \geq \varepsilon} f(x)\varphi(x)\, dx.$$

This regularization is not defined uniquely, but only up to a term $C\delta(x)$, where C is an arbitrary constant. An example of a situation in which condition (1.15) does not hold is the following: the function $\dfrac{1}{|x|}$ on $\mathbb{R}^1 \setminus \{0\}$ cannot be extended to a homogeneous distribution (of degree -1) on $\mathbb{R}^1$.

If a function $f \in C(\mathbb{R}^n \setminus \{0\})$ is homogeneous of degree $-n-k$, $k = 1, 2, \ldots$, then a necessary and sufficient condition for the existence of a homogeneous regularization $\hat{f}$ is that

$$\int\limits_{|\omega|=1} \omega^\alpha f(\omega)\, dS_\omega = 0 \tag{1.16}$$

for any multi-index α with $|\alpha| = k$. In this case the regularization is defined up to a term $\sum\limits_{|\alpha|=k} C_\alpha \delta^{(\alpha)}(x)$, where C_α are arbitrary constants.

Homogeneity considerations often make it possible to guess the result of computations with distributions. For example, consider the usual Heaviside function $H(x)$ on $\mathbb{R}$, which is homogeneous of degree 0. Then $\dfrac{d}{dx}H(x)$ is a homogeneous distribution of degree -1 with support at the point 0. It is then clear from the general form of distributions with support at the point 0 that this must be $C\delta(x)$, where C is a constant. More substantive examples will be exhibited below.

1.8. The Direct or Tensor Product of Distributions (cf. Gel'fand and Shilov 1958–1959; Shilov 1965; Vladimirov 1979; Hörmander 1983–1985). If $f(x)$ and $g(y)$ are two functions on the regions $\Omega_1 \subset \mathbb{R}^{n_1}$ and $\Omega_2 \subset \mathbb{R}^{n_2}$ respectively, their *direct* or *tensor* product is defined as the function

$$(f \otimes g)(x, y) = f(x)g(y),$$

defined on $\Omega_1 \times \Omega_2$. If $f \in L^1_{\mathrm{loc}}(\Omega_1)$ and $g \in L^1_{\mathrm{loc}}(\Omega_2)$, then the function $f \otimes g$ gives a functional on $\mathcal{D}(\Omega_1 \times \Omega_2)$ that can be defined by the formula

$$\langle f \otimes g, \varphi(x, y) \rangle = \langle f(x), \langle g(y), \varphi(x, y) \rangle \rangle, \quad \varphi \in \mathcal{D}(\Omega_1 \times \Omega_2), \qquad (1.17)$$

where the notation $\langle g(y), \varphi(x, y) \rangle$ means that the functional g is applied to $\varphi(x, \cdot)$ for fixed x, i.e.,

$$\langle g(y), \varphi(x, y) \rangle = \int g(y)\varphi(x, y)\, dy.$$

Instead of (1.17) one can also write (by Fubini's theorem)

$$\langle f \otimes g, \varphi \rangle = \langle g(y), \langle f(x), \varphi(x, y) \rangle \rangle, \quad \varphi \in \mathcal{D}(\Omega_1 \times \Omega_2). \qquad (1.17')$$

Now let $f \in \mathcal{D}'(\Omega_1)$ and $g \in \mathcal{D}'(\Omega_2)$. Then using one of the formulas (1.17) and (1.17$'$) we can define a distribution $f \otimes g \in \mathcal{D}'(\Omega_1 \times \Omega_2)$. Each of these formulas has meaning because $\langle g(y), \varphi(x, y) \rangle \in \mathcal{D}(\Omega_1)$ by virtue of the fact that $\varphi(x, y)$ can be regarded as a smooth function of $x \in \Omega_1$ with values in $\mathcal{D}(\Omega_2)$ and vanishing for $x \in \Omega_1 \setminus K$, where K is a compact subset of Ω_1. Moreover these formulas lead to the same distribution $f \otimes g$, since if $\varphi \in \mathcal{D}(\Omega_1)$ and $\psi \in \mathcal{D}(\Omega_2)$, each of the formulas (1.17) and (1.17$'$) gives

$$\langle f \otimes g, \varphi \otimes \psi \rangle = \langle f, \varphi \rangle \langle g, \psi \rangle,$$

and linear combinations of functions of the form $\varphi \otimes \psi$ are dense in $\mathcal{D}(\Omega_1 \times \Omega_2)$.

It is easy to see that if $f \in \mathcal{E}'(\Omega_1)$ and $g \in \mathcal{E}'(\Omega_2)$, then $f \otimes g \in \mathcal{E}'(\Omega_1 \times \Omega_2)$. If $f \in \mathcal{S}'(\mathbb{R}^{n_1})$ and $g \in \mathcal{S}'(\mathbb{R}^{n_2})$, then $f \otimes g \in \mathcal{S}'(\mathbb{R}^{n_1+n_2})$.

Example 2.5. $\delta(x) \otimes \delta(y) = \delta(x, y).$

Example 2.6. If we regard a function $f(x)$ on Ω_1 as a function on $\Omega_1 \times \Omega_2$ (independent of $y \in \Omega_2$), it can be thought of as $f(x) \otimes 1_y$, where 1_y is the function of y identically equal to 1. Similarly for each distribution $f \in \mathcal{D}'(\Omega_1)$ we can construct the distribution $f(x) \otimes 1_y \in \mathcal{D}'(\Omega_1 \times \Omega_2)$, which for simplicity is often denoted by $f(x)$ (and is said to be independent of y).

Example 2.7. If $\mu(x)$ is a distribution that is a measure, then $\mu(x) \otimes \delta(y)$ is also a measure on $\Omega_1 \times \Omega_2$ concentrated on the surface $y = 0$.

It is easily verified that

$$\operatorname{supp}(f \otimes g) = (\operatorname{supp} f) \times (\operatorname{supp} g).$$

Differentiation of a tensor product reduces to differentiating one of its factors. For example

$$D_x^\alpha \big(f(x) \otimes g(y) \big) = \big[D_x^\alpha f(x) \big] \otimes g(y).$$

1.9. The Convolution of Distributions (cf. Gel'fand and Shilov 1958–1959; Shilov 1965; Vladimirov 1979; Hörmander 1983–1985). The *convolution* of the ordinary functions $f, g \in L_{\mathrm{loc}}^1(\mathbb{R}^n)$ is defined to be the function

$$(f * g)(x) = \int f(x - y)g(y)\, dy = \int f(y)g(x - y)\, dy$$

(we assume that one of the integrals on the right-hand side converges absolutely for almost all x). Multiplying both sides by $\varphi(x)$, where $\varphi \in \mathcal{D}(\mathbb{R}^n)$, and integrating, in the case when the double integral so obtained, namely

$$\int f(x - y)g(y)\varphi(x)\, dy\, dx,$$

converges absolutely, we find that

$$\langle f * g, \varphi \rangle = \int f(x - y)g(y)\varphi(x)\, dy\, dx = \int f(x)g(y)\varphi(x + y)\, dy\, dx,$$

i.e.,

$$\langle f * g, \varphi \rangle = \langle (f \otimes g)(x, y), \varphi(x + y) \rangle. \tag{1.18}$$

This formula can be used as the basis of a definition of the *convolution* of two *distributions* $f, g \in \mathcal{D}'(\mathbb{R}^n)$ in the case when the right-hand side has meaning. A natural meaning can be given to the right-hand side when the set

$$(\operatorname{supp} f \times \operatorname{supp} g) \bigcap \{(x, y) \,|\, |x + y| \le R\}$$

is compact for any $R > 0$. In fact one can then consider for any $R > 0$ a function $\psi_R \in C_0^\infty(\mathbb{R}^{2n})$ equal to 1 on a neighborhood of this set and define

$$\langle (f \otimes g)(x, y), \varphi(x + y) \rangle = \langle (f \otimes g)(x, y), \psi_R(x, y)\varphi(x + y) \rangle$$

for functions $\varphi \in C_0^\infty(\mathbb{R}^n)$ with support in the ball $\{x \,|\, |x| \le R\}$ (the result will be independent of the choice of the cutoff function ψ_R). In particular the convolution $f * g$ is always defined if one of the distributions f and g has compact support or if the supports of f and g lie in a closed convex cone not containing any line. It is easy to verify that convolution is commutative and associative (the latter means that $(f * g) * h = f * (g * h)$ if both sides are naturally defined in the sense just described). Convolution is bilinear in f and g. In particular $\mathcal{E}'(\mathbb{R}^n)$ is a commutative and associative ring under the operation of convolution (and $\mathcal{D}'(\mathbb{R}^n)$ is a module over this ring). This ring has the δ-function as identity. More generally,

$$\delta * f = f * \delta = f, \quad f \in \mathcal{D}'(\mathbb{R}^n).$$

The rule for differentiating a convolution is

$$D^\alpha(f * g) = (D^\alpha f) * g = f * (D^\alpha g). \tag{1.19}$$

Convolution is continuous in each factor separately. For example, if $g_k \to g$ in $\mathcal{E}'(\mathbb{R}^n)$, then $f * g_k \to f * g$ for any distribution $f \in \mathcal{D}'(\mathbb{R}^n)$. In particular the average f_ε introduced in Sect. 1.3 for a distribution $f \in \mathcal{D}'(\mathbb{R}^n)$ is a convolution $f_\varepsilon = f * \varphi_\varepsilon$ with a δ-shaped family φ_ε. Since $\varphi_\varepsilon \to \delta$, it follows that $f_\varepsilon \to f * \delta = f$ as $\varepsilon \to +0$.

For the support of a convolution we have the *additive property*

$$\operatorname{supp}(f * g) \subset \operatorname{supp} f + \operatorname{supp} g,$$

where $A + B = \{a + b \,|\, a \in A,\, b \in B\}$ is the *arithmetic sum* of the subsets A and B of $\mathbb{R}^n$. A similar rule holds for singular support

$$\operatorname{sing\,supp}(f * g) \subset \operatorname{sing\,supp} f + \operatorname{sing\,supp} g$$

(the *singular support* $\operatorname{sing\,supp} f$ of a distribution f on Ω is the complement of the largest open set $\Omega' \subset \Omega$ for which $f|_{\Omega'} \in C^\infty(\Omega')$).

Finally if we introduce the *analytic singular support* $\operatorname{sing\,supp}_a f$ as the complement of the largest open set on which f is analytic, it will have the same property:

$$\operatorname{sing\,supp}_a(f * g) \subset \operatorname{sing\,supp}_a f + \operatorname{sing\,supp}_a g.$$

Example 2.8. The most important examples of convolutions are potentials. Thus the *Newtonian* (or *volume*) potential in $\mathbb{R}^3$

$$u(x) = -\frac{1}{4\pi} \int \frac{\rho(y)\, dy}{|x - y|},$$

where ρ is charge density, is the convoluton $\rho * \left(\dfrac{-1}{4\pi r}\right)$ of the functions ρ and $-1/4\pi r$, where $r = |x|$ (the meaning of the factor $-1/4\pi$ will become clear later).

If we allow distributions as ρ, we can write the potentials of single and double layers as the same kind of convolution. To be specific, let Γ be a compact piecewise-smooth 2-dimensional surface in $\mathbf{R}^3$. We introduce the distribution δ_Γ by the formula

$$\langle \delta_\Gamma, \varphi \rangle = \int_\Gamma \varphi(y)\, dS_y,$$

where dS_y is the element of area on the surface. The distribution δ_Γ is called the δ-function of the surface Γ and is a measure concentrated on Γ. If a piecewise-smooth function σ is defined on Γ, we can consider the distribution $\sigma\delta_\Gamma$ defined by the formula

$$\langle \sigma\delta_\Gamma, \varphi \rangle = \int_\Gamma \sigma(y)\varphi(y)\, dS_y,$$

which is a measure (in general a signed measure) or *charge* with density σ on the surface Γ. The potential of this charge (called the *single-layer potential*) has the form of a convolution $u\delta_\Gamma * \left(-1/4\pi r\right)$ and for $x \notin \Gamma$ can be written in the form

$$u(x) = -\int_\Gamma \frac{\sigma(y)\, dS_y}{4\pi|x-y|}.$$

Finally we describe the potential of a double layer with dipole density $\beta = \beta(y)$ on the surface Γ. Let the orientation of the dipoles be prescribed in the direction of the normal n_y, $y \in \Gamma$, which is chosen in an arbitrary piecewise-smooth manner (for example, as the exterior normal in the case of a closed surface Γ). We introduce the distribution $\dfrac{\partial}{\partial n}(\beta\delta_\Gamma)$ by the formula

$$\langle \frac{\partial}{\partial n}(\beta\delta_\Gamma), \varphi \rangle = -\int_\Gamma \beta(y)\frac{\partial\varphi(y)}{\partial n_y}\, dS_y.$$

Then the convolution

$$u = \frac{\partial}{\partial n}(\beta\delta_\Gamma) * \left(-\frac{1}{4\pi r}\right)$$

is the *double-layer potential*

$$u(x) = \frac{1}{4\pi}\int_\Gamma \beta(y)\left(\frac{\partial}{\partial n_y}\frac{1}{|x-y|}\right) dS_y.$$

Example 2.9. A differential operator with constant coefficients $P(D)$ can be written in the form of a convolution with the distribution $P(D)\delta$

$$P(D)u = \delta * P(D)u = P(D)\delta * u.$$

Example 2.10. The *Hilbert transform*

$$u(x) \mapsto \mathrm{PV}\left\{\int_{-\infty}^{\infty} \frac{u(y)}{x - y}\, dy\right\}$$

is a convolution operator with the distribution $\mathrm{PV}\left\{\frac{1}{x}\right\}$. A more general example is the *singular integral*

$$\mathrm{PV}\left\{\int_{\mathbb{R}^n} \frac{u(y)}{|x - y|^n} f\left(\frac{x - y}{|x - y|}\right) dy\right\},$$

where $f(\omega)$ is a continuous function on $\mathbb{S}^{n-1} = \{\omega : |\omega| = 1\}$ such that $\int_{\mathbb{S}^{n-1}} f(\omega)\, d\mathbb{S}_\omega = 0$. This integral can be represented in the form of a convolution with the homogeneous distribution (of order $-n$) that is obtained by regularizing the ordinary function $|x|^{-n} f\left(\frac{x}{|x|}\right)$ (cf. Sect. 1.7).

1.10. The Fourier Transform of Tempered Distributions (cf. Schwartz 1950–1951; Gel'fand and Shilov 1958–1959; Shilov 1965; Rudin 1973; Vladimirov 1979; Hörmander 1983–1985). The *Fourier transform* of a function $u(x)$ on $\mathbb{R}^n$ is defined by the formula

$$\tilde{u}(\xi) = (Fu)(\xi) = \int e^{-ix\cdot\xi} u(x)\, dx, \qquad (1.20)$$

where $x \cdot \xi = x_1\xi_1 + \cdots + x_n\xi_n$ is the usual inner product of the vectors $x, \xi \in \mathbb{R}^n$. The operator F is a topological isomorphism $F : S(\mathbb{R}^n) \to S(\mathbb{R}^n)$, and the inverse mapping F^{-1} is given by the formula

$$(F^{-1}v)(x) = (2\pi)^{-n}\int e^{ix\cdot\xi} v(\xi)\, d\xi. \qquad (1.20')$$

Multiplying both sides of (1.20) by $\varphi(\xi)$, where $\varphi \in S(\mathbb{R}^n)$, and integrating over ξ, we obtain (for $u \in S(\mathbb{R}^n)$):

$$\langle Fu, \varphi \rangle = \langle u, F\varphi \rangle, \qquad (1.21)$$

i.e., ${}^t F = F$. Formula (1.21) makes it possible to define the *Fourier transform* F as a mapping $F : S'(\mathbb{R}^n) \to S'(\mathbb{R}^n)$. It is the extension by continuity of the mapping $F : S(\mathbb{R}^n) \to S(\mathbb{R}^n)$ and is also a topological isomorphism. The inverse transformation F^{-1} is obtained by extending the transformation given in formula (1.20'), i.e., $F^{-1} : S'(\mathbb{R}^n) \to S'(\mathbb{R}^n)$, by continuity. Usually the Fourier transform of distributions in $S'(\mathbf{R}^n)$ is written formally in the form of the integral (1.20).

In the case when $u \in \mathcal{E}'(\mathbf{R}^n)$ the Fourier transform $\tilde{u}(\xi)$ of the distribution u can be defined by the more explicit formula

$$\tilde{u}(\xi) = \langle u(x), e^{-ix\cdot\xi} \rangle, \qquad (1.22)$$

from which it is clear immediately that $\tilde{u}(\xi) \in C^\infty(\mathbf{R}^n)$ and moreover $\tilde{u}(\xi)$ extends to an entire function of $\xi \in \mathbf{C}^n$. Using the general form of distributions $u \in \mathcal{E}'(\mathbf{R}^n)$ (cf. Sect. 1.4), it is easy to obtain the result that $\tilde{u}(\xi)$ satisfies an estimate of the form

$$|\tilde{u}(\xi)| \le C(1 + |\xi|)^N e^{a|\text{Im}\,\xi|}, \quad \xi \in \mathbf{C}^n, \qquad (1.23)$$

where a is the radius of a closed ball $\{x \mid |x| \le a\}$ containing the support of the distribution u.

Conversely if an entire function $\tilde{u}(\xi)$ is given satisfying such an estimate with some N and $a \ge 0$, then $\tilde{u}(\xi)$ is the Fourier transform of a distribution $u \in \mathcal{E}'(\mathbf{R}^n)$ with support in the ball $\{x \mid |x| \le a\}$ (the *Paley-Wiener-Schwartz Theorem*, Hörmander 1983–1985, Theorem 7.3.1).

Example 2.11. From (1.22) we find that $(F\delta)(\xi) = 1$, i.e., the Fourier transform of the δ-function is the function identically equal to 1. It follows from this that $(F^{-1}1)(x) = \delta(x)$ and $(F1)(\xi) = (2\pi)^n\delta(x)$.

Example 2.12. The Fourier transform of the Heaviside function $H(x)$ on $\mathbf{R}$ can be computed, for example, using the continuity of the Fourier transform on S':

$$\tilde{H}(\xi) = (FH)(\xi) = \lim_{\varepsilon \to +0} F\big(H(x)e^{-\varepsilon x}\big)(\xi) =$$

$$= \lim_{\varepsilon \to +0} \int_0^\infty e^{-i\xi x - \varepsilon x}dx = \lim_{\varepsilon \to +0} \frac{1}{i\xi + \varepsilon} = \frac{-i}{\xi - i0}.$$

Similarly

$$\big[FH(-x)\big](\xi) = \frac{i}{\xi + i0}.$$

It follows from these two formulas that

$$\big[F(\text{sgn}\,x)\big](\xi) = -2i\,\text{PV}\left\{\frac{1}{\xi}\right\}.$$

The connection between the Fourier transform and differentiation is given by the formula

$$F\big(D^\alpha u\big)(\xi) = \xi^\alpha (Fu)(\xi), \qquad (1.24)$$

which is easily verified for $u \in \mathcal{S}(\mathbf{R}^n)$ and true for $u \in \mathcal{S}'(\mathbf{R}^n)$, for example, by continuity. This formula means that the Fourier transform takes differentation D^α into multiplication by ξ^α. There is a more general formula that follows in an obvious way from (1.23):

$$F(P(D)u)(\xi) = P(\xi)(Fu)(\xi), \qquad (1.25)$$

where $P(\xi)$ is an arbitrary polynomial and $P(D)$ the corresponding differential operator with constant coefficients. Thus, if $u \in \mathcal{S}(\mathbf{R}^n)$ and $P(D)u = f$, then $P(\xi)\tilde{u}(\xi) = \tilde{f}(\xi)$, so that solving the equation $P(D)u = f$ in $\mathbf{R}^n$ reduces to dividing the distribution $\tilde{f}(\xi)$ by the polynomial $P(\xi)$.

Example 2.13. Using Example 2.11 and the formula (1.25), we obtain

$$F(P(D)\delta) = P(\xi).$$

The connection between the Fourier transform and convolution is given by the formula

$$F(f * g) = (Ff) \cdot (Fg), \qquad (1.26)$$

which holds, for example, if $f \in \mathcal{E}'(\mathbf{R}^n)$ and $g \in \mathcal{S}(\mathbf{R}^n)$. Thus the Fourier transform changes convolution into multiplication.

It is easy to verify that if a nonsingular linear transformation $A : \mathbf{R}^n \to \mathbf{R}^n$ is given, then

$$[F(u(Ax))](\xi) = |\det A|^{-1}(Fu)({}^tA^{-1}\xi), \qquad (1.27)$$

where ${}^tA$ is the transpose to A (with respect to the standard inner product on $\mathbf{R}^n$). In particular if A is an orthogonal transformation, so that ${}^tA^{-1} = A$, then

$$[F(u(Ax))](\xi) = (Fu)(A\xi).$$

It follows from this that the Fourier transform of a *spherically symmetric* distribution (i.e., a distribution that is invariant under all orthogonal transformations) is also a spherically symmetric distribution.

Furthermore it also follows from (1.27) that for any $t > 0$

$$[F(u(tx))](\xi) = t^{-n}(Fu)(t^{-1}\xi).$$

Therefore if u is a homogeneous distribution of degree s, then Fu is also a homogeneous distribution of degree $-s - n$.

Example 2.14. We set $v(\xi) = |\xi|^{-2}$ on $\mathbf{R}^n$ for $n \geq 3$. Then $v \in L^1_{\text{loc}}(\mathbf{R}^n)$, so that v defines a distribution $v \in \mathcal{S}'(\mathbf{R}^n)$ that is spherically symmetric and homogeneous of degree -2. We set $u = F^{-1}v$. Then $u \in \mathcal{S}'(\mathbf{R}^n)$, and u is spherically symmetric and homogeneous of degree $2 - n$. It follows from this that $u(x) = C|x|^{2-n}$, where $C \neq 0$. The constant C can be computed from the relation $-\Delta u = \delta$, since

$$[F(-\Delta u)](\xi) = |\xi|^2 (Fu)(\xi) = |\xi|^2 |\xi|^{-2} = 1.$$

A straightforward computation making use of Stokes' Theorem (cf. Sect. 2.2) reveals that $-\Delta(|x|^{2-n}) = (n-2)\sigma_{n-1}\delta(x)$, where σ_{n-1} is the surface area of the sphere of radius 1 in $\mathbb{R}^n$. Therefore $C = \sigma_{n-1}^{-1}(n-2)^{-1}$, and we obtain as a result

$$[F^{-1}(|\xi|^{-2})](x) = \frac{1}{(n-2)\sigma_{n-1}}|x|^{2-n}.$$

1.11. The Schwartz Kernel of a Linear Operator (cf. Hörmander 1983–1985, Sect. 5.2). Given an integral operator of the form

$$(Af)(x) = \int K_A(x,y)f(y)\,dy, \tag{1.28}$$

the function K_A is called its *Schwartz kernel*. Multiplying both sides of this equality by $g(x)$ and integrating, we obtain (assuming the double integral converges absolutely)

$$\langle Af, g \rangle = \langle K_A, g \otimes f \rangle. \tag{1.29}$$

This equality is the basis for the definition of a generalized kernel of an operator A. To be specific, given a linear operator $A : C_0^\infty(\Omega_2) \to \mathcal{D}'(\Omega_1)$, where Ω_1 and Ω_2 are regions in $\mathbb{R}^{n_1}$ and $\mathbb{R}^{n_2}$ respectively, the *generalized kernel* (or simply *kernel* or *Schwartz kernel*) of the operator A is defined as the distribution $K_A \in \mathcal{D}'(\Omega_1 \times \Omega_2)$ such that equality (1.29) holds for all $f \in C_0^\infty(\Omega_2)$ and $g \in C_0^\infty(\Omega_1)$.

It is easy to see that if the operator A has the kernel K_A, then it is continuous as an operator from $C_0^\infty(\Omega_2)$ into $\mathcal{D}'(\Omega_1)$ if $C_0^\infty(\Omega_2)$ has the usual topology (cf. Sect. 1.2) and $\mathcal{D}'(\Omega)$ has the weak topology (cf. Sect. 1.3). Conversely if an operator $A : C_0^\infty(\Omega_2) \to \mathcal{D}'(\Omega_1)$ is continuous in the sense just indicated, it has a Schwartz kernel (this assertion is known as the L. Schwartz kernel theorem). We note that for any kernel $K_A \in \mathcal{D}'(\Omega_1 \times \Omega_2)$ formula (1.29) defines an operator $A : C_0^\infty(\Omega_2) \to \mathcal{D}'(\Omega_1)$, and the kernel K_A is uniquely determined by the operator A, so that there is a one-to-one correspondence between continuous linear operators $A : C_0^\infty(\Omega_2) \to \mathcal{D}'(\Omega_1)$ and distributions $K_A \in \mathcal{D}'(\Omega_1 \times \Omega_2)$.

Example 2.15. The identity operator $I : C_0^\infty(\Omega) \to C_0^\infty(\Omega)$ has as its kernel the δ-function $\delta(x-y)$ defined by the formula

$$\iint \delta(x-y)\varphi(x,y)\,dy\,dx = \int \varphi(x,x)\,dx.$$

Example 2.16. The differential operator $A = a(x, D_x) = \sum\limits_{|\alpha| \leq m} a_\alpha(x) D^\alpha$ on the region $\Omega \subset \mathbb{R}^n$ has the kernel

$$K_A(x, y) = a(x, D_x)\delta(x - y) = \sum_{|\alpha| \leq m} a_\alpha(x) D_x^\alpha \delta(x - y) \in \mathcal{D}'(\Omega \times \Omega).$$

We note that the support of this kernel lies on the diagonal $\Delta = \{(x, x) | x \in \Omega\} \subset \Omega \times \Omega$, reflecting the fact that this operator is *local,* i.e., $\text{supp}\,(Au) \subset \text{supp}\,u$ for $u \in C_0^\infty(\Omega)$. It can be shown that the converse holds also (in a certain sense): every local linear operator $A : C_0^\infty(\Omega) \to C_0^\infty(\Omega)$ (not necessarily continuous *a priori*) is a differential operator with the coefficients $a_\alpha \in C^\infty(\Omega_1)$ over any region $\Omega_1 \Subset \Omega$ (Peetre 1960).

1.12. Fundamental Solutions for Operators with Constant Coefficients (cf. Shilov 1965; Hörmander 1983–1985). The distribution $E \in \mathcal{D}'(\mathbb{R}^n)$ is called a *fundamental solution* for the operator $P(D)$ (with constant coefficients) if it satisfies the equation

$$P(D)E(x) = \delta(x). \tag{1.30}$$

If E is a tempered distribution, (i.e., $E \in \mathcal{S}'(\mathbb{R}^n)$), then, passing to the Fourier transform, we can write Eq. (1.30) in the form

$$P(\xi)\tilde{E}(\xi) = 1. \tag{1.31}$$

This means in particular that $\tilde{E}(\xi)$ is a regularization of the function $1/P(\xi)$ (which may have nonintegrable singularities), i.e., $\tilde{E} = 1/P(\xi)$ on the set $\{\xi | P(\xi) \neq 0\}$. It can be proved that if $P \not\equiv 0$, then such a regularization always exists. In particular *every* nonzero operator $P(D)$ has a fundamental solution $E \in \mathcal{S}'(\mathbb{R}^n)$ (cf. Hörmander 1958). If, for example, $P(\xi) \neq 0$ for all $\xi \in \mathbb{R}^n$, then we necessarily have $\tilde{E}(\xi) = 1/P(\xi)$ everywhere and consequently $E = F^{-1}(1/P)$, so that in this case a fundamental solution belonging to $\mathcal{S}'(\mathbb{R}^n)$ is unique. In the class $\mathcal{D}'(\mathbb{R}^n)$ a fundamental solution is never unique, since it remains a fundamental solution when any solution of the equation $P(D)u = 0$ is added to it, and there are always nontrivial solutions of such an equation in $\mathcal{D}'(\mathbb{R}^n)$, for example the exponentials $u(x) = e^{ix \cdot \xi}$, where the vector $\xi \in \mathbb{C}^n$ is such that $P(\xi) = 0$.

If the function $1/P(\xi)$ is locally integrable and defines a tempered distribution, we can also take $F^{-1}(1/P)$ as a fundamental solution. Thus we obtain

Example 2.17. For $n \geq 3$ the function

$$F^{-1}\big(-1/|\xi|^2\big) = \frac{1}{(2 - n)\sigma_{n-1}} r^{2-n}$$

is a fundamental solution for the Laplacian (cf. Example 2.14, where $r = |x|$). In particular for $n = 3$ we obtain $-1/4\pi r$, which explains the presence of the factor $-1/4\pi$ in the definition of the potentials (cf. Sect. 1.9) having the form of convolutions of various distributions with a fundamental solution for the Laplacian.

Example 2.18. We shall show how to find a fundamental solution for the Laplacian Δ on $\mathbb{R}^2$. We remark first of all that since Δ commutes with rotations and the δ-function is spherically symmetric (i.e., invariant with respect to rotations), when a rotation is applied to any fundamental solution, we again obtain a fundamental solution. Averaging over all rotations, we see that there exists a spherically symmetric fundamental solution. We shall seek it in the form $E(x) = f(r)$, where $r = |x|$, and $f \in C^2$ for $r > 0$. For $x \neq 0$ we have

$$0 = \Delta E(x) = f''(r) + \frac{1}{r} f'(r),$$

whence $f(r) = C \ln r + C_1$. It is clear that C_1 is irrelevant, so that we may assume $f(r) = C \ln r$. It is easy to see that $\Delta(\ln r) = C_2 \delta(x)$. (It must be a distribution with support at the point 0, homogeneous of degree -2, since the derivatives $\dfrac{\partial}{\partial x_j} \ln r = x_j/r^2$ are locally integrable and homogeneous of degree -1.) The constant C_2 is easily computed using Green's formula (cf. Sect. 2.2) and turns out to be 2π, so that in this case $E(x) = \dfrac{1}{2\pi} \ln r$. Naturally the case $n \geq 3$ can also be handled in exactly the same way (cf. Example 2.1).

Example 2.19. For the Cauchy-Riemann operator $\dfrac{\partial}{\partial \bar{z}} = \dfrac{1}{2} \left(\dfrac{\partial}{\partial x} + i \dfrac{\partial}{\partial y} \right)$ in $\mathbb{R}^2$ a fundamental solution is the locally integrable function $1/\pi z$, where $z = x + iy$. Up to computing the constant this is also clear from homogeneity considerations.

Example 2.20. Consider an ordinary differential operator (i.e., an operator on $\mathbb{R}^1$) with constant coefficients of the form

$$P(D_t) = \frac{d^m}{dt^m} + a_1 \frac{d^{m-1}}{dt^{m-1}} + \cdots + a_{m-1} \frac{d}{dt} + a_m.$$

Let $y(t)$ be a solution of the equation $P(D_t)y = 0$ satisfying the initial conditions

$$y(0) = \dot{y}(0) = \cdots = y^{(m-2)}(0) = 0, \quad y^{(m-1)}(0) = 1.$$

Then one of the fundamental solutions for the operator $P(D_t)$ is given by the formula

$$E(t) = H(t)y(t), \tag{1.32}$$

where H is the Heaviside function. (The verification that E is a fundamental solution is easily carried out using Leibniz' formula.) For example, if $P(D_t) = \dfrac{d^2}{dt^2} + \omega^2$, then formula (1.32) gives

$$E(t) = H(t)\frac{\sin \omega t}{\omega}.$$

We note that the fundamental solution (1.32) does not necessarily belong to $\mathcal{S}'(\mathbb{R})$. Other fundamental solutions, distinguished by various additional conditions (for example, vanishing at $+\infty$ or $-\infty$ or belonging to $\mathcal{S}'(\mathbb{R})$) are often convenient. They are easy to find by combining solutions of the homogeneous equation on the semiaxes $t > 0$ and $t < 0$ or using a contour integral

$$E(t) = \frac{1}{2\pi} \int\limits_{\Gamma} \frac{e^{it\xi}}{P(\xi)} \, d\xi,$$

where Γ is a suitable contour in the complex plane enclosing the zeros of the polynomial $P(\xi)$ and traversing the real axis in a neighborhood of infinity. This way of regularizing the function $1/P(\xi)$ by allowing the variable ξ to pass into the complex plane often works and is useful in the multidimensional case.

1.13. A Fundamental Solution for the Cauchy Problem (cf. Shilov 1965, Sect. 27). We now make a general remark on the connection between a fundamental solution for an operator on $\mathbb{R}^{n+1}$ of the form

$$P = P(D_t, D_x) = D_t^m + \sum_{j=1}^{m} P_j(D_x) D_t^{m-j}$$

and the solution of a Cauchy problem for the same operator. Let $E(t, x)$ be a distribution on $\mathbb{R}^{n+1}$ equal to 0 for $t < 0$ and such that it can be regarded as a smooth function of $t \in [0, +\infty)$ with values in $\mathcal{D}'(\mathbb{R}^n)$, i.e., it defines a distribution $E(t, \cdot) \in \mathcal{D}'(\mathbb{R}^n)$ depending smoothly on $t \in [0, +\infty)$, and for $\varphi \in C_0^\infty(\mathbb{R}^{n+1})$

$$\langle E(t, x), \varphi(t, x) \rangle = \int_0^\infty \langle E(t, \cdot), \varphi(t, \cdot) \rangle \, dt$$

(by assumption $\langle E(t, \cdot), \varphi(t, \cdot) \rangle = \int E(t, x)\varphi(t, x) \, dx$ depends smoothly on $t \in [0, +\infty)$, so that the integral has meaning). Suppose for any function $\varphi(x) \in C_0^\infty(\mathbb{R}^n)$ the convolution (on the variable x)

$$u(t, x) = \int E(t, x - y)\varphi(y) \, dy, \quad t > 0, \tag{1.33}$$

is a solution of the Cauchy problem

$$Pu = 0, t > 0; \quad u\big|_{t=0} = 0, \ldots, D_t^{m-2}u\big|_{t=0} = 0, \quad D_t^{m-1}u\big|_{t=0} = \varphi. \quad (1.34)$$

As is easy to see, this is equivalent to

$$PE = 0, t > 0; \quad E\big|_{t=+0} = 0, \ldots, D_t^{m-2}E\big|_{t=+0} = 0, \quad D_t^{m-1}E\big|_{t=+0} = \delta(x). \quad (1.35)$$

But it follows from this that $PE(t, x) = \delta(t, x)$, so that $E(t, x)$ is a fundamental solution for the operator P. Conversely if E is a fundamental solution for the operator P possessing the smoothness properties described above and vanishing for $t < 0$, then it satisfies conditions (1.35) and consequently can be used to solve the Cauchy problem (1.34) from formula (1.33). Therefore such a distribution is frequently called a *fundamental solution for the Cauchy problem* for the operator P.

We note that, knowing a solution of the Cauchy problem (1.34) in the situation described above, it is easy to solve the general Cauchy problem

$$Pu = 0, t > 0; u\big|_{t=0} = \varphi_0, \ldots, D_t^{m-2}u\big|_{t=0} = \varphi_{m-2}, D_t^{m-1}u\big|_{t=0} = \varphi_{m-1}, \quad (1.36)$$

where $\varphi_j \in C_0^\infty(\mathbb{R}^n)$. To be specific, if we denote by u_φ the solution of the problem (1.34), then the solution of the problem (1.36) is given by the formula

$$u = \sum_{j=0}^{m-1} D_t^{m-1-j} u_{\varphi_j}.$$

Thus, knowing a fundamental solution for the Cauchy problem, we can find the solution of the general Cauchy problem. On the other hand, knowing the formulas that give the solution of the Cauchy problem, we know a fundamental solution for the Cauchy problem and therefore a fundamental solution for the operator P.

Example 2.21. It is clear from the Poisson formula giving the solution of the Cauchy problem for the heat equation (cf. Chapter 1, formula (1.40)) that the fundamental solution for the Cauchy problem and hence also a fundamental solution for the heat conduction operator $\frac{\partial}{\partial t} - \Delta$, where Δ is the Laplacian in $\mathbb{R}^n$, is given by the formula

$$E(t, x) = \left(2\sqrt{\pi t}\right)^{-n} H(t) \exp\left(-|x|^2/4t\right).$$

It is easy to verify that this function $E(t, x)$ is locally integrable everywhere in $\mathbb{R}^{n+1}$ and infinitely differentiable in $\mathbb{R}^{n+1} \setminus 0$ (but not analytic!).

Example 2.22. One can find a fundamental solution for the d'Alembertian operator $\square = \frac{\partial^2}{\partial t^2} - \Delta$ in $\mathbb{R}^{n+1}$ from the known formulas giving the solution of the Cauchy problem for the wave equation for $n = 1, 2, 3$ (cf. Sect. 4.5

below). To be specific, for $n = 1$, by d'Alembert's formula this fundamental solution has the form

$$E_1(t, x) = \frac{1}{2}H(t - |x|), \quad x \in \mathbb{R}^1;$$

for $n = 2$ it follows from Poisson's formula that

$$E_2(t, x) = \frac{H(t - |x|)}{2\pi\sqrt{t^2 - |x|^2}}, \quad x \in \mathbb{R}^2$$

(the functions E_1 and E_2 are locally integrable); and for $n = 3$ Kirchhoff's formula gives

$$E_3(t, x) = \frac{1}{4\pi t}\delta(|x| - t).$$

Here $\delta(|x| - t)$ can be understood, for example, as a limit

$$\delta(|x| - t) = \lim_{\varepsilon \to +0} \varphi_\varepsilon(|x| - t),$$

where $\varphi_\varepsilon(\tau)$ is a δ-shaped family of functions of $\tau \in \mathbb{R}^1$, or using the explicit formula

$$\langle \delta(|x| - t), \varphi(t, x) \rangle = \int_{-\infty}^{\infty} \left(\int_{|x|=r} \varphi(t, x) \, dS_r \right) dt,$$

where dS_r is the element of area on the sphere of radius r.

The fundamental solutions E_1, E_2, and E_3 are the only fundamental solutions for the corresponding d'Alembertian operators $\square$ having support in the half-space $\{(t, x) | t \geq 0\}$ (in fact their support even lies in the *light cone* $\{(t, x) | t \geq 0, |x| \leq t\}$). They are therefore the only fundamental solutions consistent with the principle of causality, if we keep in mind that they must describe a wave from a point source.

1.14. Fundamental Solutions and Solutions of Inhomogeneous Equations (cf. Shilov 1965; Hörmander 1983–1985). Knowing a fundamental solution $E \in \mathcal{D}'(\mathbb{R}^n)$ for the operator $P(D)$, one can find a particular solution of any inhomogeneous equation $P(D)u = f$ with a right-hand side $f \in \mathcal{E}'(\mathbb{R}^n)$ of compact support. To be specific, one must take u in the form of a convolution

$$u = E * f. \tag{1.37}$$

In fact, applying the operator $P(D)$ to both sides, we obtain

$$P(D)u = P(D)E * f = \delta * f = f.$$

Example 2.23. The potentials described in Sect. 1.9 have the form of convolutions (1.37), where E is a fundamental solution for the Laplacian operator Δ on $\mathbb{R}^3$ and consequently they satisfy the corresponding equations of the form $\Delta u = f$, understood in the sense of distributions. For example, the Newtonian potential u satisfies Poisson's equation $\Delta u = \rho$, where ρ is the charge density.

If u is a single-layer potential, the equation $\Delta u = \sigma \delta_\Gamma$ must hold, where δ_Γ is a δ-function on the surface Γ and σ is the surface charge density on Γ. This in particular means that u is a harmonic function outside the surface Γ. In addition, this leads to jump properties of the potential u near the surface itself: u and its tangent derivatives (in local coordinates near some point of the surface) are continuous on Γ, and the normal derivative $\dfrac{\partial u}{\partial n}$ has the jump σ on Γ necessary in order to obtain $\sigma \delta_\Gamma$ on taking the second derivative $\dfrac{\partial^2 u}{\partial n^2}$. Here the jump is calculated in the direction of the same normal that was chosen for the derivative $\dfrac{\partial u}{\partial n}$.

Similarly the double-layer potential with dipole density β on the surface Γ is harmonic outside Γ and has a jump β at Γ.

Example 2.24. Consider the convolution $u = f * 1/z$ on $\mathbf{R}^2$, where $z = x + iy$, x and y are coordinates on $\mathbf{R}^2$, and $f \in \mathcal{E}'(\mathbf{R}^2)$. If $f \in L^1_{\mathrm{loc}}(\mathbf{R}^2)$, this convolution can be written in the form of an integral

$$u(z) = \int \frac{f(z')}{z - z'} \, dx' \, dy',$$

where $z' = x' + iy'$. For any $f \in \mathcal{E}'(\mathbf{R}^2)$ it satisfies the equation $\partial u / \partial \bar{z} = \pi f$, since $E(z) = 1/\pi z$ is a fundamental solution for the operator $\partial / \partial \bar{z}$ (cf. Example 2.19). If $f = g \delta_\Gamma$, where Γ is a compact piecewise-smooth curve in $\mathbf{R}^2$ and g is a piecewise-smooth function on Γ, then the convolution $f * 1/z$ can be written in the form of an integral of Cauchy type

$$u(z) = \int_\Gamma \frac{h(z')}{z - z'} \, dz',$$

where the function $h(z')$ is determined from the conditions $h(z') \, dz' = g(z') \, ds$ (where ds is the element of arc length on the curve Γ). In this case the equation $\partial u / \partial \bar{z} = \pi f$ means that the function u is holomorphic outside Γ and has a jump equal to $2\pi i h(z)$ at the point z on Γ.

Example 2.25. Formula (1.32), which gives a fundamental solution for the ordinary differential operator $P(D_t)$ with leading coefficient 1, makes it possible to write a particular solution of the equation $P(D_t)u = f$, where $f \in C(\mathbf{R})$, in the form

$$u(t) = \int_{-\infty}^{t} y(t - s) f(s) \, ds,$$

if the integral converges absolutely and can be differentiated a sufficient number of times. The lower limit here can be replaced by any $t_0 \in \mathbf{R}$ and also by $+\infty$ (again under suitable convergence conditions), since this leads to the addition of a solution of the homogeneous equation $P(D_t)u = 0$ to $u(t)$. In particular for any function $f \in C(\mathbb{R})$ the formula

$$u(t) = \int_0^t y(t-s)f(s)\, ds$$

is usable. For example a particular solution of the equation $\ddot{u} + u = f$ has the form

$$u(t) = \int_0^t \sin(t-s)f(s)\, ds.$$

The lower limit of $-\infty$ (resp. $+\infty$) is convenient when we wish to find a particular solution vanishing as $t \to -\infty$ (resp. $t \to +\infty$).

1.15. Duhamel's Principle for Equations with Constant Coefficients (cf. Courant and Hilbert 1962, Chapter 3). Let $E(t, x)$ be a fundamental solution for the Cauchy problem for the evolution operator $P = P(D_t, D_x)$ of order m (cf. Sect. 1.13). Let $f = f(t, x)$ be a suffciently smooth function on $\mathbb{R}^{n+1}$ such that $f(t, x) = 0$ for $t < 0$ and the convolution $u = E * f$ is naturally defined in the sense of Sect. 1.9. In addition let f, together with E, be a smooth function of t with values in $\mathcal{D}'(\mathbb{R}^n)$, and let the convolution on the variable x

$$\int E(t, x-y)f(t', y)\, dy$$

also be naturally defined for all t and t' in $\mathbb{R}$. Consider the complete convolution $u = E * f$ that is a particular solution of the equation $Pu = f$, and write it in the form

$$u = \int_0^t v(t', t, x)\, dt' \tag{1.38}$$

where

$$v(t', t, x) = \int E(t-t', x-y)f(t', y)\, dy.$$

We remark that the distribution $v(t', t, x)$ is a distribution on x that depends smoothly on the parameters t and t' for $t' \in (0, t)$, and for $0 < t' < t$ it satisfies the conditions

$$P(D_t, D_x)v = 0, \quad v\big|_{t=t'} = 0, \dots, \frac{\partial^{m-2}v}{\partial t^{m-2}}\Big|_{t=t'} = 0, \quad \frac{\partial^{m-1}v}{\partial t^{m-1}}\Big|_{t=t'} = f(t', x). \tag{1.39}$$

For $t = 0$ the function $u(t, x)$ itself satisfies the zero Cauchy initial conditions

$$u\big|_{t=0} = 0, \dots, \frac{\partial^{m-1}u}{\partial t^{m-1}}\Big|_{t=0} = 0. \tag{1.40}$$

Formula (1.38) (in which v satisfies conditions (1.39)) is one of the variants of *Duhamel's principle*. It is clear from it that if we know how to solve the Cauchy problem for the homogeneous equation $Pu = 0$, we can obtain the solution of the Cauchy problem for the inhomogeneous equation. A straightforward computation makes it possible to verify directly from conditions (1.39), which determine v, that the equation $Pu = f$ and the initial conditions (1.40) hold. In this way it becomes clear that a principle of exactly the same kind is true for evolution equations with variable coefficients (provided the Cauchy problem for the homogeneous equation is well-posed in some natural sense).

The simplest examples of the situation just described arise in using the fundamental solutions for the heat and wave equations (cf. Examples 2.21 and 2.22).

Example 2.26. The integral

$$u(t,x) = \int_0^t \left(2\sqrt{\pi(t-t')}\right)^{-n} \left[\int_{\mathbf{R}^n} \exp\left[-|x-y|^2/4(t-t')\right] f(t',y)\, dy\right] dt',$$

(1.41)

which is an example of a *heat potential*, defines a solution of the problem

$$\frac{\partial u}{\partial t} - \Delta u = f(t,x), \quad t > 0; \quad u\big|_{t=0} = 0,$$

provided this integral converges and can be differentiated under the integral sign once on t and twice on x (however, it suffices to require instead that it define a convolution of the distributions $\left(2\sqrt{\pi t}\right)^{-n} \exp\left[-|x|^2/4t\right]$ and $H(t)f(t,x)$ as was explained above).

Example 2.27. Let $f = f(t,x)$ be a sufficiently smooth function on $\mathbf{R}^4$ vanishing for $t < 0$. Then the function

$$u(t,x) = \int_0^t \frac{dt'}{4\pi(t-t')} \int_{|y-x|=t-t'} f(t',y)\, dS_y,$$

(1.42)

called a *retarded potential*, is a solution of the Cauchy problem

$$\Box u(t,x) = f(t,x), \quad t > 0; \quad u\big|_{t=0} = 0, \quad \frac{\partial u}{\partial t}\Big|_{t=0} = 0.$$

The formulas analogous to (1.42) in the case of two or one spatial variables have the respective forms:

$$u(t,x) = \frac{1}{2\pi} \int_0^t \left[\int_{|y-x|\le t-t'} \frac{f(t',y)\, dy}{\sqrt{(t-t')^2 - |y-x|^2}}\right] dt', \quad x,y \in \mathbf{R}^2,$$

and

$$u(t,x) = \frac{1}{2} \int_0^t \left[\int_{-(t-t')}^{t-t'} f(t',y)\,dy \right] dt', \quad x,y \in \mathbb{R}^1.$$

Essentially all these formulas are the mathematical expression of the principle of superposition of waves, which is a consequence of the linearity of the wave equation.

1.16. The Fundamental Solution and the Behavior of Solutions at Infinity. As we have just seen, completely definite particular solutions of the equation $P(D)u = f$ can sometimes be written in the form of a convolution $u = E * f$ (for example, the solution of the Cauchy problem with zero initial conditions). We give another example of such a situation. If $u \in \mathcal{E}'(\mathbb{R}^n)$ (i.e., u has compact support), $P(D)u = f$, and $E = E(x)$ is any fundamental solution for the operator $P(D)$, then $u = E * f$. Indeed in this case we have

$$E * f = E * P(D)u = \big[P(D)E\big] * u = \delta * u = u.$$

Thus a solution u of the equation $P(D)u = f$ having compact support can always be recovered from the right-hand side f in the form of a canonical convolution of the right-hand side with a fundamental solution.

We now give another example of a simple situation where similar reasoning can be applied. Let $u = u(x)$ be a harmonic function defined for $|x| > R$ in $\mathbb{R}^3$ and such that $u(x) \to 0$ as $|x| \to \infty$. We shall obtain an integral representation for it. To do this we consider a cutoff function $\chi \in C^\infty(\mathbb{R}^3)$ equal to 1 for $|x| > R + 2$ and 0 for $|x| < R + 1$, and we set $f = \Delta(\chi u)$, so that $f \in C_0^\infty(\mathbb{R}^3)$. Consider the convolution

$$v(x) = -\frac{1}{4\pi} \int \frac{f(y)}{|x-y|}\,dy \in C^\infty(\mathbb{R}^3),$$

which also satisfies the equation $\Delta v = f$ and, as is clear from the way the function v is written, tends to zero as $|x| \to +\infty$. But then $w = \chi u - v$ is a harmonic function everywhere in $\mathbb{R}^3$ and $w(x) \to 0$ as $|x| \to \infty$. By the maximum modulus principle or Liouville's theorem it follows from this that $w \equiv 0$, i.e., $v = \chi u$ and, in particular

$$u(x) = -\frac{1}{4\pi} \int \frac{f(y)}{|x-y|}\,dy, \quad |x| > R + 2.$$

It follows from this, for example, that $|u(x)| \le C/|x|$ for large $|x|$. Starting from this integral representation, we can write the complete asymptotics of $u(x)$ as $|x| \to \infty$. Such reasoning can also be applied to a variety of other equations.

1.17. Local Properties of Solutions of Homogeneous Equations with Constant Coefficients. Hypoellipticity and Ellipticity (cf. Shilov 1965; Hörmander 1983–1985).

A. The local properties of solutions of the equation $P(D)u = 0$ are connected with the local properties of the fundamental solution $E(x)$ for the operator $P(D)$ because it is possible to write an "integral" representation of an arbitrary solution $u \in \mathcal{D}'(\Omega)$, as follows. Let $x_0 \in \Omega$, and suppose we wish to study the local properties of the solution u near the point x_0. Let $\varphi \in C_0^\infty(\Omega)$ and $\varphi \equiv 1$ in a neighborhood of the point x_0. We set $f = P(D)(\varphi u)$. Then $f \in \mathcal{E}'(\Omega)$, $f \equiv 0$ in a neighborhood of the point x_0, and

$$u = E * f = E * [P(D)(\varphi u)] \tag{1.43}$$

(cf. Sect. 1.16). Essentially this means that the solution $u(x)$ can be written in the form of a superposition (sum) of translates of the fundamental solution $E(x - y)$ in a neighborhood of the point x_0 (and the translations are by vectors y separated from x_0, so that these translates are indeed solutions of the equation $P(D)u = 0$ in a neighborhood of the point x_0).

In order to establish the simplest facts about the regularity of the function u using formula (1.43), it often suffices to use the additivity properties of the singular support and the analytic singular support (cf. Sect. 1.9). It follows from them, for example, that if $E \in C^\infty(\mathbb{R}^n \setminus \{0\})$, i.e., the only singularity of the fundamental solution is at the point 0 (so that $\operatorname{sing\,supp} E = \{0\}$), then $\operatorname{sing\,supp} u = \operatorname{sing\,supp} [P(D)(\varphi u)]$, and in particular $u \in C^\infty$ in a neighborhood of the point x_0. Since the point x_0 can be arbitrary, it follows that $u \in C^\infty(\Omega)$.

Conversely, if any solution $u \in \mathcal{D}'(\Omega)$ of the equation $P(D)u = 0$ is infinitely differentiable, then of course $E \in C^\infty(\mathbb{R}^n \setminus \{0\})$ for any fundamental solution E of the operator $P(D)$. Thus the following assertions are equivalent:

a) there exists a fundamental solution $E \in \mathcal{D}'(\mathbb{R}^n)$ for the operator $P(D)$ such that $E \in C^\infty(\mathbb{R}^n \setminus \{0\})$;

b) every solution $u \in \mathcal{D}'(\Omega)$ of the equation $P(D)u = 0$ is infinitely differentiable.

When these conditions are satisfied, the operator $P(D)$ is called *hypoelliptic*. Examples of hypoelliptic operators are the Laplacian Δ in $\mathbb{R}^n$, the heat operator $\frac{\partial}{\partial t} - \Delta$ in $\mathbb{R}^{n+1}$, and the Cauchy-Riemann operator $\frac{\partial}{\partial \bar{z}}$ in $\mathbb{R}^2$. The hypoellipticity of these operators follows from the fact pointed out in Sect. 1.7 that their fundamental solutions are infinitely differentiable except at the origin. Thereby any generalized solutions of the corresponding homogeneous equations are infinitely differentiable. An example of a nonhypoelliptic operator is the d'Alembertian $\square$ in $\mathbb{R}^{n+1}$ for any $n \geq 1$.

Similarly it can be established that the following conditions are equivalent:

a') there exists a fundamental solution E for the operator $P(D)$ that is analytic except at the origin;

b') all solutions $u \in \mathcal{D}'(\Omega)$ of the equation $P(D)u = 0$ are analytic in Ω.

Conditions a') and b') are fulfilled for the Laplacian and Cauchy-Riemann operators, but not for the heat-conduction operator In particular every generalized solution of Laplace's equation or the Cauchy-Riemann equation is analytic.

It turns out that conditions a') and b') are equivalent to the condition that the operator $P(D)$ be *elliptic*. Thus every elliptic operator with constant coefficients has a fundamental solution that is analytic except at the origin.

B. We shall exhibit explicit and comparatively easily verified conditions for the operator $P(D)$ to be hypoelliptic, and also a method by which they may be obtained.

Let the operator $P(D)$ be hypoelliptic, Ω a region in $\mathbb{R}^n$, Ω' a subregion of Ω such that $\overline{\Omega}'$ is a compact subset of Ω. The set of solutions $u \in L^1(\Omega)$ of the equation $P(D)u = 0$ is closed in $L^1(\Omega)$ and consequently is a Banach space (with the norm of the space $L^1(\Omega)$). The restriction operator $u \mapsto u|_{\Omega'}$ from this space into $C^k(\Omega')$ (which is defined by virtue of the hypoellipticity of the operator $P(D)$) has a closed graph and is consequently continuous. In particular

$$\|u\|_{C^k(\Omega')} \le C\|u\|_{L^1(\Omega)},$$

where the constant C is independent of u. Applying this estimate in the case of concentric balls $\Omega' \subset \Omega$ with center at the origin and for the exponentials $u(x) = e^{ix\cdot\zeta}$, where $\zeta \in \mathbb{C}^n$ is such that $P(\zeta) = 0$, we find that the roots $\zeta = \xi + i\eta$ of the polynomial $P(\xi)$ must satisfy the condition

$$|\eta| \ge A \ln |\xi| - B, \tag{1.44}$$

where A and B are positive constants independent of the choice of the root ζ. Thus estimate (1.44) for the roots $\zeta = \xi + i\eta$ of the polynomial $P(\zeta)$ is a necessary condition for the operator $P(D)$ to be hypoelliptic. It turns out that this estimate is also sufficient, as can be established, for example, by an explicit construction of a fundamental solution using an extension into the complex plane.

In fact one can strengthen estimate (1.44) by applying the Seidenberg-Tarski Theorem (cf., for example, Hörmander 1983–1985, Appendix A to Vol. 2), which asserts that a condition for a real system of polynomial equations depending polynomially on parameters to be solvable can be written in the form of a finite system of polynomial equations and inequalities. (or, more briefly, that the linear projection of a real algebraic manifold is semialgebraic). Because of this theorem condition (1.44) is equivalent to the stronger condition

$$|\eta| \ge A|\xi|^{1/\beta} - B, \tag{1.45}$$

where $\beta > 0$.

Further, we can give simple necessary and sufficient conditions for estimates (1.44) and (1.45) to hold in terms of the polynomial $P(\xi)$ directly,

not in terms of its complex roots. In particular in this way we find that a
necessary and sufficient condition for hypoellipticity is the following limiting
relation

$$\frac{|\nabla P(\xi)|}{|P(\xi)|} \to 0 \quad \text{as } |\xi| \to +\infty \tag{1.46}$$

(here $\nabla P = \left(\dfrac{\partial P}{\partial \xi_1}, \ldots, \dfrac{\partial P}{\partial \xi_n}\right)$ is the gradient of the polynomial ξ). It follows
in particular from relation (1.46) that

$$|P(\xi)| \to +\infty \quad \text{as } |\xi| \to \infty, \tag{1.47}$$

but the converse assertion is false (for example, the polynomial $P(\xi_1, \xi_2) = \xi_1^2 \xi_2^2 + \xi_1^2 + \xi_2^2 + 1$ satisfies (1.47) but not (1.46)).

Along with hypoellipticity and ellipticity, we can consider variants of Gevrey hypoellipticity intermediate between them. The Gevrey class to which the fundamental solution (and all solutions of the homogeneous equation) belongs is determined by the index β in (1.45) (cf. Shilov 1965 or Hörmander 1983–1985).

1.18. Liouville's Theorem for Equations with Constant Coefficients. The general Liouville theorem for equations with constant coefficients can be given a natural statement and proof in the context of distributions. We shall exhibit one variant of it (cf. Vladimirov 1967, Sect. 24).

Theorem 2.28. *Let an operator $P(D)$ with constant coefficients be given in $\mathbb{R}^n$ such that $P(\xi) \neq 0$ for $\xi \neq 0$. Then if $u \in S'(\mathbb{R}^n)$ and $P(D)u = 0$, it follows that u is a polynomial in x.*

To prove this it suffices to apply the Fourier transform to the equation $P(D)u = 0$. We then obtain $P(\xi)\tilde{u}(\xi) = 0$, from which it follows that the tempered distribution $\tilde{u}(\xi)$ is concentrated at the point $0 \in \mathbb{R}_\xi^n$ and consequently is equal to $\sum\limits_{|\alpha| \leq N} c_\alpha \delta^{(\alpha)}(\xi)$, i.e., $u(x) = \sum\limits_{|\alpha| \leq N} c'_\alpha x^\alpha$, where c_α and c'_α are constants.

If $P(\xi) \neq 0$ for all $\xi \in \mathbb{R}^n$, then the equation $P(D)u = 0$ has no nontrivial tempered distribution solutions.

The condition $P(\xi) \neq 0$ for $\xi \neq 0$ is obviously necessary for the conclusion of the theorem to hold since if $P(\xi_0) = 0$, then the equation $P(D)u = 0$ has the solution $u(x) = e^{i\xi_0 \cdot x}$.

Example 2.29. The Laplacian operator Δ satisfies the hypotheses of the theorem. In particular, every solution that is bounded throughout $\mathbb{R}^n$ is a polynomial and therefore constant.

Example 2.30. The heat operator $\dfrac{\partial}{\partial t} - \Delta$ in $\mathbf{R}^{n+1}$ has the form $P(D_t, D_x)$, where $P(\tau, \xi) = i\tau + |\xi|^2$, so that it also satisfies the conditions for Liouville's theorem to be applicable.

1.19. Isolated Singularities of Solutions of Hypoelliptic Equations. Let Ω be a region in $\mathbf{R}^n$ containing the point 0, and let $u = u(x)$ be a solution of the hypoelliptic equation $P(D)u = 0$, defined on $\Omega \setminus \{0\}$. Assume that near the point 0 the solution u has at most polynomial growth, i.e., there exist C, N, and $\varepsilon > 0$ such that

$$|u(x)| \le C|x|^{-N} \quad \text{for } 0 < |x| < \varepsilon. \tag{1.48}$$

In this case we can describe the behavior of $u(x)$ near the point 0 using some fundamental solution $E(x)$ for the operator $P(D)$. To be specific, let $\hat{u}(x)$ be an extension of $u(x)$ to a distribution $\hat{u} \in \mathcal{D}'(\Omega)$. It is easy to see that such an extension always exists because of condition (1.48). For example, we can set

$$\langle \hat{u}, \varphi \rangle = \int\limits_{|x| \le \varepsilon} u(x) \left[\varphi(x) - \sum_{|\alpha| \le N} \frac{\varphi^{(\alpha)}(0)}{\alpha!} x^\alpha \right] dx + \int\limits_{|x| > \varepsilon} u(x)\varphi(x)\, dx.$$

Then $P(D)\hat{u} = \sum\limits_{|\alpha| \le k} c_\alpha \delta^{(\alpha)}(x)$, where c_α are some constants. Now consider the distribution $v(x) = \sum\limits_{|\alpha| \le k} c_\alpha E^{(\alpha)}(x)$, which is obviously a particular solution of the equation $P(D)v = \sum\limits_{|\alpha| \le k} c_\alpha \delta^{(\alpha)}(x)$. But then $P(D)(\hat{u} - v) = 0$, whence $\hat{u} - v \in C^\infty(\Omega)$ by the hypoellipticity of the operator $P(D)$. Thus we find as a result that

$$u(x) = \sum_{|\alpha| \le k} c_\alpha E^{(\alpha)}(x) + w(x), \quad w \in C^\infty(\Omega), \quad P(D)w = 0. \tag{1.49}$$

This means that all the singularities of polynomial growth are like those of linear combinations of derivatives of the fundamental solution. This fact can be used in particular to establish theorems of *removable singularity type*.

Example 2.31. Let $n = 2$, let $\mathbf{R}^2$ be identified with $\mathbf{C}$, and let the function $u(z)$ be holomorphic in $\Omega \setminus \{0\}$, where Ω is a region in $\mathbf{C}$ containing 0. Then taking $P(D) = \partial/\partial\bar{z}$, we see that if u satisfies the polynomial growth condition (1.48) near the point 0, then

$$u(z) = \sum_{k=1}^{N} \frac{c_k}{z^k} + v(z),$$

where $v(z)$ is holomorphic in Ω (of course this is a well-known proposition on the Laurent expansion).

Example 2.32. Let $P(D) = \Delta$ and $n \geq 3$. We find from the discussion above that if u is harmonic in $\Omega \setminus 0$, where $0 \in \Omega$, and u satisfies (1.48), then

$$u(x) = \sum_{|\alpha| \leq k} c_\alpha D^\alpha \frac{1}{r^{n-2}} + v(x),$$

where v is harmonic in Ω. Assuming in particular that u is bounded near the point 0 or even the weaker condition $u(x) = o(|x|^{2-n})$ as $|x| \to 0$, we find that $u \equiv v$, which gives the classical removable singularity theorem for harmonic functions. For $n = 2$ and $P(D) = \Delta$ the representation (1.49) assumes the form

$$u(x) = c_0 \ln |x| + \sum_{|\alpha| \leq k} \left[a_\alpha D^\alpha \frac{x_1}{|x|^2} + b_\alpha D^\alpha \frac{x_2}{|x|^2} \right] + v(x),$$

where c_0, a_α, and b_α are constants and $v(x)$ is harmonic in Ω. In particular if u is harmonic in $\Omega \setminus \{0\}$, and $u(x) = o\left(\ln \frac{1}{|x|} \right)$ as $|x| \to 0$, then $u \equiv v$ and we obtain a removable singularity theorem for this case.

§2. Elliptic Equations and Boundary-Value Problems

2.1. The Definition of Ellipticity. The Laplace and Poisson Equations. We recall (cf. Sect. 1.3) that an *elliptic equation* is an equation of the form $Au = f$, where A is an *elliptic* differential operator, i.e.

$$A = \sum_{|\alpha| \leq m} a_\alpha(x)D^\alpha, \tag{2.1}$$

and for all x in the region Ω in which the equation is being studied and all $\xi \neq 0$ the principal symbol of the operator A

$$a_m(x, \xi) = \sum_{|\alpha|=m} a_\alpha(x)\xi^\alpha \tag{2.2}$$

does not vanish. The unknown function u and the right-hand side f can be vector-valued functions, so that the equation $Au = f$ is actually a system of equations. In this case the principal symbol is a matrix-valued function and ellipticity (more precisely, *Petrovskij ellipticity*) means that the matrix $a_m(x, \xi)$ is invertible for $\xi \neq 0$.

The simplest of all elliptic equations are the Cauchy-Riemann equation (cf. Sect. 1.1) and second-order elliptic equations of the form

$$\sum_{i,j=1}^{n} a_{ij}(x)\frac{\partial^2 u}{\partial x_i \partial x_j} + \sum_{j=1}^{n} b_j(x)\frac{\partial u}{\partial x_j} + c(x)u = f. \qquad (2.3)$$

The principal part of such an equation can be written in *divergent form*, so that the equation assumes the form

$$\sum_{i,j=1}^{n} \frac{\partial}{\partial x_i}\left(a_{ij}(x)\frac{\partial u}{\partial x_j}\right) + \sum_{j=1}^{n} b_j(x)\frac{\partial u}{\partial x_j} + c(x)u = f \qquad (2.3')$$

(with different functions b_j). It is clear that if $a_{ij} \in C^1(\Omega)$, then we can pass from the form (2.3) to the form (2.3') and back again. For the equations (2.3) or (2.3') ellipticity means that the quadratic form

$$\sum_{i,j=1}^{n} a_{ij}(x)\xi_i\xi_j, \qquad (2.4)$$

which differs only in sign from the principal symbol, is either positive-definite or negative-definite.

Elliptic equations usually describe stationary situations in which x is a set of spatial variables, so that there are no distinguished variables (of time type). The statement of the simplest boundary-value problems for these equations, for example the Dirichlet and Neumann problems, is connected with this fact (cf. Sect. 1.1).

The simplest second-order elliptic equations are Laplace's equation

$$\Delta u = 0 \qquad (2.5)$$

and Poisson's equation

$$\Delta u = f. \qquad (2.5')$$

A solution of Laplace's equation, i.e., a function $u \in C^2(\Omega)$ satisfying the equation in a region $\Omega \subset \mathbb{R}^n$, is called a *harmonic function* (in the region Ω). Many properties of solutions of the equations of Laplace and Poisson generalize to solutions of the second-order elliptic equations (2.3) and (2.3') with various modifications. We shall now exhibit the most basic of these properties.

2.2. A Fundamental Solution for the Laplacian Operator. Green's Formula (cf. Petrovskij 1961; Shilov 1965; Vladimirov 1967; Hörmander 1983–1985). The most important example of a harmonic function is the fundamental solution for the Laplacian operator, which from the physical point of view, is the potential of a unit point charge located at the point 0. It can be obtained by seeking harmonic functions of the form $E(x) = f(r)$, where $r = |x|$. In doing this we arrive at the equation

$$f''(r) + \frac{n-1}{r} f'(r) = 0,$$

which has the general solution

$$f(r) = C_1 r^{2-n} + C_2, \quad n \geq 3,$$
$$f(r) = C_1 \ln r + C_2, \quad n = 2.$$

The constant C_2 plays no role here and can be omitted. We choose the constant C_1 as follows:

$$E(x) = -\frac{1}{(n-2)\sigma_{n-1}} r^{2-n}, \quad n \geq 3, \tag{2.6}$$

where σ_{n-1} is the area of the unit sphere in $\mathbf{R}^n$ and

$$E(x) = \frac{1}{2\pi} \ln r, \quad n = 2. \tag{2.6'}$$

This is connected with the fact that E must satisfy the equation (cf. Sect. 2.1)

$$\Delta E = \delta(x), \tag{2.7}$$

where $\delta(x)$ is the Dirac δ-function. This equation means that if $\varphi \in C_0^\infty(\mathbf{R}^n)$ (i.e., $\varphi \in C^\infty(\mathbf{R}^n)$ and $\varphi(x) = 0$ for large $|x|$), then

$$\int E(x)\Delta\varphi(x)\,dx = \varphi(0). \tag{2.7'}$$

Let us apply *Green's formula*

$$\int_\Omega (u\Delta v - v\Delta u)\,dx = \int_{\partial\Omega} \left(u\frac{\partial v}{\partial n} - v\frac{\partial u}{\partial n} \right) dS, \quad u, v \in C^2(\bar{\Omega}) \tag{2.8}$$

(here Ω is a bounded region with a smooth boundary and n is the exterior normal to $\partial\Omega$), which follows from the general Stokes' theorem via the observation that $u\Delta v - v\Delta u = \operatorname{div}(u\operatorname{grad} v - v\operatorname{grad} u)$. To be specific, let us take $\Omega = \{x : \varepsilon \leq |x| \leq R\}$, where $\varepsilon > 0$ is very small and $R > 0$ very large, and let $u = E$ and $v = \varphi$. Then by passing to the limit as $\varepsilon \to +0$ we find that (2.7') is equivalent to the relation

$$\lim_{\varepsilon \to +0} \int_{|x|=\varepsilon} \frac{\partial E}{\partial r}\,dS = 1,$$

from which the form of the constant in formulas (2.6) and (2.6') follows. We remark in passing that the same reasoning leads to what is called *Green's second formula*

$$u(x_0) = \int\limits_{\Omega} E(x - x_0)\Delta u(x)\, dx +$$ (2.9)

$$+ \int\limits_{\partial\Omega} \left[u(x)\frac{\partial E(x - x_0)}{\partial n_x} - E(x - x_0)\frac{\partial u(x)}{\partial n_x} \right] dS_x,$$

where Ω is a bounded region with smooth boundary, $u \in C^2(\bar{\Omega})$, and $x_0 \in \Omega$ (if $x_0 \notin \bar{\Omega}$, the right-hand side of (2.9) is 0). In particular if u is a harmonic function in Ω, then we obtain

$$u(x_0) = \int\limits_{\partial\Omega} \left[u(x)\frac{\partial E(x - x_0)}{\partial n_x} - E(x - x_0)\frac{\partial u(x)}{\partial n_x} \right] dS_x, \quad x_0 \in \Omega. \quad (2.10)$$

It follows from this formula in particular that the function u is analytic in Ω, since $E(x)$ has this property for $x \neq 0$.

2.3. Mean-Value Theorems for Harmonic Functions (cf. Petrovskij 1961; Courant and Hilbert 1962; Vladimirov 1967). A consequence of (2.10) is a *mean-value theorem* for harmonic functions. To see this we first remark that if we take $v = 1$ in (2.8), we obtain for any harmonic function $u \in C^2(\bar{\Omega})$

$$\int\limits_{\partial\Omega} \frac{\partial u(x)}{\partial n_x}\, dS_x = 0.$$ (2.11)

We now apply formula (2.10), taking as Ω the ball $\{x : |x - x_0| \leq R\}$. We then obtain the formula

$$u(x_0) = \frac{1}{\sigma_{n-1}R^{n-1}} \int\limits_{|x-x_0|=R} u(x)\, dS_x,$$ (2.12)

which is the content of the mean-value theorem. Another variant of this theorem is obtained if we multiply both sides of (2.12) by R^{n-1} and integrate on R from 0 to R:

$$u(x_0) = \frac{1}{V_n R^n} \int\limits_{|x-x_0|\leq R} u(x)\, dx,$$ (2.13)

where V_n is the volume of the unit ball in $\mathbb{R}^n$.

2.4. The Maximum Principle for Harmonic Functions and the Normal Derivative Lemma (Landkof 1966; Landis 1971). An immediate consequence of the mean-value theorem (2.13) is the *maximum principle* for harmonic functions: if u is a harmonic function in Ω, then u cannot have local maxima or minima in Ω. More precisely, if the value of u at the point x_0 is such that $u(x_0) \geq u(x)$ for $|x - x_0| < \varepsilon$, or $u(x_0) \leq u(x)$ for $|x - x_0| < \varepsilon$, where $\varepsilon > 0$, then $u(x) = $ const in the component of the set Ω containing x_0. In particular if Ω is a bounded region in $\mathbb{R}^n$, $u \in C(\bar{\Omega})$, and u is harmonic in Ω, then for $x \in \Omega$

$$\min_{x \in \partial\Omega} u(x) \leq u(x) \leq \max_{x \in \partial\Omega} u(x). \tag{2.14}$$

The maximum principle can also be proved by considering the behavior of the function u and its derivatives at an extremum. For example, if x_0 is a strict maximum and lies in the interior of Ω, then $\dfrac{\partial^2 u}{\partial x_j^2}\Big|_{x=x_0} \leq 0$. This would be impossible if u satisfied the condition $\Delta u > 0$ instead of the equation $\Delta u = 0$. But everything reduces to this case if we replace u by $u + \varepsilon x_1^2$, where $\varepsilon > 0$ is very small. This proof can be extended to general second-order elliptic equations (cf. Sect. 2.16 below).

We note also the following property of harmonic functions at a point $x_0 \in \partial\Omega$ where the function u attains its maximum on Ω.

Lemma 2.33 (The normal derivative lemma). *Let u be harmonic in the region $\Omega \subset \mathbb{R}^n$ and continuous in $\bar{\Omega}$, and suppose it attains its maximum at a point $x_0 \in \partial\Omega$ such that the boundary $\partial\Omega$ has a tangent plane at this point. Further suppose the derivative $\dfrac{\partial u}{\partial \nu}\Big|_{x=x_0} = \lim\limits_{\varepsilon \to +0} \varepsilon^{-1}\big(u(x_0) - u(x_0 - \varepsilon\nu)\big)$ exists, where ν is the unit external normal to $\partial\Omega$ at the point x_0. Then if $u \not\equiv$ const, it follows that $\dfrac{\partial u}{\partial \nu}\Big|_{x=x_0} > 0$.*

The exterior normal ν here can be replaced by any direction forming an acute angle with ν.

It suffices to prove the normal derivative lemma in the case when Ω is a ball of radius R with center at the origin. Let $w(x) = |x|^{2-n} - R^{2-n}$ for $n > 2$ and $w(x) = \ln R - \ln|x|$ for $n = 2$. The function $v(x) = u(x) + \varepsilon w(x)$ is harmonic for $R/2 < |x| < R$ and therefore attains its maximum on the boundary of this region. On the other hand, by the maximum principle, if $u(x) \not\equiv C$, then $\max\limits_{|x|=R/2} u(x) < u(x_0)$. Consequently $v(x) < u(x_0)$ for $|x| = R/2$ if $\varepsilon > 0$ is sufficiently small. Since $v(x) = u(x)$ for $|x| = R$, the function $v(x)$ attains a maximum at the point $x = x_0$, and therefore $\dfrac{\partial v(x_0)}{\partial \nu} \geq 0$, so that $\dfrac{\partial u(x_0)}{\partial \nu} \geq -\varepsilon \dfrac{\partial w(x_0)}{\partial \nu} > 0$.

2.5. Uniqueness of the Classical Solutions of the Dirichlet and Neumann Problems for Laplace's Equation (cf. Petrovskij 1961; Vladimirov 1967). The maximum principle and the normal derivative lemma guarantee the uniqueness of the classical solutions of the *Dirichlet problem*

$$\Delta u = 0, \quad \text{in } \Omega, \quad u\big|_{\partial\Omega} = \varphi \qquad (2.15)$$

and the *Neumann problem*

$$\Delta u = 0 \quad \text{in } \Omega, \quad \frac{\partial u}{\partial n}\bigg|_{\partial\Omega} = \psi \qquad (2.16)$$

in a bounded region Ω. (The solution of the Neumann problem is unique up to an additive constant). Here the term *classical solution* for the Dirichlet problem is interpreted to mean a solution $u \in C^2(\Omega) \cap C(\bar{\Omega})$ and for the Neumann problem a solution $u \in C^2(\Omega) \cap C(\bar{\Omega})$ having a normal derivative $\frac{\partial u}{\partial n}$ at each point of the boundary (in the case of the Neumann problem it is necessary to assume that the boundary has a tangent plane at each point). Obviously the solution of both problems is unique for Poisson's equation $\Delta u = f$ as well as Laplace's equation $\Delta u = 0$.

It also follows from the maximum principle that the solution u of the Dirichlet problem (2.15) depends continuously on the boundary function φ if both u and φ are given the sup-norm (i.e., φ is considered to be in the space $C(\partial\Omega)$ and the solution in the space $C(\bar{\Omega})$). The corresponding fact for the Neumann problem (if we consider solutions normalized in some way, for example solutions equal to zero at some point $x_0 \in \Omega$) requires other norms and more delicate considerations. We shall return to this question later. We remark in passing that that (2.11) gives a necessary condition for the Neumann problem (2.16) to have a solution in the form

$$\int_{\partial\Omega} \psi \, dS_x = 0 \qquad (2.17)$$

in the case of regions with sufficiently smooth boundary. As will be explained in more detail below, this condition is also sufficient, while the Dirichlet problem is solvable in such regions with any continuous function φ on $\partial\Omega$.

2.6. Internal A Priori Estimates for Harmonic Functions. Harnack's Theorem (cf. Petrovskij 1961; Vladimirov 1967; Mikhlin 1977; Mikhailov 1983). From formula (2.10), which gives an integral representation of a harmonic function $u(x)$ in Ω in terms of the values of $u\big|_{\partial\Omega}$ and $\frac{\partial u}{\partial n}\big|_{\partial\Omega}$, it follows in an obvious manner that if K is a compact subset of Ω, then for any multi-index α

$$\sup_K |D^\alpha u(x)| \le C\Big(\sup_\Omega |u(x)| + \sum_{j=1}^n \sup_\Omega \Big| \frac{\partial u}{\partial x_j} \Big| \Big),$$

where C depends on α and K, but not on u. This inequality is an example of an *internal a priori estimate*. It can easily be sharpened by starting from some other, more convenient, integral representation of the harmonic function. For example, if the function $\varphi \in C_0^\infty(\mathbf{R}^n)$ vanishes for $|x| > \varepsilon$, where $\varepsilon > \rho(K, \partial\Omega)$ (here $\rho(K, \partial\Omega)$ is the distance from K to $\partial\Omega$) and φ is spherically symmetric, i.e., $\varphi(x) = f(r)$, where $r = |x|$ and $\int \varphi(x) \, dx = 1$, then it follows from the mean-value theorem that for $x \in K$

$$u(x) = \int \varphi(x - y) u(y) \, dy.$$

Differentiating this relation, we obtain a sharper internal a priori estimate

$$|D^\alpha u(x)| \le C_{\alpha, K} \int_\Omega |u(x)| \, dx, \quad x \in K. \tag{2.18}$$

A particular consequence is *Harnack's theorem*, which asserts that if a sequence of harmonic functions $\{u_k\}_{k=1}^\infty$ defined in Ω converges uniformly on each compact subset K of Ω to a function u, then u is harmonic in Ω. (In fact it is clear from (2.18) that it suffices for the sequence u_k to converge in $L^1(K)$ for every compact set K. Moreover even weaker convergence will suffice – weak convergence in the space of distributions in Ω – cf. Sect. 2.1.)

It is simple to sharpen the dependence of the constants $C_{\alpha, K}$ on α and K in estimates of type (2.18). For example we have the estimates

$$|D^\alpha u(x)| \le \Big(\frac{n|\alpha|}{d} \Big)^{|\alpha|} \max_{\bar\Omega} |u(x)|, \quad d = \rho(K, \partial\Omega), \tag{2.19}$$

from which it follows that the function u is analytic (cf. Mikhailov 1983, Sect. 4.3).

2.7. The Green's Function of the Dirichlet Problem for Laplace's Equation (cf. Vladimirov 1967; Tikhonov and Samarskij 1977; Smirnov 1981). In the study of the Dirichlet problem for the equations of Laplace and Poisson an important role is played by the *Green's function*, sometimes called a *source function*. If Ω is a bounded region in $\mathbf{R}^n$, the Green's function of this region is defined as a function $G(x, y)$ on $\Omega \times \Omega$ having the form

$$G(x, y) = E(x - y) + v(x, y) \tag{2.20}$$

where $v(x, y)$ satisfies the equation $\Delta_x v(x, y) = 0$, i.e., is harmonic on x for each fixed y, and $G(x, y)$ satisfies the boundary condition

$$G(x, y)\big|_{x \in \partial\Omega} = 0. \tag{2.21}$$

The physical interpretation of the Green's function is obvious from (2.20) and (2.21): it is the potential at the point x due to a point charge located at the point y inside a grounded conducting surface $\partial\Omega$. Instead of (2.20) and (2.21) we can write more briefly

$$\Delta_x G(x, y) = \delta(x - y), \quad G(x, y)\big|_{x \in \partial\Omega} = 0. \tag{2.22}$$

In view of the uniqueness of the solution of the Dirichlet problem it is clear that the Green's function of a bounded region is unique. It is also clear that if the solution of the Dirichlet problem (2.15) exists for any function $\varphi \in C(\partial\Omega)$, then there exists a Green's function for this region (equal to the solution of this problem with boundary values $\varphi(x) = -E(\dot{x} - y)$). Conversely, given a Green's function, one can find the solution of the Dirichlet problem for arbitrary boundary values φ. To be specific, in the case of a region with a smooth boundary, using Green's formula (2.8) with $v(x) = G(x, y)$, we find that if there exists a solution u of the Dirichlet problem

$$\Delta u = f, \quad u\big|_{\partial\Omega} = \varphi, \tag{2.23}$$

then it is given by the formula

$$u(x) = \int_\Omega G(y, x) f(y)\, dy + \int_{\partial\Omega} \frac{\partial G(y, x)}{\partial n_y} \varphi(y)\, dS_y. \tag{2.24}$$

In particular for the solution of the problem (2.15) the formula

$$u(x) = \int_{\partial\Omega} \frac{\partial G(y, x)}{\partial n_y} \varphi(y)\, dS_y \tag{2.25}$$

holds. We now indicate the operator significance of the Green's function. If we introduce the operator A with domain of definition

$$D_A = \{u : u \in C^\infty(\bar\Omega),\ u\big|_{\partial\Omega} = 0\},$$

equal to Δ on D_A and assume that it is invertible as an operator $A : D_A \to C^\infty(\bar\Omega)$, then $G(y, x)$ will be the kernel for the inverse operator, since

$$(A^{-1}f)(x) = \int_\Omega G(y, x) f(y)\, dy$$

by virtue of (2.24). We note that because the operator A is symmetric, i.e. the identity

$$(Au, v) = (u, Av), \quad u, v \in D_A,$$

holds, (here $(\cdot, \cdot)$ is the inner product on $L^2(\Omega)$), the operator A^{-1} is also symmetric, from which it follows easily that the Green's function is symmetric:

$$G(x, y) = G(y, x). \tag{2.26}$$

In particular we can replace $G(y, x)$ by $G(x, y)$ in formulas (2.24) and (2.25).

It is often convenient to use the Green's function for unbounded regions Ω as well. In this case additional conditions besides (2.22) are usually imposed on G at infinity to guarantee the uniqueness and optimal behavior of G as $|x| \to \infty$. For example if Ω is the exterior of the bounded region $\mathbf{R}^n \setminus \bar{\Omega}$, it suffices to require that

$$\lim_{|x| \to \infty} G(x, y) = 0, \quad n \geq 3; \tag{2.27}$$

$$|G(x, y)| = O(1) \quad \text{as } |x| \to \infty, \quad n = 2. \tag{2.28}$$

2.8. The Green's Function and the Solution of the Dirichlet Problem for a Ball and a Half-Space. The Reflection Principle (cf. Vladimirov 1967; Tikhonov and Samarskij 1977; Smirnov 1981). By writing out the Green's function explicitly for a specific region it is often possible to use it to prove the existence of a solution of the Dirichlet problem for Laplace's equation, i.e., problem (2.15), by directly verifying that formula (2.25) gives such a solution. This can be done, for example, in the case of a ball and a half-space. The more general problem (2.23) (the Dirichlet problem for Poisson's equation) could be solved similarly, but this problem reduces easily to the problem (2.15) if one subtracts from a hypothetical solution u a particular solution of the equation $\Delta u = f$ having the form

$$u_1(x) = \int_\Omega E(x - y) f(y) \, dy. \tag{2.29}$$

That u_1 is indeed a solution of the equation $\Delta u = f$ can be verified directly for $f \in C^1(\bar{\Omega})$ using Green's formula or the theory of distributions. If we know only that $f \in C(\bar{\Omega})$, then u_1 is a solution in the weaker sense of distribution theory (cf. Sect. 1.9). The Green's functions of the ball and the half-space are found by the *reflection principle*. To be specific, in the case of a half-space

$$\Omega = \{(x', x_n) : x_n > 0\} \subset \mathbf{R}^n,$$

where $x' = (x_1, \ldots, x_{n-1})$, the Green's function has the form

$$G(x, y) = E(x - y) - E(x - \bar{y}) \tag{2.30}$$

where $\bar{y} \in \mathbf{R}^n$ is the point symmetric to y with respect to the plane $x_n = 0$, i.e., if $y = (y', y_n)$, then $\bar{y} = (y', -y_n)$. It is easy to see that for any $n \geq 2$

$$\lim_{|x| \to \infty} G(x, y) = 0,$$

and the Green's function possessing this property is unique. Elementary calculations show that formula (2.25) assumes the form

$$u(x) = \frac{2x_n}{\sigma_{n-1}} \int_{\mathbf{R}^{n-1}} \frac{\varphi(y') \, dy'}{\left(|x' - y'|^2 + x_n^2\right)^{n/2}}, \quad x_n > 0. \tag{2.31}$$

For example for $n = 2$ we obtain

$$u(x_1, x_2) = \frac{x_2}{\pi} \int\limits_{\mathbf{R}} \frac{\varphi(y) \, dy}{(x_1 - y)^2 + x_2^2}. \tag{2.31'}$$

It is easy to verify that formula (2.31) actually defines a bounded solution of the Dirichlet problem in the half-space for any bounded continuous function φ on $\mathbf{R}^{n-1}$. If $\varphi(y) \to 0$ as $|y| \to \infty$, then $u(x) \to 0$ as $|x| \to \infty$.

For the ball $\Omega = \{x : |x| < R\}$ the geometric concept of inversion suggests that the point symmetric to x should be taken to be the point $\bar{x} = \dfrac{R^2 x}{|x|^2}$. Instead of (2.30) we write

$$G(x, y) = E(x - y) - c(y)E(x - \bar{y}) - c_1(y),$$

where $c(y)$ and $c_1(y)$ must be chosen so that the boundary condition (2.21) is satisfied. From this it follows that

$$G(x, y) = E(x - y) - \frac{R^{n-2}}{|y|^{n-2}} E(x - \bar{y}). \tag{2.32}$$

Applying formula (2.25), we find that the solution of the Dirichlet problem for the ball must have the form

$$u(x) = \frac{R^2 - |x|^2}{\sigma_{n-1} R} \int\limits_{|y|=R} \frac{\varphi(y) \, dS_y}{|x - y|^n}. \tag{2.33}$$

This formula is often called *Poisson's formula*. For $n = 2$ it can be rewritten in the form

$$u(x) = \frac{1}{2\pi R} \int_0^{2\pi} \frac{\varphi(R \cos \theta, R \sin \theta)(R^2 - r^2) \, d\theta}{\sqrt{R^2 + r^2 - 2Rr \cos \theta}}, \qquad r = |x|. \tag{2.33'}$$

As in the case of the half-space, it can be verified that for any function φ continuous on the sphere $\{x : |x| = R\}$ formula (2.33) indeed gives a solution of the Dirichlet problem.

2.9. Harnack's Inequality and Liouville's Theorem (cf. Petrovskij 1961; Landis 1971). A consequence of formula (2.33), which gives the solution of the Dirichlet problem in a ball is *Harnack's inequality* for a nonnegative harmonic function $u \in C(\bar{\Omega}) \cap C^2(\Omega)$, where $\Omega = \{x : |x| < R\}$ is the ball:

$$\frac{R^{n-2}(R - |x|)}{(R + |x|)^{n-1}} u(0) \le u(x) \le \frac{R^{n-2}(R + |x|)}{(R - |x|)^{n-1}} u(0), \tag{2.34}$$

(the proof uses the estimates $|y| - |x| \le |x - y| \le |y| + |x|$ and the spherical mean-value theorem). Hence it follows in particular that if u is harmonic everywhere in $\mathbf{R}^n$ and $u \geq 0$, then $u = $ const. (We pass to the limit as $R \to \infty$ in (2.34).) An obvious consequence is the following theorem.

Theorem 2.34 (Liouville). *If the function u is harmonic in $\mathbf{R}^n$ and bounded below (i.e., if $u(x) \geq -C$ for some $C > 0$ and all $x \in \mathbf{R}^n$), then $u(x) = $ const.*

It can also be shown that if u is harmonic in $\mathbf{R}^n$ and

$$u(x) \geq -C(1 + |x|)^m,$$

then $u(x)$ is a polynomial of degree at most m. For information on theorems of Liouville type for more general equations see Sect. 1, Ch. 2.

2.10. The Removable Singularities Theorem (cf. Vladimirov 1967, Sect. 24). It is natural to pose the question of the minimal irremovable singularity that a function harmonic in $\mathcal{U} \setminus \{x_0\}$ can have at the point x_0. Here $\mathcal{U}$ is some neighborhood of the point x_0. An example of a function with an irremovable singularity is the translated Green's function $E(x - x_0)$. It turns out that this singularity is the minimum possible, as the following theorem asserts.

Theorem 2.35 (Removable singularities theorem). *Let $u(x)$ be harmonic in $\mathcal{U} \setminus \{x_0\}$, where $\mathcal{U}$ is a neighborhood of the point x_0 and*

$$u(x) = o\big(|E(x - x_0)|\big) \quad \text{as } x \to x_0. \tag{2.35}$$

Then u can be extended to a harmonic function in $\mathcal{U}$.

To prove this we take $r_0 > 0$ such that the closure of the ball $B_{r_0}(x_0) = \{x : |x - x_0| < r_0\}$ is contained in $\mathcal{U}$. We then subtract from the function u the harmonic function v in $B_{r_0}(x_0)$ equal to u on the boundary of the ball. Now, using condition (2.35), we see that for any $\varepsilon > 0$ there exists $\delta > 0$ such that $|u(x) - v(x)| \leq \varepsilon |E(x - x_0)|$ on the boundary of the strip $\{x : \delta \leq |x - x_0| \leq r_0\}$, and so, by the maximum principle, throughout this strip, whence it follows that $u \equiv v$ and $B_{r_0}(x_0) \setminus \{x_0\}$, which gives the required assertion. Another explanation for the appearance of a condition of the form (2.35) and a proposition of removable-singularities type was given in Sect. 1, Ch. 2.

2.11. The Kelvin Transform and the Statement of Exterior Boundary-Value Problems for Laplace's Equation (cf. Vladimirov 1967, Sect. 28; Mikhailov 1983, Chapt. 40). The *Kelvin transform* is defined as the transformation taking a function $u(x)$ defined in the region $\Omega \subset \mathbf{R}^n$ into the function $v(y) = |y|^{2-n} u\big(\frac{y}{|y|^2}\big)$. Direct computation shows that if the function u is harmonic in Ω, then v is harmonic in the region Ω' obtained by inversion from Ω, i.e., $\Omega' = \{y : y \in \mathbb{R}^n \setminus \{0\},\ \dfrac{y}{|y|^2} \in \Omega\}$. It is easy to see that the Kelvin transform is involutive, i.e., applying it to the function $v(y) = |y|^{2-n} u\!\left(\dfrac{y}{|y|^2}\right)$, we again obtain the function $u(x)$. It is natural to extend the definition of the inversion $x \mapsto \dfrac{x}{|x|^2}$ to an involutive homeomorphism of the sphere $\mathbb{S}^n$, which is the one-point compactification of the space $\mathbb{R}^n$, so that $\mathbb{S}^n = \mathbb{R}^n \cup \{\infty\}$. Then under inversion a region containing the point 0 maps to a region containing ∞ and vice versa. This makes the following definition natural: the function $u(x)$ defined and harmonic outside some sphere is called *harmonic at infinity* if the Kelvin transform maps it to a function having a removable singularity at 0. It is clear that a necessary and sufficient condition for this to happen is

$$u(x) = O\!\left(\frac{1}{|x|^{n-2}}\right) \quad \text{as } |x| \to \infty. \tag{2.36}$$

By the removable-singularities theorem the following weaker condition is sufficient:

$$u(x) \to 0 \quad \text{as } |x| \to \infty, \quad n \geq 3; \tag{2.37}$$
$$u(x) = o\!\left(\ln |x|\right) \quad \text{as } |x| \to \infty, \quad n = 2. \tag{2.37'}$$

Corresponding to this we have the statement of the exterior Dirichlet and Neumann problems for Laplace's equation. These problems are posed in a region $\Omega \subset \mathbb{R}^n$ such that the complement $\mathbb{R}^n \setminus \Omega$ is bounded. In the case of the Neumann problem one must also assume that the boundary $\partial\Omega$ has a tangent plane at each point. The problems themselves have the same form as (2.15) and (2.16), but it is additionally assumed that the function $u(x)$ is harmonic at the point ∞, i.e., that either condition (2.36) or one of conditions (2.37) and (2.37') holds, depending on n. Thus these problems in case $n \geq 3$ have the form

$$\Delta u = 0 \quad \text{in } \Omega;\ u|_{\partial\Omega} = \varphi;\ u(x) = O\!\left(\frac{1}{|x|^{n-2}}\right) \quad \text{as } |x| \to \infty \tag{2.38}$$

(*the exterior Dirichlet problem*);

$$\Delta u = 0 \quad \text{in } \Omega;\ \frac{\partial u}{\partial n}\Big|_{\partial\Omega} = \varphi;\ u(x) = O\!\left(\frac{1}{|x|^{n-2}}\right) \quad \text{as } |x| \to \infty \tag{2.39}$$

(*the exterior Neumann problem*).

We note that the Kelvin transform essentially reduces the exterior Dirichlet problem to the interior problem. For $n = 2$ it also reduces the exterior Neumann problem to the interior problem. For $n \geq 3$ the exterior Neumann problem yields the interior problem with a boundary condition of the form

$$\left(\frac{\partial u(y)}{\partial n_y} + \gamma(y) u(y)\right)\Bigg|_{y \in \partial\Omega} = \varphi(y), \tag{2.40}$$

where $\gamma(y)$ is a given function on $\partial\Omega$ (depending only on the shape of the boundary $\partial\Omega$). Therefore existence and uniqueness for the exterior Dirichlet problem for any n and the exterior Neumann problem for $n = 2$ are equivalent to the corresponding facts for the interior problems. The properties of the exterior Neumann problem with $n \geq 3$ differ somewhat from the properties of the interior problem. For example, it follows from the maximal principle applied to the region $\Omega \cap \{x : |x| \leq R\}$ for sufficiently large R and the normal derivative lemma (cf. Sect. 2.4) that for $n \geq 3$ the exterior Neumann problem has at most one solution, while the solution of the interior problem is determined only up to an additive constant.

2.12. Potentials (cf. Petrovskij 1961; Vladimirov 1967; Mikhlin 1977; Smirnov 1981). We have encountered potentials as examples of convolutions in Sects. 1.9 and 1.14. We now consider the more general case and discuss the properties of potentials in more detail. Potentials are defined as the following integrals of special form, constructed using a fundamental solution $E(x)$ for the Laplacian operator:

$$u(x) = \int_\Omega E(x - y)\rho(y)\,dy \quad \text{(Newtonian potential);} \quad (2.41)$$

$$v(x) = \int_\Gamma E(x - y)\sigma(y)dS_y \quad \text{(single-layer potential);} \quad (2.42)$$

$$w(x) = -\int_\Gamma \frac{\partial E(x - y)}{\partial n_y}\beta(y)\,dS_y \quad \text{(double-layer potential).} \quad (2.43)$$

Here Ω is a region in $\mathbb{R}^n$ and Γ is a surface of dimension $(n - 1)$ in $\mathbb{R}^n$ (not necessarily closed). The potential is often written in a slightly different form, replacing $E(x)$ in formulas (2.41)–(2.43) by $|x|^{2-n}$ for $n \geq 3$ and by $\ln \frac{1}{|x|}$ for $n = 2$, leading to definitions that differ by numerical factors from those given here. Potentials have an obvious physical significance: the Newtonian potential is the potential of charges distributed with density $\rho(y)$ in the region Ω; the single-layer potential is the potential of charges distributed over the surface Γ with density $\sigma(y)$; the double-layer potential is the potential of dipoles distributed over the surface with density $\beta(y)$ and oriented in the direction of the chosen normal n_y.

In what follows we shall assume for simplicity that the region Ω is bounded and has a smooth boundary and that the function ρ is continuous on $\bar{\Omega}$. Similarly we shall assume that Γ is a piecewise smooth compact hypersurface (possibly with boundary) and the densities σ and β are continuous functions on Γ. It is easy to see that the integrals (2.41)–(2.43) converge in this case (the latter two for $x \notin \Gamma$).

If $\rho \in C^1(\Omega)$, then, as we have already pointed out in Sect. 2.8, for $x \notin \partial\Omega$ the Newtonian potential is a solution of Poisson's equation $\Delta u = \chi_\Omega \rho$, where χ_Ω is the characteristic function of the region Ω (equal to 1 on Ω and 0 outside Ω). In particular u is a harmonic function outside Ω. If $\rho \in C^1(\bar{\Omega})$, then $u \in C^1(\mathbb{R}^n)$ and the equation $\Delta u = \chi_\Omega \rho$ holds everywhere in $\mathbb{R}^n$ in the sense of distributions (cf. Sect. 1, Ch. 2). The latter is true even for a function $\rho \in L^1(\Omega)$.

The single- and double-layer potentials v and w are obviously harmonic functions outside Γ. Their behavior in approaching Γ is described by *jump theorems* (cf. Example 2.23). We shall state these theorems for points of the surface Γ where the surface is smooth.

The surface Γ locally divides the space into two parts. To be specific, each point of the surface has a neighborhood $\mathcal{U}$ that can be represented as the following disjoint union:

$$\mathcal{U} = \mathcal{U}^- \cup \Gamma_\mathcal{U} \cup \mathcal{U}^+,$$

where $\Gamma_\mathcal{U} = \Gamma \cap \mathcal{U}$, and $\mathcal{U}^\pm$ are nonempty connected open subsets of $\mathcal{U}$. It is possible to choose coordinates $x = (x_1, \ldots, x_n)$ in $\mathcal{U}$ such that $\Gamma_\mathcal{U} = \{x : x_n = 0\}$, $\mathcal{U}^- = \{x : x_n < 0\}$, and $\mathcal{U}^+ = \{x : x_n > 0\}$. In doing this we shall always assume that the normal direction n on Γ is chosen so that it is directed from $\mathcal{U}^-$ into $\mathcal{U}^+$ (in the coordinates just displayed the normal must have a positive last component). If u is a function on $\mathcal{U}$ or on $\mathcal{U} \setminus \Gamma$, we denote by u^+ and u^- the restrictions of u to $\mathcal{U}^+$ and $\mathcal{U}^-$ respectively. If the functions u^+ and u^- have limiting values on Γ (i.e., are continuous on $\overline{\mathcal{U}^+}$ and $\overline{\mathcal{U}^-}$ respectively), then these limiting values will also be denoted $u^+|_\Gamma$ and $u^-|_\Gamma$.

With this notation the theorem on potential jumps assumes the following form:

$$(v^+ - v^-)|_\Gamma = 0, \quad \left(\frac{\partial v^+}{\partial n} - \frac{\partial v^-}{\partial n} \right)\Big|_\Gamma = \sigma \qquad (2.44)$$

(the *jump theorem for the single-layer potential*); and

$$(w^+ - w^-)|_\Gamma = \beta \qquad (2.45)$$

(the *jump theorem for the double-layer potential*). For sufficiently smooth surface Γ and densities σ, β the jumps of any derivatives of the potentials v and w can be calculated (not just normal, but also tangent and mixed). However it is natural to do this using distribution theory by applying an equation satisfied by the potentials throughout $\mathbb{R}^n$ (cf. Sect. 1, Ch. 2). The jump theorems themselves (2.44) and (2.45) can also be deduced from this equation or proved immediately by analyzing the behavior of the potentials near Γ.

It is easy to verify that the single-layer potential $v(x)$ is defined at points of the surface Γ itself by the same integral, which converges absolutely. Extending the definition to Γ in this way, we obtain a continuous function throughout $\mathbb{R}^n$ (this is natural in view of the first of relations (2.44)). The integral that gives the double-layer potential $w(x)$ is actually absolutely convergent for $x \in \Gamma$ also. One can also demonstrate the relation

$$w(x) = \frac{1}{2}(w^+(x) + w^-(x)), \quad x \in \Gamma, \qquad (2.46)$$

which is a useful supplement to the jump theorems. A similar relation holds
for the normal derivative of the single-layer potential also, provided the
derivative is understood for $x \in \Gamma$ as the integral

$$\frac{\partial v(x)}{\partial n} = \int_{\Gamma} \frac{\partial E(x - y)}{\partial n_x} \sigma(y) \, dS_y$$

(this integral also converges absolutely). To be specific,

$$\frac{\partial v(x)}{\partial n} = \frac{1}{2}\left(\frac{\partial v^+(x)}{\partial n} + \frac{\partial v^-(x)}{\partial n}\right), \quad x \in \Gamma. \tag{2.47}$$

Sometimes the potentials can be computed explicitly using the equations they
satisfy and the jump theorems together with various symmetry considerations
and the behavior as $|x| \to \infty$. For example, suppose the surface Γ is a sphere
$\Gamma = \{x : |x| = R\}$. Then it is clear from symmetry considerations that
the single- and double-layer potentials $v(x)$ and $w(x)$ with constant densities
$\sigma(x) \equiv \sigma_0$ and $\beta(x) \equiv \beta_0$ are spherically symmetric, i.e., depend only on
$r = |x|$. Therefore both of them are constant for $|x| < R$, since any spherically
symmetric harmonic function in the ball $|x| < R$ is constant in this ball. This
constant value can be calculated by setting $x = 0$, from which we find for the
single-layer potential

$$v(x)\big|_{|x|<R} = v(0) = \sigma_{n-1} R^{n-1} \sigma_0 E(x)\big|_{|x|=R} = q_0 E(x)\big|_{|x|=R}. \tag{2.48}$$

(Here $q_0 = \sigma_{n-1} R^{n-1} \sigma_0$ is the total charge on the sphere $|x| = R$). Thus

$$v(x)\big|_{|x|<R} = v(0) = -\frac{R\sigma_0}{(n-2)}, \quad n \geq 3,$$

$$v(x)\big|_{|x|<R} = v(0) = R\sigma_0 \ln R, \quad n = 2.$$

Similarly for the double-layer potential we find that

$$w(x)\big|_{|x|<R} = w(0) = -\beta_0, \quad n \geq 2. \tag{2.49}$$

For $|x| > R$ we find in exactly the same way that the potentials $v(x)$ and $w(x)$
must have the form $C_1 + C_2 E(x)$. Using the jump theorems or finding the
asymptotic behavior of the desired potentials as $|x| \to \infty$, we easily determine
the constants C_1 and C_2. The result is

$$v(x) = q_0 E(x), \quad |x| > R, \tag{2.50}$$

$$w(x) = 0, \quad |x| > R. \tag{2.51}$$

Formulas (2.48) and (2.50) mean that a uniformly charged sphere creates no
field inside itself and the field it creates outside itself is the same as the field
of a point charge equal to the total charge of the sphere and located at its
center (a fact first proved by Newton). Formulas (2.49) and (2.51) mean, in
particular, that a uniform distribution of dipoles on the sphere creates no

field either inside the sphere or outside it (however, this was clear from the preceding, since the double layer can be represented as a pair of infinitely close single layers with charge densities of equal intensity but opposite sign).

2.13. Application of Potentials to the Solution of Boundary-Value Problems (cf. Petrovskij 1961; Vladimirov 1967; Mikhlin 1977; Smirnov 1981). Green's formula (2.10) essentially means that every function $u \in C^1(\bar{\Omega})$ that is harmonic in a bounded region Ω with smooth boundary can be represented in the form of a sum of single- and double-layer potentials (with densities $\sigma = -\frac{\partial u}{\partial n}\big|_{\partial\Omega}$ and $\beta = -u\big|_{\partial\Omega}$ respectively). Therefore one can look for the solution of any boundary-value problem for Laplace's equation in Ω in the form of a sum of such potentials with unknown densities σ and β. This is inconvenient, however, since arbitrary densities cannot be uniquely recovered from the sum of the corresponding potentials and an underdetermined system will be obtained (a single equation in the pair of densities σ and β). For this reason we look for the solution of the Dirichlet and Neumann boundary-value problems in the form of just one of the potentials. To be specific, the solution of Dirichlet problem (interior or exterior) can be conveniently sought in the form of a double-layer potential $w(x)$ and the solution of the Neumann problem (again interior or exterior) in the form of a single-layer potential $v(x)$. Now, using the theorems on potential jumps, it is easy to obtain integral equations for the desired densities equivalent to the boundary conditions. For example, using formulas (2.45) and (2.46), we find that the Dirichlet condition for the interior problem, which in this case has the form $w^-(x) = \varphi(x)$, $x \in \Gamma = \partial\Omega$, is equivalent to $w - \frac{\beta}{2} = \varphi$ on Γ, which in turn can be written in the form of the integral equation

$$(D_i) : -\frac{1}{2}\beta(x) - \int_\Gamma \frac{\partial E(x-y)}{\partial n_y}\beta(y)\,dS_y = \varphi(x), \quad x \in \Gamma, \qquad (2.52)$$

where the symbol (D_i) means that this is the integral equation for the interior Dirichlet problem. Similarly the condition for the exterior Dirichlet problem $w^+(x) = \varphi(x)$, $x \in \Gamma$, for a double-layer potential w with density β is equivalent to the integral equation

$$(D_e) : \frac{1}{2}\beta(x) - \int_\Gamma \frac{\partial E(x-y)}{\partial n_y}\beta(y)\,dS_y = \varphi(x), \quad x \in \Gamma. \qquad (2.53)$$

In analogy with (2.44) and (2.47) we find that the density $\sigma(x)$ of the single-layer potential $v(x)$ that gives the solution of the interior or exterior Neumann problem ($v^-(x) = \psi(x)$ or $v^+(x) = \psi(x)$ respectively) must satisfy respectively the integral equations

$$(\mathrm{N}_i) : -\frac{1}{2}\sigma(x) + \int_{\Gamma} \frac{\partial E(x-y)}{\partial n_x}\sigma(y)\,dS_y = \psi(x), \quad x \in \Gamma; \qquad (2.54)$$

$$(\mathrm{N}_e) : \frac{1}{2}\sigma(x) + \int_{\Gamma} \frac{\partial E(x-y)}{\partial n_x}\sigma(y)\,dS_y = \psi(x), \quad x \in \Gamma. \qquad (2.55)$$

All these integral equations are *Fredholm equations of second kind*. To be specific, the second terms in (2.52)–(2.55) are the result of applying to the densities β and σ the integral operator whose kernel $K(x,y)$ is smooth outside the diagonal on $\Gamma \times \Gamma$, and on the diagonal itself (as $x \to y$) has a so-called *weak singularity*, i.e., is integrable on y for fixed x; more precisely

$$|K(x,y)| \le C_\varepsilon |x-y|^{-n+1-\varepsilon}, \qquad (2.56)$$

with sufficiently small $\varepsilon > 0$. It can be proved that every such operator is compact in $L^2(\Gamma)$, and its image lies in $C(\Gamma)$. It follows from this that the known Fredholm theorems are applicable to Eqs. (2.52)–(2.55), and every solution of them, which a priori belongs only to $L^2(\Gamma)$, is actually continuous on Γ (provided the right-hand side is continuous). If we change the signs on the left-hand side of Eq. (D$_i$), it becomes adjoint to the equation (N$_e$). Similarly Eq. (D$_e$) becomes adjoint to Eq. (N$_i$) when the sign is changed on the left-hand side of one of these equations. Thus we are dealing with two pairs of adjoint equations.

According to general theorems of functional analysis, to prove the unique solvability of a pair of adjoint integral equations of this type (such unique solvability holds simultaneously for an equation and its adjoint), it suffices to verify that one of the homogeneous equations (with right-hand side zero) has a unique solution. This is easy to do, for example, for the equation (N$_e$) for $n \ge 3$. To be specific, for a single-layer potential $v(x)$ with density $\sigma(x)$ satisfying the equation (N$_e$) with $\psi \equiv 0$ we have obviously $\left.\dfrac{\partial v^+}{\partial n}\right|_{\Gamma} \equiv 0$, whence $v(x) \equiv 0$ on $\mathbf{R}^n \setminus \Omega$ by the uniqueness of the solution of the exterior Neumann problem. But then, by the continuity of v and the uniqueness of the solution of the interior Dirichlet problem, we find that $v(x) \equiv 0$ in Ω, whence, by the jump theorem for the normal derivative of a single-layer potential we obtain $\sigma \equiv 0$, as required. Thus the interior Dirichlet problem and the exterior Neumann problem for $n \ge 3$ have a unique solution.

The integral equations (D$_e$) and (N$_i$) are studied similarly. However the study leads to slightly different results. First of all the equation (D$_e$) with $\varphi(x) \equiv 0$ has a nontrivial solution $\beta_0 = 1$. Therefore the equation adjoint to it (N$_i$) is solvable if and only if its right-hand side ψ is orthogonal to β_0, i.e., satisfies condition (2.17), which is a necessary condition for solvability of the interior Neumann problem. Thus the interior Neumann problem is solvable if and only if the necessary condition (2.17) holds, and then the solution is determined only up to an arbitrary additive constant. For $n = 2$, using inversion (the Kelvin transform), we deduce the same result for the exterior problem: the exterior Neumann problem with $n = 2$ is solvable if and only if condition (2.17) holds, and then the solution is determined only up to an arbitrary additive constant.

By what has been said above the equation (D_e) is not solvable in every case, but only when an orthogonality condition on the right-hand side φ is satisfied. The Kelvin transform, however, reduces the exterior Dirichlet problem to the interior problem, whence the exterior Dirichlet problem is uniquely solvable for any function $\varphi \in C(\Gamma)$. Thus the reason the integral equation (D_e) is not solvable is that not every solution of the exterior Dirichlet problem can be represented in the form of a double-layer potential (it can be shown that the solutions for which the condition $u(x) = O(|x|^{1-n})$ holds at infinity are representable in this form and that no other solutions are so representable. This condition is stronger than the usual condition (2.36)).

Thus using potentials we can establish that the Dirichlet and Neumann problems for Laplace's equation are solvable in a region with smooth boundary. The problem with more general (mixed) boundary conditions (2.40) reduces to a Fredholm integral equation of second kind by a similar device. The more precise information that can be obtained about the smoothness of the solution will be discussed below in a more general context.

2.14. Boundary-Value Problems for Poisson's Equation in Hölder Spaces. Schauder Estimates (cf. Courant and Hilbert 1962; Miranda 1970; Ladyzhenskaya and Ural'tseva 1973). As already noted, boundary-value problems for Poisson's equation (for example the Dirichlet problem (2.23)) can be reduced to the corresponding boundary-value problems for Laplace's equation by subtracting the particular solution (2.29) of Poisson's equation. However the important question of the precise connection of the smoothness properties of the right-hand side f and the boundary values φ with the smoothness properties of the solution u was left open. We shall show here how this question is solved using Hölder spaces (we shall discuss another solution, using Sobolev spaces, in Sect. 2.6).

Let Ω be a bounded region with a C^∞ boundary, m a nonnegative integer, and $0 < \gamma < 1$. The *Hölder space* $C^{m+\gamma}(\Omega)$ consists of the functions $u \in C^m(\bar{\Omega})$ for which the following norm is finite.

$$\|u\|_{(m+\gamma)} = \sum_{|\alpha| \le m} \sup_{\Omega} |D^\alpha u(x)| + \sum_{\substack{|\alpha|=m}} \sup_{\substack{|x-y| \le 1 \\ x \in \Omega, y \in \Omega}} \frac{|D^\alpha u(x) - D^\alpha u(y)|}{|x - y|^\gamma}. \quad (2.57)$$

The space $C^{m+\gamma}(\Gamma)$ is defined similarly for a compact hypersurface Γ in $\mathbb{R}^n$.

We now consider the Dirichlet problem. It is natural to associate with it the transformation

$$C^{m+\gamma}(\Omega) \to C^{m-2+\gamma}(\Omega) \times C^{m+\gamma}(\partial\Omega), \quad u \mapsto (\Delta u, u|_{\partial\Omega}), \quad (2.58)$$

where $m \ge 2$. It is clear that this is a continuous linear transformation. It turns out that it is a *topological isomorphism* of Banach spaces (this can be verified by analyzing the potentials that give the solution). This means, first of all, that the relation $u \in C^{m+\gamma}(\Omega)$ for the solution of the problem (2.23) is equivalent to the two relations $\Delta u = f \in C^{m-2+\gamma}(\Omega)$ and $u|_{\partial\Omega} = \varphi \in C^{m+\gamma}(\partial\Omega)$, and second that, besides the fact that the transformation (2.58) is continuous, the a priori estimate

$$\|u\|_{(m+\gamma)} \le C(\|\Delta u\|_{(m-2+\gamma)} + \|u|_{\partial\Omega}\|_{(m+\gamma)}), \quad u \in C^{m+\gamma}(\Omega), \quad (2.59)$$

holds. This estimate is equivalent to the continuity of the inverse transformation.

We could have taken $\gamma = 1$ in (2.57), and for $\gamma = 0$ we could have considered $C^{m+\gamma} = C^m$. However for such γ estimate (2.59) and the analogous estimates stated below no longer hold, and the entire theory becomes more complicated.

Estimate (2.59) is an example of a *Schauder estimate*. We note that it implies that the solution of the Dirichlet problem for the equations of Laplace and Poisson is unique. Therefore the analogous estimate for the Neumann problem cannot be true. However, we have instead the estimate

$$\|u\|_{m+\gamma} \le C(\|\Delta u\|_{(m-2+\gamma)} + \left\|\frac{\partial u}{\partial n}\Big|_{\partial\Omega}\right\|_{(m-1+\gamma)} + \|u\|_{C(\bar{\Omega})}), \quad u \in C^{m+\gamma}(\Omega),$$
$$(2.60)$$

where $\|u\|_{C(\bar{\Omega})} = \sup_{\bar{\Omega}} |u(x)|$. The last term in (2.60) can be replaced by the norm of the function u in $L^2(\Omega)$ or in general by any norm of the function u that has meaning. Estimate (2.60) is then a typical Schauder estimate. The same kind of estimate for the mixed boundary condition (2.40) (i.e., with $\frac{\partial u}{\partial n}\Big|_{\partial\Omega}$ replaced by $\left(\frac{\partial u(y)}{\partial n_y} + \gamma(y)u(y)\right)\Big|_{\partial\Omega}$ also holds. Estimate (2.60) means in particular that the relation $u \in C^{m+\gamma}(\Omega)$ is equivalent to the two relations $\Delta u \in C^{m-2+\gamma}(\Omega)$ and $\frac{\partial u}{\partial n}\Big|_{\partial\Omega} \in C^{m-1+\gamma}(\partial\Omega)$, and the estimate corresponding to the mixed condition has an analogous meaning. But the transformation

$$C^{m+\gamma}(\Omega) \to C^{m-2+\gamma}(\Omega) \times C^{m-1+\gamma}(\partial\Omega), \quad u \mapsto \left(\Delta u, \frac{\partial u}{\partial n}\Big|_{\partial\Omega}\right), \quad (2.61)$$

although it is not invertible, is nevertheless a Fredholm transformation, i.e., has a finite-dimensional kernel and cokernel (actually one-dimensional). In this situation it will have a continuous inverse if we replace the left- and right-hand sides of (2.61) by suitable subspaces of codimension 1.

2.15. Capacity (cf. Brelot 1959; Landkof 1966; Landis 1971; Maz'ya 1985). The concept of capacity is a mathematical formalization and generalization of the concept of capacitance of a condensor in electrostatics. Let K be a compact subset of $\mathbb{R}^n$. For any nonnegative Borel measure μ on K we introduce its potential using the formula

$$u_\mu(x) = \int_K E(x - y)\, d\mu(y), \tag{2.62}$$

where $E(x)$ is a fundamental solution for the Laplacian in $\mathbb{R}^n$. In particular if the measure μ has density $\rho(x)$ with respect to Lebesgue measure, then u_μ is the usual Newtonian potential (cf. (2.41)). We remark that if $n \geq 3$, then $u_\mu(x) \leq 0$ for all $x \in \mathbb{R}^n$, and for $n = 2$ the same inequality holds at points x near to K for compact sets K of diameter less than 1. We can now define the *capacity* of the compact set K by the formula

$$C(K) = \sup\big\{\mu(K) : -u_\mu(x) \leq 1 \quad \text{for all } x \in \mathbb{R}^n\big\}. \tag{2.63}$$

In particular the set of measures μ for which $u_\mu(x)$ is bounded below may consist of only the zero measure, and in that case $C(K) = 0$. In particular the capacity of a point is zero. For simplicity we shall assume for the time being that $n \geq 3$. Capacity can be extended from compact sets to more general sets by the standard extension procedure (just as measures are extended). For an open set $E \subset \mathbb{R}^n$ we must set $C(E) = \sup\limits_{K \subset E} C(K)$, where K is an arbitrary compact set contained in E; then we must define the *inner* and *outer* capacities of an arbitrary set $E \subset \mathbb{R}^n$ by the formulas

$$\underline{C}(E) = \sup_{K \subset E} C(K), \quad K \text{ compact};$$

$$\overline{C}(E) = \inf_{G \supset E} C(G), \quad G \text{ an open subset of } \mathbb{R}^n.$$

It can be shown that all Borel sets $E \subset \mathbb{R}^n$ are C-measurable, i.e., for them $\underline{C}(E) = \overline{C}(E)$. Here the common value of the inner and outer capacities will be called simply the *capacity* and denoted $C(E)$.

For compact subsets of sufficiently simple structure one can find the capacity using the solution of the exterior Dirichlet problem. For example, let the compact set K have a smooth boundary $\partial K = K \setminus \text{Int}\, K$, where $\text{Int}\, K$ is the set of interior points of K. Let Ω_e be the unbounded connected component of the open set $\mathbb{R}^n \setminus K$, and let Γ_e be the boundary of Ω_e, so that Γ_e is a smooth closed hypersurface in $\mathbb{R}^n$, $\Gamma_e \subset K$. Consider the solution v of the exterior Dirichlet problem in Ω_e with boundary values $v\big|_{\Gamma_e} = -1$. Extend v to $\mathbb{R}^n$ by setting $v\big|_{\mathbb{R}^n \setminus \Omega_e} = -1$. Then v can be represented in the form of a single-layer potential with density σ on Γ_e. The upper bound in (2.63) is attained at the measure $d\mu(x) = \sigma(x)\, dS_x$, which is concentrated on Γ_e, so that in this case

$$C(K) = \int_{\Gamma_e} \sigma(x)\, dS_x = \int_{\Gamma_e} \frac{\partial v(x)}{\partial n_x}\, dS_x, \tag{2.64}$$

where n_x is the normal to Γ_e at the point x directed into Ω_e, and the derivative is taken over points lying in Ω_e (the last equality in (2.64) follows from the jump theorem for the normal derivative of a single-layer potential).

In particular, using (2.64) it is easy to find the capacity of a ball, a sphere, or any compact set contained in a closed ball and containing the boundary of the ball. To be specific, if the radius of the sphere under consideration is R, then $v(x) = -R^{n-2}r^{2-n}$, where $r = |x|$, and the capacity of interest to us is $(n-2)\sigma_{n-1}R^{n-2}$.

Here are some simple properties of capacity. If $k > 0$ and kE is the set obtained from E by a dilation with coefficient k, then $C(kE) = k^{n-2}C(E)$. If $E = \bigcup_{j=1}^{\infty} E_j$, then $C(E) \leq \sum_{j=1}^{\infty} C(E_j)$.

Sets of zero capacity play an important role. We shall describe some of their properties and applications. An example of a set of zero capacity is the union of a finite number of smooth $(n-2)$-dimensional submanifolds (together with their boundaries). A necessary condition for the equality $C(E) = 0$ is the vanishing of the $(n-1)$-dimensional Lebesgue measure of the projection of E onto any given hyperplane. In particular pieces of hypersurfaces having positive $(n-1)$-dimensional Lebesgue measure also have positive capacity, as do sets having an interior point.

The removable singularities theorem can be sharpened using the concept of capacity as follows: if a function u is harmonic and bounded in a neighborhood of a set E of capacity 0, then it is also harmonic at all points of the set E.

The theorem that asserts the uniqueness of the solution of the Dirichlet problem is strengthened as follows: if two bounded harmonic functions u and v in a bounded region Ω have the same limiting values at all points of the boundary $\partial\Omega$ except the points of a set of capacity zero, then they coincide everywhere in Ω.

2.16. The Dirichlet Problem in the Case of Arbitrary Regions (The Method of Balayage). Regularity of a Boundary Point. The Wiener Regularity Criterion (cf. Brelot 1959; Landkof 1966). Let Ω be an arbitrary bounded region in $\mathbb{R}^n$. Consider the Dirichlet problem (2.15) in the region with data function $\varphi \in C(\partial\Omega)$. We shall try to find a classical solution of this problem $u \in C^2(\Omega) \cap C(\bar{\Omega})$. Such a solution does not always exist for a region with nonsmooth boundary. However, one can construct a harmonic function u in Ω that is an optimal solution of the problem (2.15) in a certain sense. This is done using the following method proposed by Poincaré and now known as the method of *balayage*.

Extend φ to a function $\Phi \in C(\bar{\Omega})$ such that $\Phi|_{\partial\Omega} = \varphi$. Now consider a sequence of regions Ω_k, $k = 1, 2, \ldots$, contained in Ω such that

a) Ω_k is a region with smooth boundary;

b) $\bar{\Omega}_k \subset \Omega_{k+1}$, $k = 1, 2, \ldots$; $\Omega = \bigcup_{k=1}^{\infty} \Omega_k$.

Solve the Dirichlet problem in Ω_k

$$\Delta u_k = 0, \quad u_k\big|_{\partial\Omega_k} = \varphi_k = \Phi\big|_{\partial\Omega_k}.$$

We obtain a sequence of functions $u_1, u_2, u_3, \ldots$ defined at each point $x \in \Omega$ from some index on, since $x \in \Omega_k$ for all $k \geq k_0$ if $k = k_0(x)$ is sufficiently large. It follows from the internal a priori estimate (2.18) that the functions $u_{k+1}, u_{k+2}, \ldots$ are uniformly bounded and uniformly continuous on $\bar\Omega_k$. But then by the theorem of Arzelà we can choose a subsequence of the sequence $\{u_k\}$ that is uniformly convergent on each compact set $K \subset \Omega$. We denote the limit of this sequence by u. Then $u \in C^\infty(\Omega)$ and u is harmonic and bounded in Ω. From the maximum principle it is easy to deduce that this function u is independent of the choice of the extension Φ of the boundary function φ and independent of the choice of the sequence of regions $\{\Omega_k\}$ satisfying a) and b). If the Dirichlet problem under consideration has a classical solution, then u will coincide with this solution (this is clear, for example, from the fact that in this case we can take the function Φ equal to the classical solution).

A boundary point $x_0 \in \partial\Omega$ is called *regular* if for any function $\varphi \in C(\partial\Omega)$ and for the function u constructed from φ by the method of balayage the relation

$$\lim_{x \to x_0} u(x) = \varphi(x_0)$$

holds, i.e., u actually assumes the value $\varphi(x_0)$ at the point x_0.

It is clear that the Dirichlet problem (in the classical sense) is solvable for any function $\varphi \in C(\partial\Omega)$ if and only if all the boundary points are regular. It turns out that the regularity of a point $x_0 \in \partial\Omega$ depends only on the local structure of Ω in a neighborhood of the point. The solution can be stated as follows using the concept of capacity.

Theorem 2.36 (Wiener's Criterion). *A necessary and sufficient condition for a point $x_0 \in \partial\Omega$ to be regular is that the series*

$$\sum_{k=1}^{\infty} 4^{k(n-2)} C\big(\{x : x \in \mathbb{R}^n \setminus \Omega, \quad |x - x_0| < 4^{-k}\}\big) \qquad (2.65)$$

diverge.

We shall give examples of the application of this theorem. If a point $x_0 \in \partial\Omega$ is isolated in $\partial\Omega$, it is nonregular since all the terms of the series (2.65) vanish. This agrees with the removable singularities theorem. If there exists a closed cone with vertex at the point x_0 having interior points and lying in $\mathbb{R}^n \setminus \Omega$ near the point x_0, then the point x_0 is regular. For $n = 2$ a sufficient condition for regularity of the point x_0 is that it can be included to a nontrivial connected compact set (not coinciding with x_0) lying in $\mathbb{R}^2 \setminus \Omega$. In particular all the points of the boundary of a simply connected region in $\mathbb{R}^2$ are regular. For $n \geq 3$ it is easy using Wiener's criterion to construct an example of a nonregular boundary point of a region Ω for which $\bar\Omega$ is homeomorphic to a ball. To be specific one must take a region Ω whose complement has the form $\left\{x : 0 \le x_n \le e^{-(x_1^2 + \cdots + x_{n-1}^2)^{-1}}\right\}$ near the origin. Then 0 will be a nonregular point of $\partial \Omega$.

Important information about the structure of the set of nonregular points is contained in the following theorem.

Theorem 2.37 (Kellog). *For any bounded region $\Omega \subset \mathbb{R}^n$ the set of nonregular points of its boundary has capacity zero.*

In particular the Dirichlet problem in a bounded region Ω has a unique bounded solution assuming given boundary values at all regular boundary points.

Everything that has been said in this section about the interior Dirichlet problem is true also for the exterior problem, since the criterion for regularity of a point $x_0 \in \partial \Omega$ is local.

2.17. General Second-Order Elliptic Equations. Eigenvalues and Eigenfunctions of Elliptic Operators (cf. Courant and Hilbert 1931, 1962; Friedman 1964; Miranda 1970; Landis 1971; Ladyzhenskaya and Ural'tseva 1973). Now consider the general elliptic equation (2.3). The basic facts relating to the equations of Laplace and Poisson carry over to this more general case, although with certain stipulations. We shall give the corresponding statements here. For definiteness we shall always assume that the quadratic form (2.4) is positive definite (the opposite case reduces to this by a change of sign in the equation).

The *maximum principle* for the general equation (2.3) has the following form: if $c(x) \le 0$ and $f(x) \le 0$, then a solution $u \in C^2(\Omega) \cap C(\bar{\Omega})$ defined in a bounded region attains a negative minimum on the boundary $\partial \Omega$ of the region Ω; if $c(x) \le 0$ and $f(x) \ge 0$, then the solution $u(x)$ attains a positive maximum on the boundary $\partial \Omega$. In particular if $c(x) \le 0$ and $f(x) \equiv 0$, then any extremal value is necessarily attained on $\partial \Omega$.

It follows from the maximal principle that the Dirichlet problem for Eq. (2.3) with $c(x) \le 0$ cannot have more than one solution. Under the same assumption, in the case when the coefficients of the equation and the boundary are smooth, the Dirichlet problem is solvable. Schauder estimates also hold, namely estimate (2.59) for $c(x) \le 0$ with Δ replaced by A, where

$$A = \sum_{i,j=1}^{n} a_{ij}(x) \frac{\partial^2}{\partial x_i \partial x_j} + \sum_{j=1}^{n} b_j(x) \frac{\partial}{\partial x_j} + c(x), \qquad (2.66)$$

as well as estimate (2.60) (again with Δ replaced by A), now without any restrictions on the sign of the coefficient $c(x)$. (In this case the Dirichlet problem is a Fredholm problem – cf. Sect. 2.13; the same applies to the Neumann problem and the problem with mixed boundary condition (2.40).)

We note that in the case of an arbitrary coefficient c the Dirichlet problem may have more than one solution (and may fail to have even one solution). In particular if an operator A of the form (2.66) has coefficients that are smooth in $(\bar{\Omega})$ and is symmetric with the Dirichlet conditions (in the sense explained in Sect. 2.7), then there exists a countable sequence of eigenvalues of the operator A with the Dirichlet conditions, i.e., numbers $\{\lambda_k\}_{k=1}^{\infty}$ such that the problem

$$A\psi_k = \lambda_k\psi_k, \quad \psi_k\big|_{\partial\Omega} = 0 \qquad (2.67)$$

has a nontrivial solution. In particular this means that there is more than one solution of the Dirichlet problem for Eq. (2.3) with $c(x)$ replaced by $c(x) - \lambda_k$. This problem also may have no solution, since in this case if $(A - \lambda_k)u = f$ and $u\big|_{\partial\Omega} = 0$, then $f \perp \psi_k$ in $L^2(\Omega)$.

The eigenvalues and eigenfunctions of selfadjoint elliptic operators play an important role in solving boundary-value problems for hyperbolic and parabolic equations. For that reason the study of the properties of eigenvalues and eigenfunctions is important. We shall give only the most fundamental properties here, assuming both the coefficients and the boundary are smooth.

First of all $\lambda_k \to -\infty$ as $k \to \infty$. The eigenfunctions ψ_k belong to $C^{\infty}(\bar{\Omega})$ (and are analytic in $\bar{\Omega}$ in the case when the coefficients themselves are analytic in $\bar{\Omega}$). If the eigenvalues are arranged in decreasing order (counting multiplicities): $\lambda_1 \geq \lambda_2 \geq \lambda_3 \geq \ldots$, then λ_1 is simple (i.e., $\lambda_1 > \lambda_2$) and the corresponding eigenfunction ψ_1 is of constant sign in Ω. Eigenfunctions corresponding to distinct eigenvalues are orthogonal in $L^2(\Omega)$. The eigenfunctions constitute a complete orthogonal system in $L^2(\Omega)$. The same facts hold when the Dirichlet condition is replaced by the Neumann condition.

For general equations of the form (2.3) with smooth coefficients the removable singularities theorem holds in the same form as for Laplace's equation (cf. Sect. 2.10).

Boundary-value problems for the general equation (2.3) can be reduced to integral equations using special potentials analogous to the procedure followed above in the case of Laplace's equation.

Finally, we point out that in the case of an arbitrary bounded region Ω the Dirichlet problem for Eq. (2.3) with $f(x) \equiv 0$ and $c(x) \leq 0$ may be treated just like Laplace's equation in Sect. 2.16. In particular Wiener's criterion for regularity of a boundary point holds (in the same form) for general equations of this form. Thus a boundary point $x_0 \in \partial\Omega$ is regular for the general second-order equation if and only if it is regular for Laplace's equation.

2.18. Higher-Order Elliptic Equations and General Elliptic Boundary-Value Problems. The Shapiro-Lopatinskij Condition (cf. Agmon, Douglas, and Nirenberg 1959; Hörmander 1963; Lions and Magenes 1968; Hörmander 1983–1985). In a bounded region Ω with smooth boundary $\partial\Omega$ consider an operator A of the form (2.1) with smooth coefficients $a_\alpha \in C^{\infty}(\bar{\Omega})$. We shall assume that the operator A is elliptic for $x \in \bar{\Omega}$. For $n \geq 3$ the order of the elliptic operator A is necessarily an even number m, and if ξ and η are two linearly independent vectors of $\mathbb{R}^n$ the equation

$$a_m(x, \xi + \tau\eta) = 0 \qquad (2.68)$$

(here a_m is the principal symbol of the operator A defined by formula (2.2)) has exactly $m/2$ roots with respect to τ that lie in the upper half-plane $\operatorname{Im}\tau > 0$; and therefore exactly $m/2$ roots lying in the lower half-plane (Eq. (2.68) has no real roots). This is proved by elementary topological reasoning: if ξ changes continuously into $-\xi$ and in such a way that the linear independence of ξ and η is preserved, then the root τ_j changes into $-\tau_j$, from which it follows that the number of roots τ in the upper half-plane is the same as the number in the lower half-plane. The same holds for $n = 2$ for an operator with real coefficients. In the general case for $n = 2$ we require for simplicity that the number m be even and that the number of roots of Eq. (2.68) lying in each of the two half-planes be $m/2$. We shall call operators satisfying this condition *properly elliptic*.

Consider the boundary-value problem

$$\begin{cases} Au = f, \\ B_j u\big|_{\partial\Omega} = \varphi_j, \quad j = 1, 2, \ldots, m/2, \end{cases} \qquad (2.69)$$

where $B_j = B_j(x, D)$ are differential operators of orders $m_j < m$ defined in a neighborhood of $\partial\Omega$ and having smooth coefficients.

We now choose any point $x_0 \in \partial\Omega$, and introduce coordinates $x_1, \ldots, x_n$ in a neighborhood of it such that the boundary $\partial\Omega$ assumes the form $\{x : x_n = 0\}$ and the region Ω is given as the set $\{x : x_n > 0\}$. We then replace the coefficients of the operators A and B_j by constant coefficients equal to the values of the corresponding coefficients at the point x_0 (we freeze the coefficients of the operators A and B_j at the point x_0, as it is said). In addition we keep only the leading parts of the operators (i.e., only the derivatives of order m in the operator A and the derivatives of order m_j in the operator B_j). Then instead of problem (2.69) we obtain a model problem in a half-space

$$\begin{cases} A_{(x_0)}u(x) = f(x), & x_n > 0, \\ B_{j(x_0)}u\big|_{x_n=0} = \varphi_j(x'), & j = 1, 2, \ldots, m/2, \end{cases} \qquad (2.70)$$

where $A_{(x_0)}$ and $B_{j(x_0)}$ are the leading parts of the operators A and B_j in the new coordinates $x = (x', x_n)$ with coefficients frozen at the point x_0. In problem (2.70) we take the Fourier transform on x'. We then obtain the problem

$$\begin{cases} A_{(x_0)}(\xi', D_n)\tilde{u}(\xi', x_n) = \tilde{f}(\xi', x_n), & x_n > 0, \\ B_{j(x_0)}(\xi', D_n)\tilde{u}(\xi', x_n)\big|_{x_n=0} = \tilde{\varphi}_j(\xi'), & j = 1, 2, \ldots, m/2, \end{cases}$$

where the tilde denotes the Fourier transform on x'. We now have a problem for ordinary differential equations depending on ξ as a parameter. Fixing $\xi' \in \mathbb{R}^n \setminus \{0\}$, we obtain a problem on a half-line

$$\begin{cases} A_{(x_0)}(\xi', D_n)v(x_n) = g(x_n), & x_n > 0, \\ B_{j(x_0)}(\xi', D_n)v(x_n)\big|_{x_n=0} = \psi_j, & j = 1, 2, \ldots, m/2. \end{cases} \tag{2.71}$$

We wish to solve this problem in the class of functions that are decreasing on x_n; more precisely, in the class $\mathcal{S}(\overline{\mathbf{R}}_+)$ consisting of functions $v \in C^\infty(\overline{\mathbf{R}}_+)$ for which $\sup\limits_{x_n>0} |x_n^k D_n^p v(x_n)| < \infty$ for any integers $k \geq 0$ and $p \geq 0$. In conjunction with this problem there is a fundamental condition, which we now state:

(Ell) (*the ellipticity condition for the boundary-value problem*, or *the complementarity condition*, or *the covering condition*, or *the Shapiro-Lopatinskij condition*): for any $x_0 \in \partial\Omega$ and $\xi' \in \mathbf{R}^{n-1} \setminus \{0\}$ the problem (2.71) has a unique solution $v \in \mathcal{S}(\overline{\mathbf{R}}_+)$ for any $g \in \mathcal{S}(\overline{\mathbf{R}}_+)$ and $\psi_1, \psi_2, \ldots, \psi_{m/2} \in \mathbf{C}$.

This condition can be simplified if we note that the equation

$$A_{(x_0)}(\xi', D_n)v(x_n) = g(x_n), \quad x_n > 0,$$

always has a particular solution $v \in \mathcal{S}(\overline{\mathbf{R}}_+)$ for $g \in \mathcal{S}(\overline{\mathbf{R}}_+)$. Subtracting this particular solution, we see that the problem reduces to the problem (2.71) with $g(x_n) \equiv 0$. This means that our condition (Ell) is equivalent to the following condition:

(Ell$_0$): for any $x_0 \in \partial\Omega$ and $\xi' \in \mathbf{R}^{n-1} \setminus \{0\}$ the problem

$$\begin{cases} A_{(x_0)}(\xi', D_n)v(x_n) = 0, & v \in \mathcal{S}(\overline{\mathbf{R}}_+), \\ B_{j(x_0)}(\xi', D_n)v(x_n)\big|_{x_n=0} = \psi_j, & j = 1, 2, \ldots, m/2 \end{cases} \tag{2.72}$$

has a unique solution. Problem (2.72) is considerably easier to solve since it is a problem for homogeneous equations (with zero right-hand side) with constant coefficients. Such equations, as is known, can be solved explicitly and their solutions have the form of linear combinations of functions of the form $x_n^p e^{i\tau_j(\xi')x_n}$, where $\tau_j(\xi')$ is a root of the equation $A_{(x_0)}(\xi', \tau) = 0$ and p is a nonnegative integer less than the multiplicity of the root $\tau_j(\xi')$. In particular if we assume that the equation $A_{(x_0)}(\xi', \tau) = 0$ has $m/2$ distinct roots $\tau_1(\xi'), \ldots, \tau_{m/2}(\xi')$ lying in the upper half-plane $\mathrm{Im}\, \tau > 0$ (these are the only roots corresponding to solutions $e^{i\tau_j(\xi')x_n}$ that tend to zero as $x_n \to +\infty$), then the condition (Ell$_0$) becomes a condition for unique solvability of the system of linear equations

$$\sum_{k=1}^{m/2} B_{j(x_0)}(\xi', \tau_k(\xi'))c_k = \psi_j, \quad j = 1, \ldots, m/2, \tag{2.73}$$

with respect to the unknowns c_k. This condition means that

$$\det \left\| B_{j(x_0)}(\xi', \tau_k(\xi')) \right\|_{j,k=1}^{m/2} \neq 0 \quad \text{for all } \xi' \neq 0. \tag{2.74}$$

In the general form it is possible to write conditions analogous to (2.74) if instead of the system of exponentials $e^{i\tau_j(\xi')x_n}$, $j = 1, \ldots, m/2$, we choose some basis of the space of solutions of the equation $A_{(x_0)}(\xi', D_n)v(x_n) = 0$ consisting of functions that tend to zero as $x_n \to +\infty$. Without doing this explicitly we can nevertheless use the fact that the problem reduces to unique solvability of a system of linear equations, which is equivalent to the condition that that homogeneous system (with zero right-hand sides) has no nontrivial solutions. The latter leads to the following restatement of conditions (Ell) and (Ell$_0$):

(Ell$_1$) for any $x_0 \in \partial\Omega$ and $\xi' \in \mathbf{R}^{n-1} \setminus \{0\}$ the problem

$$\begin{cases} A_{(x_0)}(\xi', D_n)v(x_n) = 0, \\ B_{j(x_0)}(\xi', D_N)v(x_n)\big|_{x_n=0} = 0, \end{cases} \tag{2.75}$$

has no nontrivial solutions $\dot{v} \in \mathcal{S}(\overline{\mathbf{R}}_+)$.

Taking account of the structure of the solutions of the equation with constant coefficients and the fact that the equation $A_{(x_0)}(\xi', \tau) = 0$ has no real roots, we see that instead of the condition $v \in \mathcal{S}(\overline{\mathbf{R}}_+)$ it suffices to require that $v(x_n) \to 0$ as $x_n \to +\infty$, or even that $v(x_n)$ be bounded for $x_n \geq 0$.

We now give another equivalent algebraic formulation of the ellipticity condition for a boundary-value problem. To do this we set

$$A^+_{(x_0)}(\xi', \tau) = \prod_{j=1}^{m/2} (\tau - \tau_j(\xi')), \tag{2.76}$$

where $\tau_1(\xi'), \dots, \tau_{m/2}(\xi')$ are all the roots of the polynomial $A_{(x_0)}(\xi', \tau)$ in the upper half-plane (counting multiplicities). Then the ellipticity condition is equivalent to the following:

(Ell$_2$): for any $x_0 \in \partial\Omega$ and $\xi' \in \mathbf{R}^{n-1} \setminus \{0\}$ the following polynomials, when regarded as polynomials in τ, are linearly independent modulo the polynomial $A^+_{(x_0)}(\xi', \tau)$:

$$B_{1(x_0)}(\xi', \tau), \dots, B_{m/2(x_0)}(\xi', \tau).$$

It is easy to see that the ellipticity condition for the boundary-value problem is stable: under a small perturbation of the coefficients of the operator and boundary conditions an elliptic problem remains elliptic. Moreover it is clear from the definition itself that ellipticity depends only on the leading terms in the operator and the boundary conditions.

Example 2.38. The following boundary-value problem for an operator A of order m is called the *Dirichlet problem*:

$$\begin{cases} Au = f, \\ \dfrac{\partial^j u}{\partial n^j}\bigg|_{\partial\Omega} = \varphi, \quad j = 0, 1, \dots, \dfrac{m}{2} - 1. \end{cases} \tag{2.77}$$

It is easy to verify that it is elliptic for any properly elliptic operator A.

Example 2.39. (The oblique derivative problem.) Let A be a second-order properly elliptic operator. Let a vector-valued function (with values in $\mathbb{R}^n$) be defined on $\partial\Omega$ and denoted $\nu = \nu(x)$. Then the problem

$$\begin{cases} Au = f, \\ \dfrac{\partial u}{\partial \nu}\Big|_{\partial\Omega} = \varphi, \end{cases} \tag{2.78}$$

is called an *oblique derivative problem*. It is easy to verify that for $n \geq 3$ this problem is elliptic if and only if the vector field $\nu(x)$ is not tangent to $\partial\Omega$ at any point $x \in \partial\Omega$, and for $n = 2$ it is elliptic if and only if for all $x \in \partial\Omega$ the relation $\nu(x) \neq 0$ holds.

We now state one of the basic facts relating to elliptic boundary-value problems: *a theorem on Fredholm operators, regularity of solutions, and an a priori estimate.*

Theorem 2.40. *Suppose the boundary-value problem* (2.69) *is elliptic. Then*

1) *for any noninteger $\gamma > m$ the operator*

$$\mathfrak{A} : C^\gamma(\Omega) \to C^{\gamma-m}(\Omega) \times \prod_{j=1}^{m/2} C^{\gamma-m_j}(\partial\Omega),$$

$$\tag{2.79}$$

$$u \mapsto \left\{ Au, B_1 u\big|_{\partial\Omega}, \ldots, B_{m/2} u\big|_{\partial\Omega} \right\}$$

is a Fredholm operator, i.e., has a finite-dimensional kernel and a closed image (set of values) of finite codimension;

2) *the relation $u \in C^\gamma(\Omega)$ is equivalent to the set of conditions $Au = f \in C^{\gamma-m}(\Omega)$ and $B_j u\big|_{\partial\Omega} = \varphi_j \in C^{\gamma-m_j}(\partial\Omega)$, $j = 1, 2, \ldots, m/2$; if the boundary $\partial\Omega$, the coefficients of the operators A and B_j, and the data f and φ_j are analytic, then the solution u is also analytic;*

3) *for any noninteger $\gamma > m$ the a priori estimate*

$$\|u\|_{(\gamma)} \leq C\left(\|Au\|_{(\gamma-m)} + \sum_{j=1}^{m/2} \left\|B_j u\big|_{\partial\Omega}\right\|_{(\gamma-m_j)} + \|u\|_{C(\bar\Omega)} \right) \tag{2.80}$$

holds, where the constant C is positive and independent of u.

We note that all the assertions of the theorem are closely connected with one another (for example, estimate (2.80) is easily deduced from the Fredholm property of the operator (2.79) using the closed graph theorem); however, we have presented all three of them for the sake of completeness.

The proof of the theorem can be based on the reduction of the problem to integral equations using potentials obtained from the solution of the model problems (2.70). The significance of the hypothesis that the solutions tend to zero on x_n is that if a root $\tau_j(\xi')$ is chosen with $\operatorname{Im}\tau_j(\xi') < 0$ on a sufficiently large set of values of ξ', the corresponding exponential $e^{i\tau_j(\xi')x_n}$ will increase rapidly on ξ' for $x_n > 0$, and its inverse Fourier transform on ξ' will not belong to the usual function spaces.

We now discuss several generalizations. First of all, we can dispense with the hypothesis of proper ellipticity; when this is done, the number of boundary conditions B_j must become equal to the number of roots τ_j of the polynomial $A_{(x_0)}(\xi', \tau)$ lying in the upper half-plane $\mathrm{Im}\,\tau > 0$. In addition, the entire theory extends naturally to elliptic systems. In this extension the condition (Ell$_2$) no longer has meaning (the conditions (Ell), (Ell$_0$), and (Ell$_1$) can be stated in exactly the same way if the principal parts of the operators A and B_j are suitably interpreted). We note, however, that the Dirichlet problem for elliptic systems is no longer necessarily elliptic, and there exist elliptic systems having no elliptic boundary-value problems.

Finally, elliptic boundary-value problems can be studied not only in Hölder spaces, but also in other appropriate spaces. In particular, analogous results are obtained when these problems are studied in the Sobolev spaces, which will be discussed below in Sect. 3.

2.19. The Index of an Elliptic Boundary-Value Problem (cf. Fedosov 1974; Rempel and Schulze 1982). The question arises: is it possible to give verifiable conditions for an elliptic boundary-value problem to have a unique solution, i.e., conditions for an operator $\mathfrak{A}$ of the form (2.79) to be invertible? If we ask for necessary and sufficient conditions, the answer to this question is negative in the general situation. The reason for this is that a noninvertible operator of the form (2.79) may become invertible under a small perturbation of the coefficients. The simplest example of such a situation is the operator for the Neumann problem

$$\mathfrak{A}_0 : C^\gamma(\Omega) \to C^{\gamma-2}(\Omega) \times C^{\gamma-1}(\partial\Omega), \quad u \mapsto \left\{\Delta u, \frac{\partial u}{\partial n}\Big|_{\partial\Omega}\right\}, \qquad (2.81)$$

which is noninvertible – the kernel $\mathrm{Ker}\,\mathfrak{A}_0$ is nontrivial since it contains all the constant functions – but becomes invertible if Δ is replaced by $\Delta - \varepsilon I$, where $\varepsilon > 0$ may be taken arbitrarily small and I is the identity operator. We thus see that a noninvertible operator in an elliptic boundary-value problem may become invertible under an arbitrarily small perturbation of the nonleading terms of the operator. (This can also be achieved by an analogous perturbation of the boundary condition.) An important quantity that does not change under these perturbations (and in general under any homotopy-deformations in the class of elliptic boundary-value problems) is the *index*

$$\mathrm{ind}\,\mathfrak{A} = \dim\mathrm{Ker}\,\mathfrak{A} - \dim\mathrm{Coker}\,\mathfrak{A} \qquad (2.82)$$

(we recall that for any linear operator $\mathfrak{A} : E_1 \to E_2$, where E_1 and E_2 are vector spaces, $\mathrm{Ker}\,\mathfrak{A} = \{x : x \in E_1, \mathfrak{A}x = 0\}$ is the kernel of the operator $\mathfrak{A}$, $\mathrm{Coker}\,\mathfrak{A} = E_1/\mathrm{Im}\,\mathfrak{A}$ is the cokernel of the operator $\mathfrak{A}$, and $\mathrm{Im}\,\mathfrak{A} = \mathfrak{A}E_1$ is the image of the operator $\mathfrak{A}$). The index of an operator $\mathfrak{A}$ of the form

(2.79) is independent of the choice of the noninteger $\gamma > m$ and independent of the lower-order terms of the operator A and the boundary operators B_j; therefore it is often called simply the *index* of the boundary-value problem. The index of a boundary-value problem that has a unique solution is obviously zero. Therefore the index of any elliptic boundary-value problem that is homotopic to a uniquely solvable problem is also zero. In particular, since the perturbation of the Neumann problem given above changes the problem into a uniquely solvable problem, the index of the Neumann problem is zero.

We shall also consider the oblique derivative problem for the Laplacian (cf. Example 2.39), which we shall regard as an elliptic problem. For $n \geq 3$ its index is zero, since in that case ellipticity means that the vector field $\nu(x)$ is not tangent to the boundary and therefore can be deformed into a normal field, so that the problem becomes the Neumann problem. For $n = 2$ the index of the problem is $2 - 2p$, where p is the index of the vector field $\nu(x)$ (the winding number of the vector $\nu(x)$ as the contour $\partial\Omega$ is traversed counterclockwise). Knowing the index makes it possible to find one of the dimensions occurring in (2.82) when the other is known. For example, if $\operatorname{ind}\mathfrak{A} = 0$ and $\dim\operatorname{Ker}\mathfrak{A} = 0$, then $\dim\operatorname{Coker}\mathfrak{A} = 0$, i.e., the problem has a unique solution for any data f and φ_j in the corresponding spaces. If it is known that $\dim\operatorname{Coker}\mathfrak{A} = 0$ (i.e., the problem is solvable for any f and φ_j), then $\dim\operatorname{Ker}\mathfrak{A} = \operatorname{ind}\mathfrak{A}$, which sometimes makes it possible to find $\dim\operatorname{Ker}\mathfrak{A}$.

The index of an elliptic boundary-value problem can often be computed using a homotopy from the given problem to a simpler one. To calculate it one can also apply the general topological formula for the index or analytic formulas (Fedosov 1974, Rempel and Schulze 1982).

2.20. Ellipticity with a Parameter and Unique Solvability of Elliptic Boundary-Value Problems (cf. Agranovich and Vishik 1964). For problems depending polynomially on a parameter one can give an easily verified sufficient condition for unique solvability of the problem for values of the parameter that are sufficiently large in absolute value. We shall discuss here the simplest example of such a situation. In a bounded region Ω with a smooth boundary $\partial\Omega$ consider the boundary-value problem

$$\begin{cases} (A - \lambda)u = f, \\ B_j u\big|_{\partial\Omega} = \varphi_j, \quad j = 1, 2, \ldots, m/2, \end{cases} \tag{2.83}$$

where A and B_j are as in (2.69) and λ is a complex parameter. As in Sect. 2.18, we shall assume the operator A is properly elliptic. We choose the point $x_0 \in \partial\Omega$, straighten the boundary in a neighborhood of this point (so that Ω will be given locally as the set $\{x : x_n > 0\}$), freeze the coefficients of the operators A and B_j at the point x_0, and keep only the leading terms of these operators $A_{(x_0)}$ and $B_{j(x_0)}$. We now consider the problem analogous to the problem (2.71), only with the parameter λ:

$$\begin{cases} (A_{(x_0)}(\xi', D_n) - \lambda)v(x_n) = 0, & v \in \mathcal{S}(\overline{\mathbb{R}}_+), \\ B_{j(x_0)}(\xi', D_n)v(x_n)\big|_{x_n=0} = \psi_j, & j = 1, 2, \ldots, m/2. \end{cases} \quad (2.84)$$

Suppose the parameter λ varies in some closed angle $\Lambda \subset \mathbb{C}$ with vertex at the point 0 (we do not exclude the possibility that the sides of the angle Λ coincide, so that Λ may be only a ray). We now state the fundamental condition.

(Ell$_\Lambda$) (*ellipticity condition with the parameter* $\lambda \in \Lambda$ or *the Agranovich-Vishik condition* or *the Agmon condition*).

a) if $a_m(x, \xi)$ is the principal symbol of the operator A, then $a_m(x, \xi) - \lambda \neq 0$ (or $\det(a_m(x, \xi) - \lambda) \neq 0$ in the matrix-valued case) for all $(x, \xi) \in \overline{\Omega} \times \mathbb{R}^n$ and $\lambda \in \Lambda$ such that $|\xi| + |\lambda| \neq 0$;

b) for any $x_0 \in \partial\Omega$, $\xi' \in \mathbb{R}^{n-1}$, $\lambda \in \Lambda$, and $\psi_1, \ldots, \psi_{m/2} \in \mathbb{C}$, for $|\lambda| + |\xi'| \neq 0$ the problem (2.84) has a unique solution $v \in \mathcal{S}(\overline{\mathbb{R}}_+)$.

We note that a) implies that the operator A is elliptic throughout $\overline{\Omega}$. The condition (Ell$_\Lambda$) is satisfied, for example, in the case of the Dirichlet and Neumann problems for the Laplacian if Λ is taken as any angle not containing the ray $(-\infty, 0]$. In fact in this case for $v(x_n)$ we obtain the equation

$$\left(-|\xi'|^2 - \lambda + \frac{\partial^2}{\partial x_n^2} \right)v(x_n) = 0,$$

which has the solutions

$$v(x_n) = C_1 e^{-\tau(\xi', \lambda)x_n} + C_2 e^{\tau(\xi', \lambda)x_n}, \quad \tau(\xi', \lambda) = \sqrt{|\xi'|^2 + \lambda},$$

where the branch of the radical is taken with a cut along the ray $(-\infty, 0]$ such that $\sqrt{\mu} > 0$ for $\mu > 0$; it follows from this that $\mathrm{Re}\,\tau(\xi', \lambda) > 0$ for $\xi' \in \mathbb{R}^{n-1}$, $\lambda \in \Lambda$, $|\xi'| + |\lambda| \neq 0$. A solution that tends to zero as $x_n \to +\infty$ has the form $v(x_n) = C_1 e^{-\tau(\xi', \lambda)x_n}$. The boundary condition in (2.84) assumes the form of the equation $C_1 = \psi_1$ in the case of the Dirichlet condition and $-C_1\tau(\xi', \lambda) = \psi_1$ in the case of the Neumann condition, from which it is obvious that the condition (Ell$_\Lambda$) is satisfied.

For the case $n = 2$ and the oblique derivative problem for the Laplacian (cf. Example 2.39) the condition for ellipticity with parameter does not hold at the points $x_0 \in \partial\Omega$ where the vector field ν is tangent to the boundary. In fact, in the notation just introduced, at such a point x_0 the boundary condition for the solution that tends to zero assumes the form $|\nu(x_0)|i\xi'C_1 = \psi_1$, whence for $\xi' = 0$ (and $\lambda \neq 0$) problem (2.84) does not have a unique solution. This example shows the difference between the condition for ordinary ellipticity (in which the value $\xi' = 0$ is forbidden) and the condition for ellipticity with parameter.

Theorem 2.41. *When the condition* (Ell$_\Lambda$) *holds, there exists $R > 0$ such that for $|\lambda| > R$ and $\lambda \in \Lambda$ the problem (2.83) under consideration has a unique solution in the class C^∞. More precisely, for any noninteger $\gamma > m$ the operator*

$$\mathfrak{A}_\lambda : C^\gamma(\Omega) \to C^{\gamma-m}(\Omega) \times \prod_{j=1}^{m/2} C^{\gamma-m_j}(\partial\Omega),$$

(2.85)

$$u \mapsto \left\{ (A - \lambda)u, B_1 u \big|_{\partial\Omega}, \ldots, B_{m/2} u \big|_{\partial\Omega} \right\}$$

is invertible and the inverse operator is also continuous.

In particular it is clear from this that if the problem (2.83) is elliptic with parameter, then for any fixed $\lambda \in \mathbf{C}$ its index is 0.

Theorem 2.41 can be proved by the same method as Theorem 2.40. It plays an important role in spectral theory, where λ has the interpretation of the spectral parameter.

§3. Sobolev Spaces and Generalized Solutions of Boundary-Value Problems

3.1. The Fundamental Spaces (cf. Sobolev 1950; Eskin 1961; Hörmander 1963; Bers, John, and Schechter 1964; Palais 1965; Ladyzhenskaya, Solonnikov, and Ural'tseva 1967; Lions and Magenes 1968; Nikol'skij 1969; Miranda 1970; Ladyzhenskaya 1973; Birman and Solomyak 1974; Sobolev 1974; Besov, Il'in and Nikol'skij 1975; Mikhlin 1977; Triebel 1978; Hörmander 1983–1985; Mikhailov 1983; Triebel 1983; Maz'ya 1985). As we have seen above in Sect. 2.1, the study of distributions that are solutions of differential equations (generalized solutions) is quite useful in answering a variety of questions in the theory of partial differential equations with constant coefficients. It is very often useful, however, especially in the theory of boundary-value problems, to make use of the Hilbert-space structure, or at least the Banach-space structure, of the function spaces from which the solutions are taken. Doing so makes it possible to apply the methods of functional analysis in the theory of linear and nonlinear partial differential equations. In many cases these methods are the only ones possible; in other cases they lead to more definitive results.

Function spaces adapted to the study of partial differential equations were first introduced by S. L. Sobolev and later called Sobolev spaces in his honor. We shall describe the simplest Sobolev spaces in the region Ω that are Hilbert spaces.

Definition 2.42. Let $s \in \mathbf{Z}$, $s \geq 0$. The *Sobolev space* $H^s(\Omega)$ consists of the functions (or rather, equivalence classes of functions) $u \in L^2(\Omega)$ such that $D^\alpha u \in L^2(\Omega)$ for any multi-index α with $|\alpha| \leq s$.

Here $L^2(\Omega)$ is the usual Hilbert space of (equivalence classes of) measurable functions u that are square-integrable in Ω with the norm given by

$$\|u\|^2 = \int_\Omega |u(x)|^2 \, dx,$$

and the derivatives $D^\alpha u$ are understood in the sense of distributions. Using the usual definition of the inner product in $L^2(\Omega)$, i.e., the formula

$$(u, v) = \int_\Omega u(x)\overline{v(x)} \, dx,$$

we introduce an inner product in $H^s(\Omega)$ by the formula

$$(u, v)_s = \sum_{|\alpha| \le s} (D^\alpha u, D^\alpha v),$$

and let $\|\cdot\|_s$ be the corresponding norm, i.e., $\|u\|_s = \sqrt{(u, u)_s}$. It is easy to see that $H^s(\Omega)$ is a Hilbert space. Its completeness follows easily from the completeness of the space $L^2(\Omega)$ and the fact that convergence in $L^2(\Omega)$ implies convergence in the space $\mathcal{D}'(\Omega)$ of distributions on Ω, hence convergence in $\mathcal{D}'(\Omega)$ of all its derivatives (cf. Sect. 1). We note that $H^0(\Omega) = L^2(\Omega)$, and for $s > s'$ there is a continuous imbedding $H^s(\Omega) \subset H^{s'}(\Omega)$. The separability of the space $L^2(\Omega)$ implies the separability of the space $H^s(\Omega)$ for any $s \ge 0$, since the mapping $u \mapsto \{D^\alpha u : |\alpha| \le s\}$ defines an isometry of $H^s(\Omega)$ onto a subspace of $(L^2(\Omega))^{N(s)}$, where $N(s)$ is the collection of distinct n-dimensional multi-indices α with $|\alpha| \le s$ (of which there are $\binom{n + s}{n} = \frac{(n + s)!}{n! s!}$).

Example 2.43. Let $\Omega = \{x : |x| < 1\}$ be the unit ball in $\mathbb{R}^n$, $u(x) = |x|^\gamma$, where $\gamma \in \mathbb{R}$. If $\gamma \in 2\mathbb{Z}_+$ (i.e., γ is an even nonnegative integer), then $u \in C^\infty(\bar\Omega)$ and consequently $u \in H^s(\Omega)$ for any s. But if $s \notin 2\mathbb{Z}_+$, then the condition $u \in H^s(\Omega)$, as is easily verified, is equivalent to the inequality $\gamma - s > -n/2$, i.e., $\gamma + n/2 > s$, since the derivative $D^\alpha u$ in this case is homogeneous in x of degree $\gamma - |\alpha|$ in $\mathbb{R}^n \setminus \{0\}$.

Consider the case $\Omega = \mathbb{R}^n$. The space $H^s(\mathbb{R}^n)$ is easily described using the Fourier transform. To be specific, it follows from properties of the Fourier transform of tempered distributions (cf. Sect. 2.5) that the relation $u \in H^s(\mathbb{R}^n)$ is equivalent to the relations

$$\xi^\alpha \tilde{u}(\xi) \in L^2(\mathbb{R}^n), \quad |\alpha| \le s,$$

for the Fourier transform $\tilde{u}(\xi)$. The entire set of these relations is equivalent to

$$(1 + |\xi|^2)^{s/2} \tilde{u}(\xi) \in L^2(\mathbb{R}^n).$$

By Plancherel's theorem the norm $\|u\|_s$ can be written in the form

$$\|u\|_s^2 = (2\pi)^{-n} \sum_{|\alpha| \le s} \int |\xi^\alpha \tilde{u}(\xi)|^2 \, d\xi = (2\pi)^{-n} \int \left(\sum_{|\alpha| \le s} |\xi^\alpha|^2 \right) |\tilde{u}(\xi)|^2 \, d\xi,$$

from which it is clear that this norm is equivalent to the norm given by the formula

$$\|u\|_s^2 = (2\pi)^{-n} \int (1 + |\xi|^2)^{s/2} |\tilde{u}(\xi)|^2 \, d\xi \qquad (3.1)$$

and denoted just like the preceding norm. (In the questions of interest to us no danger of confusion arises since the difference between equivalent norms is of no importance.)

Applying the representation of the space $H^s(\mathbb{R}^n)$ using the Fourier transform, it is easy to verify that $S(\mathbb{R}^n)$ is dense in $H^s(\mathbb{R}^n)$; it follows easily from this that $C_0^\infty(\mathbb{R}^n)$ is dense in $H^s(\mathbb{R}^n)$ for any integer $s \geq 0$. Therefore $H^s(\mathbb{R}^n)$ can be defined as the completion of $C_0^\infty(\mathbb{R}^n)$ in the norm $\| \cdot \|_s$.

Using the Fourier transform one can define the spaces $H^s(\mathbb{R}^n)$ for any $s \in \mathbb{R}^n$.

Definition 2.44. Let $s \in \mathbb{R}$; the space $H^s(\mathbb{R}^n)$ consists of the distributions $u \in S'(\mathbb{R}^n)$ such that

$$(1 + |\xi|^2)^{s/2} \tilde{u}(\xi) \in L^2(\mathbb{R}^n).$$

Equivalently $H^s(\mathbb{R}^n)$ is the completion of $C_0^\infty(\mathbb{R}^n)$ in the norm $\| \cdot \|_s$ defined by formula (3.1).

It is easily verified that $H^s(\mathbb{R}^n)$ is a Hilbert space for any $s \in \mathbb{R}$, and that for $s > s'$ there is a continuous imbedding $H^s(\mathbb{R}^n) \subset H^{s'}(\mathbb{R}^n)$.

Sometimes the "local" Sobolev spaces are also useful. Let Ω be a region in $\mathbb{R}^n$. For any $s \in \mathbb{R}$ we define the space $H^s_{\text{loc}}(\Omega)$, which consists of the distributions $u \in \mathcal{D}'(\Omega)$ such that $\varphi u \in H^s(\mathbb{R}^n)$ for any function $\varphi \in C_0^\infty(\Omega)$. It is easy to see that when one does this, instead of the whole set of functions φ, it suffices to consider only a subset of them $\Phi \subset C_0^\infty(\Omega)$ possessing the property that for any point $x_0 \in \Omega$ there exists a function $\varphi \in \Phi$ for which $\varphi(x_0) \neq 0$. The space $H^s_{\text{loc}}(\Omega)$ acquires the structure of a Fréchet space if we introduce on it the topology defined by the seminorms

$$\|u\|_{s,\varphi} = \|\varphi u\|_s. \qquad (3.2)$$

In doing this we may restrict ourselves to a countable system of such seminorms generated by the functions $\varphi \in \Phi$, where Φ is a countable subset of $C_0^\infty(\Omega)$ possessing the property described above.

The spaces $H^s_{\text{loc}}(\Omega)$ are invariant under diffeomorphisms: given a diffeomorphism $f : \Omega \to \Omega'$ of class C^∞ (here Ω and Ω' are regions in $\mathbb{R}^n$), with $f^* : \mathcal{D}'(\Omega') \to \mathcal{D}'(\Omega)$ the corresponding distribution mapping (the extension by continuity of the usual change of variable mapping $f^* : C^\infty(\Omega') \to C^\infty(\Omega)$), then f^* induces a mapping $f^* : H^s_{\text{loc}}(\Omega') \to H^s_{\text{loc}}(\Omega)$. This mapping makes it possible to define the spaces $H^s_{\text{loc}}(X)$ on a manifold X. To be specific, suppose a measure $d\mu$ is defined on X with a smooth positive density with respect to Lebesgue measure in any local coordinates (such a measure is

easily constructed using a partition of unity). Then, defining distributions on
the manifold as continuous linear functionals on $C_0^\infty(X)$, we can construct
an imbedding of the ordinary functions (in $L_{\text{loc}}^1(X)$) into the distributions,
assigning a functional on $C_0^\infty(X)$ to each ordinary function $u \in L_{\text{loc}}^1(X)$ by
the formula

$$\langle u, \varphi \rangle = \int_X u\varphi \, d\mu. \tag{3.3}$$

We can now define the space $H_{\text{loc}}^s(X)$ by saying $u \in H_{\text{loc}}^s(X)$ if $u \in \mathcal{D}'(X)$
and for any coordinate neighborhood $\Omega \subset X$ we have $u|_\Omega \in H_{\text{loc}}^s(\Omega)$ in local
coordinates on Ω. By the invariance shown above this relation is independent
of the choice of local coordinates on Ω, and in verifying the inclusion $u \in$
$H_{\text{loc}}^s(X)$ we can restrict ourselves to a system of coordinate neighborhoods
that form a covering of the manifold X. The seminorms of all the restrictions
$u|_\Omega$ induce the structure of a Fréchet space on $H_{\text{loc}}^s(X)$.

We can also introduce the space $H_{\text{comp}}^s(X)$ consisting of the $u \in H_{\text{loc}}^s(X)$
having compact support, i.e, $H_{\text{comp}}^s(X) = H_{\text{loc}}^s(X) \cap \mathcal{E}'(X)$. For any compact
set $K \subset X$ we also denote by $H^s(K)$ the set of $u \in H_{\text{loc}}^s(X)$ such that
$\operatorname{supp} u \subset K$. It is clear that $H_{\text{comp}}^s(X) = \bigcup_{K \in X} H^s(K)$. Since $H^s(K)$ is a
closed subspace of $H_{\text{loc}}^s(X)$, the structure of a Fréchet space is induced on
$H^s(K)$. It is easy to verify, however, that in fact $H^s(K)$ is a Hilbert space,
since we can restrict ourselves to a finite set of the seminorms in the space
$H_{\text{loc}}^s(X)$ that are used to define the topology in $H^s(K)$. We can now introduce
the inductive limit topology in $H_{\text{comp}}^s(X)$, in which a balanced convex set
$\mathcal{U}$ is a neighborhood of zero if and only if the intersection $\mathcal{U} \cap H^s(K)$ is a
neighborhood of zero for each compact set $K \subset X$. The fundamental property
of this topology is that a linear mapping $A : H_{\text{comp}}^s(X) \to E$, where E is
any locally convex space, is continuous if and only if all of the restrictions
$A|_{H^s(K)} : H^s(K) \to E$ are continuous. In particular a linear functional $u :$
$H_{\text{comp}}^s(X) \to \mathbb{C}$ is continuous if and only if all of its restrictions to the spaces
$H^s(K)$ are continuous.

If the manifold X is compact, then $H_{\text{loc}}^s(X) = H_{\text{comp}}^s(X) = H^s(X)$, so
that in this case $H^s(X)$ is a Hilbert space, although there is no canonical
inner product in it.

We note the following important fact: for any $s \in \mathbb{R}$ the spaces $H_{\text{loc}}^s(X)$
and $H_{\text{comp}}^{-s}(X)$ are dual with respect to the bilinear form (3.3). This means
that the form (3.3) extends by continuity from $C^\infty(X) \times C_0^\infty(X)$ to a bilinear
mapping $\langle \cdot, \cdot \rangle : H_{\text{loc}}^s(X) \times H_{\text{comp}}^{-s}(X) \to \mathbb{C}$ that is continuous in each variable
separately, and any continuous linear functional on $H_{\text{comp}}^{-s}(X)$ can be written
uniquely in the form $u(\varphi) = \langle u, \varphi \rangle$, where $u \in H_{\text{loc}}^s(X)$, and any continuous
linear functional φ on $H_{\text{loc}}^s(X)$ can be written uniquely in the form $\varphi(u) =$
$\langle u, \varphi \rangle$, where $\varphi \in H_{\text{comp}}^{-s}(X)$. In particular, in the case of a compact X the
Hilbert spaces $H^s(X)$ and $H^{-s}(X)$ are dual with respect to the bilinear form
(3.3). This duality is topological, i.e., the mapping $u \mapsto \langle u, \cdot \rangle$, which assigns to

each $u \in H^s(X)$ a linear functional on $H^{-s}(X)$ is a topological isomorphism. All these facts about duality are obtained from the elementary fact of the topological duality of the spaces $H^s(\mathbf{R}^n)$ and $H^{-s}(\mathbf{R}^n)$ with respect to the standard form

$$\langle u, \varphi \rangle = \int_{\mathbf{R}^n} u(x)\varphi(x)\, dx. \tag{3.4}$$

We now return to the Sobolev spaces $H^s(\Omega)$, where Ω is a region in $\mathbf{R}^n$ and define these spaces for any $s \in \mathbf{R}$. To do this we begin by considering the case when $\Omega = \mathbf{R}^n_+ = \{x : x = (x', x_n),\ x_n > 0\}$. If $s \in \mathbf{Z}_+$ and $u \in H^s(\mathbf{R}^n_+)$, then there exists $\hat{u} \in H^s(\mathbf{R}^n)$ such that $\hat{u}\big|_{\mathbf{R}^n_+} = u$. Moreover in this case there exists a continuous linear extension operator $l : H^s(\mathbf{R}^n_+) \to H^s(\mathbf{R}^n)$, i.e., a continuous linear operator possessing the property that if $lu = \hat{u}$, then $\hat{u}\big|_{\mathbf{R}^n_+} = u$.

Such an operator can be constructed, for example, from the formula

$$lu(x) = \begin{cases} u(x), & x_n > 0, \\ \sum_{j=1}^{N} a_j u(x', -jx_n), & x_n < 0, \end{cases} \tag{3.5}$$

where the coefficients a_j satisfy the system of linear equations

$$\sum_{j=1}^{N} (-j)^k a_j = 1, \quad k = 0, 1, \ldots, N-1,$$

whose determinant is nonzero, and the number N must be taken to be large ($N \geq s + 1$). We can now define the space $H^s(\mathbf{R}^n_+)$ for any $s \in \mathbf{R}$ to be the space of distributions that admit an extension to $\mathbf{R}^n$ as a distribution $\hat{u} \in H^s(\mathbf{R}^n)$. In other words

$$H^s(\mathbf{R}^n_+) = H^s(\mathbf{R}^n) \big/ \{u : u \in H^s(\mathbf{R}^n),\ u\big|_{\mathbf{R}^n_+} = 0\}.$$

In doing this we must define the topology on $H^s(\mathbf{R}^n_+)$ as the quotient topology, i.e., the norm in $H^s(\mathbf{R}^n_+)$ is given by the formula

$$\|u\|_s = \inf\{\|\hat{u}\|_s : \hat{u} \in H^s(\mathbf{R}^n),\ \hat{u}\big|_{\mathbf{R}^n_+} = u\}. \tag{3.6}$$

The spaces $H^s(\Omega)$ for $s \in \mathbf{R}$ and any region Ω with a smooth boundary $\partial\Omega$ are defined analogously. To be specific, we set

$$H^s(\Omega) = H^s(\mathbf{R}^n) \big/ \{u : u \in H^s(\mathbf{R}^n),\ u\big|_{\Omega} = 0\}$$

and define the norm in $H^s(\Omega)$ by the formula

$$\|u\|_s = \inf\{\|\hat{u}\|_s : \hat{u} \in H^s(\mathbf{R}^n),\ \hat{u}\big|_{\Omega} = u\}. \tag{3.7}$$

The spaces

$$\overset{\circ}{H}{}^s(\bar{\Omega}) = \{u : u \in H^s(\mathbb{R}^n), \ \operatorname{supp} u \subset \bar{\Omega}\}$$

are also sometimes useful.

For $s > 1/2$ the space $\overset{\circ}{H}{}^s(\bar{\Omega})$ contains no distributions concentrated on $\partial\Omega$ and instead of $\overset{\circ}{H}{}^s(\bar{\Omega})$ we may write $\overset{\circ}{H}{}^s(\Omega)$. It can be shown that the spaces $H^s(\Omega)$ and $\overset{\circ}{H}{}^{-s}(\bar{\Omega})$ are dual to each other in the sense described above. Also $\overset{\circ}{H}{}^s(\bar{\Omega}) = H^s(\Omega)$ for $|s| < \frac{1}{2}$, $\overset{\circ}{H}{}^s(\bar{\Omega}) \subset H^s(\Omega)$ for $s \geq 0$, and $\overset{\circ}{H}{}^s(\bar{\Omega}) \supset H^s(\Omega)$ for $s \leq 0$. The space $C_0^\infty(\Omega)$ is dense in $\overset{\circ}{H}{}^s(\bar{\Omega})$ for $s > -\frac{1}{2}$, so that in this case $\overset{\circ}{H}{}^s(\Omega)$ can be regarded as the completion of $C_0^\infty(\Omega)$ in the norm $\| \cdot \|_s$. The space $H^s(\Omega)$ can be defined by this method for any region Ω (with any kind of boundary).

We shall now give a direct description of the spaces $H^s(\Omega)$ for $s \geq 0$ without using extensions or the Fourier transform. Let $s = k + \lambda$, where $k \in \mathbb{Z}_+$ and $0 < \lambda < 1$. Let the region Ω be bounded and have a smooth boundary. Then the condition $u \in H^s(\Omega)$ is equivalent to the condition that u belongs to $H^k(\Omega)$ and the integral

$$\mathcal{F}_s(u) = \iint\limits_{\Omega\ \Omega} \sum_{|\alpha|=k} \frac{|D^\alpha u(y) - D^\alpha u(x)|^2}{|y - x|^{n+2\lambda}} \, dx\, dy,$$

converges. A norm equivalent to the norm (3.7) can be introduced in $H^s(\Omega)$ by the formula

$$\|u\|^2 = \|u\|_k^2 + \mathcal{F}_s(u). \tag{3.8}$$

All the spaces $H^s(\Omega)$ and $\overset{\circ}{H}{}^s(\bar{\Omega})$ are Hilbert spaces.

The spaces $H^s(\Omega)$ are often denoted $W_2^s(\Omega)$. In many problems the spaces $W_p^s(\Omega)$ are also useful. These spaces are Banach spaces defined for integers $s \geq 0$ by the formula

$$W_p^s(\Omega) = \{u : u \in L^p(\Omega), \ D^\alpha u \in L^p(\Omega) \text{ for } |\alpha| \leq s\},$$

and for nonintegers $s \geq 0$ using the integral

$$\mathcal{F}_{s,p}(u) = \iint\limits_{\Omega\ \Omega} \sum_{|\alpha|=k} \frac{|D^\alpha u(y) - D^\alpha u(x)|^p}{|y - x|^{k+\lambda p}} \, dx\, dy,$$

where $s = k + \lambda$ for $k \in \mathbb{Z}_+$ and $0 < \lambda < 1$. To be specific, for a bounded region $\Omega \subset \mathbb{R}^n$ the space $W_p^s(\Omega)$ consists of the $u \in W_p^k(\Omega)$ such that the integral $\mathcal{F}_{s,p}(u)$ is finite. The norm in $W_p^s(\Omega)$ can be defined by the formula

$$\|u\|_{s,p}^p = \|u\|_{k,p}^p + \mathcal{F}_{s,p}(u), \tag{3.9}$$

where

$$\|u\|_{k,p}^p = \sum_{|\alpha|\leq k} \int_\Omega |D^\alpha u(x)|^p \, dx. \tag{3.10}$$

The spaces $W_p^s(\Omega)$ can also be defined on any compact manifold with boundary in analogy with the procedure we followed above for $p = 2$.

3.2. Imbedding and Trace Theorems (cf. references to Sect. 3.1). *Imbedding Theorems* describe the imbedding of Sobolev spaces into other spaces (in particular, other Sobolev spaces, the spaces L_p and C^k, Hölder spaces, and the like), as well as the properties of imbedding transformations. *Trace theorems* described the restrictions of functions of Sobolev spaces to submanifolds.

The simplest propositions of imbedding theorem type were mentioned above: $H^s(\Omega) \subset H^{s'}(\Omega)$ for integers s and s' with $s > s'$ (it is easy to see that this is also true for any real s and s'). Similarly $W_p^s(\Omega) \subset W_p^{s'}(\Omega)$ for $s \geq s'$ and any fixed $p \in [1, +\infty)$. If the region Ω is bounded, then $L_p(\Omega) \subset L'_p(\Omega)$ for $p \geq p'$, from which it follows that in this case $W_p^s(\Omega) \subset W_{p'}^{s'}(\Omega)$ whenever $p \geq p'$ and $s \geq s'$. The question arises whether an increase in the index of smoothness s will bring about an increase in the index of integrability. The answer is contained in the following theorem.

Theorem 2.45. *Let Ω be a bounded region with a piecewise-smooth boundary. Then there exist a compact (completely continuous) imbedding*

$$W_p^s(\Omega) \subset C_b(\Omega) \quad \text{for } s > n/p, \tag{3.11}$$

where $C_b(\Omega)$ is the space of bounded continuous functions on Ω with the sup-norm, and also a more general compact imbedding

$$W_p^s(\Omega) \subset C_b^k(\Omega) \quad \text{for } s > n/p + k, \tag{3.11'}$$

where $C_b^k(\Omega)$ is the space of functions belonging to $C^k(\Omega)$ and having bounded derivatives of orders $\leq k$ (here $k \in \mathbb{Z}_+$).
There is a continuous imbedding

$$W_p^s(\Omega) \subset L_q(\Omega) \quad \text{for } s \geq 0, \ s \geq n\left(\frac{1}{p} - \frac{1}{q}\right), \tag{3.12}$$

and a more general imbedding (also continuous)

$$W_p^s(\Omega) \subset W_q^r(\Omega) \quad \text{for } s - r \geq 0, \ s - r \geq n\left(\frac{1}{p} - \frac{1}{q}\right). \tag{3.12'}$$

In the case $ps < n$ the imbeddings become compact if both of the signs $\geq$ are changed to $>$ in (3.12) or (3.12').

The proof of Theorem 2.45 can be based, for example, on various integral formulas that express a function u in terms of its derivatives (for integer s) or in terms of expressions like the integrals $\mathcal{F}_{s,p}(u)$ occurring in the norms of the spaces $W_p^s(\Omega)$. When this is done, in the proof of the existence and continuity of the imbedding one has only to verify the estimate for the norm (continuity of the corresponding imbedding operator) for sufficiently smooth functions, since all of the spaces $W_p^s(\Omega)$ are completions of spaces of smooth functions. In this way the compactness of the imbedding is obtained from the theorem of Arzelà in the case of (3.11) (we note that (3.11') follows immediately from (3.11)) and from the analogues of this theorem in which the sup-norm is replaced by the L^p-norm (a set $A \subset L^p(\Omega)$ in the case of a bounded region Ω is compact if and only if it is bounded in L^p-norm and uniformly equicontinuous in that norm with respect to translations by vectors of $\mathbb{R}^n$).

If $C^k(\Omega)$ denotes the Hölder space introduced in Sect. 2.14 for nonintegral $k > 0$, the same imbedding theorem holds as for integer k, i.e., (3.11') holds also for nonintegral $k > 0$ and the imbedding is compact.

All the imbedding theorems described above are *sharp* in the sense that there is no such imbedding if the indices do not satisfy the stated inequalities.

Imbedding theorems for the local spaces can be obtained in an obvious way from these imbedding theorems. For example, it follows from the imbedding (3.11') with $p = 2$ that

$$H_{\text{loc}}^s(\Omega) \subset C^k(\Omega), \quad s > \frac{n}{2} + k.$$

We now turn to the discussion of trace theorems. Consider any piecewise-smooth hypersurface Γ (possibly with boundary) contained in $\bar{\Omega}$ (in particular Γ may coincide with the boundary $\partial\Omega$ of the region Ω). We introduce the restriction operator

$$\gamma : C^\infty(\bar{\Omega}) \to C^\infty(\Gamma), \quad \gamma(u) = u|_\Gamma.$$

We wish to discuss when this operator can be extended to a continuous operator on some Sobolev spaces. The answer is contained in the following theorem.

Theorem 2.46. *Suppose the hypersurface Γ is either compact or a portion of a hyperplane. Then for $s > 1/2$ the operator γ can be extended to a continuous operator*

$$\gamma : H^s(\Omega) \to H^{s-1/2}(\Gamma). \tag{3.13}$$

If the hypersurface Γ is smooth (in particular, if Γ is part of a hyperplane), this extension is an epimorphism.

Here the space $H^r(\Gamma)$ for a piecewise-smooth surface Γ is understood as the space of functions belonging to $H^r(\Gamma_0)$ on each smooth piece Γ_0 of the surface Γ.

This theorem makes it possible to give an unambiguous definition of the trace $u|_\Gamma$ for a function $u \in H^s(\Omega)$ for $s > 1/2$. This trace is an element of the space $H^{s-1/2}(\Gamma)$ depending linearly and continuously on u. It also depends continuously on Γ in a natural sense.

In the case of the space $W_p^s(\Omega)$ for arbitrary $p > 1$ the trace can be defined in the case when $s > 1/p$, but a precise description of the space in which it lies is more complicated (this is the so-called Besov space $B_p^{s-1/p}(\Gamma)$, which we shall not define here). In any case γ extends to a continuous linear mapping

$$\gamma : W_p^s(\Omega) \to W_p^{s-1/p-\varepsilon}(\Gamma), \quad s > 1/p, \quad \varepsilon > 0, \tag{3.14}$$

where $\varepsilon > 0$ can be chosen arbitrarily small. This mapping, of course, is no longer an epimorphism for any nonempty hypersurface Γ.

We also point out that if Γ is a piecewise-smooth submanifold of codimension k in Ω, compact or coinciding with a portion of a $(n - k)$-dimensional plane, then the mapping γ can be extended to a continuous linear mapping

$$\gamma : H^s(\Omega) \to H^{s-k/2}(\Gamma), \quad s > k/2. \tag{3.15}$$

This assertion can be obtained immediately from Theorem 2.46 by constructing the following submanifold flag in Ω: the sequence $\Omega \supset \Gamma_1 \supset \Gamma_2 \supset \cdots \supset \Gamma_k = \Gamma$, where Γ_j is a piecewise-smooth submanifold of codimension 1 in Γ_{j-1}, $j = 1, 2, \ldots, k$ (we take $\Gamma_0 = \Omega$). Then it is necessary to pass successively from Ω to Γ_1, from Γ_1 to Γ_2, $\ldots$, from Γ_{k-1} to $\Gamma_k = \Gamma$, each time taking the trace as in Theorem 2.46. The desired flag obviously exists for a smooth Γ and can be constructed locally in the general case, which is quite sufficient.

We note that for $k = n$ the operator (3.15) is continuous if and only if there is an imbedding $H^s(\Omega) \subset C(\Omega)$ (in this case Γ is zero-dimensional, i.e., a collection of points), so that (3.15) is consistent with Theorem 2.45 (the imbedding (3.11) with $p = 2$).

The proof of Theorem 2.46 is carried out most simply if an estimate for the norms equivalent to the continuity of the operator (3.13) is carried out in terms of the Fourier transform (when the situation is localized, one can choose local coordinates so that Γ becomes a hyperplane), i.e., using expressions of the form (3.1) for the corresponding norms. When this is done the estimate can be obtained by an elementary application of the Cauchy-Bunyakovskij inequality.

It is often important to take not only the trace $u|_\Gamma$ of the function u itself, but also the trace of some of its derivatives, for example, $D^\alpha u|_\Gamma$ or $\dfrac{\partial^j u}{\partial n^j}\Big|_\Gamma$, where n is a fixed normal to Γ chosen continuously on some smooth piece of the hypersurface Γ. But this obviously reduces to Theorem 2.46, since if $u \in H^s(\Omega)$, then $D^\alpha u \in H^{s-|\alpha|}(\Omega)$, and so the trace of $D^\alpha u|_\Gamma$ is defined for $s - |\alpha| > 1/2$ and belongs to the space $H^{s-|\alpha|-1/2}(\Omega)$. To establish the existence of the trace of $\dfrac{\partial^j u}{\partial n^j}\Big|_\Gamma$ we can choose local coordinates $x = (x_1, x_2, \ldots, x_n)$ in a neighborhood of a smooth piece Γ_0 of the hypersurface Γ so that Γ_0 becomes a piece of the hyperplane $\{x : x_n = 0\}$ and $n = (0, 0, \ldots, 0, 1)$. But then $\dfrac{\partial^j u}{\partial n^j}\Big|_\Gamma = \dfrac{\partial^j u}{\partial x_n^j}\Big|_\Gamma$ and the problem again reduces to

Theorem 2.46, since $\dfrac{\partial^j u}{\partial x_n^j} \in H^{s-j}(\mathcal{U})$ for $u \in H^s(\mathcal{U})$ (here $\mathcal{U}$ is a neighborhood of the piece Γ_0 in which the local coordinates under consideration are defined).

We note further that the concept of the trace makes it possible to describe how the subspaces $\overset{\circ}{H}{}^s(\Omega)$ are distinguished among the spaces $H^s(\Omega)$ for $s > -1/2$ and $s - 1/2$ nonintegral, at least in the case of bounded regions Ω with a smooth boundary $\partial\Omega$ or in the case when $\partial\Omega$ is a piece of a hyperplane in $\mathbf{R}^n$. To be specific, in this case the inclusion $u \in \overset{\circ}{H}{}^s(\Omega)$ for $s > -1/2$ and nonintegral $s - 1/2$ is equivalent to the set of conditions $u \in H^s(\Omega)$ and $\dfrac{\partial^j u}{\partial n^j}\Big|_{\partial\Omega} = 0$ for all $j = 0, 1, \ldots, [s - 1/2]$ (here $[x]$ denotes the integer part of the number x). In particular, if $s \in \mathbf{Z}_+$, then the inclusion $u \in \overset{\circ}{H}{}^s(\Omega)$ is equivalent to $u\big|_{\partial\Omega} = 0$, $\dfrac{\partial u}{\partial n}\Big|_{\partial\Omega} = 0, \ldots, \dfrac{\partial^{s-1} u}{\partial n^{s-1}}\Big|_{\partial\Omega} = 0$, and also to the conditions $D^\alpha u\big|_{\partial\Omega} = 0$ for $|\alpha| \le s - 1$.

3.3. Generalized Solutions of Elliptic Boundary-Value Problems and Eigenvalue Problems (cf. Hörmander 1963; Bers, John, and Schechter 1964; Lions and Magenes 1968; Ladyzhenskaya 1973; Mizohata 1973; Mikhlin 1977; Mikhailov 1983).

A. The Dirichlet Problem. We consider the simplest elliptic boundary-value problem, the Dirichlet problem for Laplace's equation or Poisson's equation, and give a generalized statement of it. We begin by discussing the problem for Poisson's equation with zero boundary conditions:

$$\begin{cases} \Delta u(x) = f(x), & x \in \Omega, \\ u\big|_{\partial\Omega} = 0. \end{cases} \tag{3.16}$$

Instead of the boundary condition $u\big|_{\partial\Omega} = 0$ we write $u \in \overset{\circ}{H}{}^1(\Omega)$ (as we have pointed out earlier, in the case of bounded regions with a smooth boundary this inclusion is equivalent to the conditions $u \in H^1(\Omega)$ and $u\big|_{\partial\Omega} = 0$). Next, multiplying both sides of the equation $\Delta u = f$ by $\bar{v}(x)$, where $v \in C_0^\infty(\Omega)$ and integrating by parts, we obtain

$$[u, v] = -(f, v), \tag{3.17}$$

where $(\cdot, \cdot)$ denotes the inner product on $L^2(\Omega)$ and

$$[u, v] = \int_\Omega \sum_{j=1}^n \frac{\partial u}{\partial x_j} \frac{\partial \bar{v}}{\partial x_i} \, dx, \tag{3.18}$$

so that $[\cdot, \cdot]$ is a hermitian form, continuous on the space $H^1(\Omega)$, i.e.,

$$|[u, v]| \le C\|u\|_1 \|v\|_1,$$

where the constant $C > 0$ is independent of u and v.

The quantity

$$\mathcal{D}(u) = [u, u] = \int_\Omega |\nabla u(x)|^2 \, dx = \int_\Omega \sum_{j=1}^n \left| \frac{\partial u(x)}{\partial x_j} \right|^2 \, dx \qquad (3.18')$$

is called a *Dirichlet integral* and has the physical interpretation of potential energy of a vibrating medium (for example, a membrane when $n = 2$, a gas or elastic solid when $n = 3$).

Equality (3.17) has meaning for any functions $u, v \in H^1(\Omega)$ and for $f \in L^2(\Omega)$. It will be our replacement for the equation $\Delta u = f$. In doing this we shall take only functions v such that $v \in \overset{o}{H}{}^1(\Omega)$. In the case of the classical solution u (i.e., a solution $u \in C^2(\bar{\Omega})$ of problem (3.16)) equality (3.17) is obtained by the procedure just described for $v \in C_0^\infty(\Omega)$ and then for $v \in \overset{o}{H}{}^1(\Omega)$ by passing to the limit. If Ω is a bounded region with smooth boundary, one can take $v \in C^\infty(\bar{\Omega})$ from the outset, and then (3.17) can be obtained (in the case of the classical solution u) for any $v \in H^1(\Omega)$. We do not wish to restrict ourselves to the case of regions with smooth boundary, however; and besides, as we shall see below, equality (3.17), given only for all $v \in \overset{o}{H}{}^1(\Omega)$ (or even only for $v \in C_0^\infty(\Omega)$) determines a function $u \in \overset{o}{H}{}^1(\Omega)$ uniquely).

Thus we arrive at the following *generalized statement of problem* (3.16):

(Π): Given a function $f \in L^2(\Omega)$, find a function $u \in \overset{o}{H}{}^1(\Omega)$ such that (3.17) holds for any function $v \in C_0^\infty(\Omega)$.

As already pointed out, one can write $v \in \overset{o}{H}{}^1(\Omega)$ instead of $v \in C_0^\infty(\Omega)$, leading to an equivalent statement. In addition, transferring the derivatives from v to u through integration by parts, we find that (3.17) is equivalent to the equation $\Delta u = f$, understood in the sense of distributions, so that the problem (Π) is equivalent to the following problem:

(Π'): Given a function $f \in L^2(\Omega)$, find a function $u \in \overset{o}{H}{}^1(\Omega)$ such that $\Delta u = f$ in the sense of distributions.

Every solution u of problem (Π) (or (Π')) will be called a *generalized* or *weak* solution (in contrast to the classical solution, which has meaning when $f \in C(\bar{\Omega})$). As we have seen above, every classical solution $u \in C^2(\bar{\Omega})$ is a generalized solution.

Theorem 2.47. *If Ω is any bounded region in $\mathbf{R}^n$ and $f \in L^2(\Omega)$, then there exists a unique generalized solution of the problem (Π) (or (Π')).*

For the proof one must first of all remark that $[\cdot, \cdot]$ can be regarded as an inner product on the space $\overset{o}{H}{}^1(\Omega)$. This is equivalent to the condition that the expression $\|u\|_1' = \sqrt{\mathcal{D}(u)} = [u, u]^{1/2}$ is a norm equivalent to the norm $\|\cdot\|_1$ on C_0^∞. In view of the obvious relation

$$\|u\|_1^2 = \|u\|^2 + \mathcal{D}(u),$$

the equivalence of the norms $\| \cdot \|_1$ and $\| \cdot \|_1'$ follows from the so-called *Friedrichs inequality*

$$\|u\|^2 \leq C\mathcal{D}(u), \quad u \in C_0^\infty(\Omega), \tag{3.19}$$

where $C > 0$ is independent of u and $\| \cdot \|$ is the L^2-norm. This inequality can be obtained by assuming that $\Omega \subset \{x : 0 < x_n < a\}$ and writing

$$u(x) = \int_0^{x_n} \frac{\partial u}{\partial x_n}(x', t)\, dt,$$

from which, by the Cauchy-Bunyakovskij inequality, for $x \in \Omega$,

$$|u(x)|^2 \leq |x_n| \int_0^{x_n} \left| \frac{\partial u}{\partial x_n}(x', t) \right|^2 dt \leq a \int_0^a \left| \frac{\partial u(x', x_n)}{\partial x_n} \right|^2 dx_n$$

and then, integrating over $x' \in \mathbb{R}^{n-1}$ and over $x_n \in [0, a]$, we obtain (3.19) with $C = a^2$.

We now remark that the functional $l(v) = -(f, v)$ is conjugate-linear and continuous on $\overset{\circ}{H}{}^1(\Omega)$. Therefore by the Riesz representation theorem, it can be written (and moreover uniquely) in the form $l(v) = [u, v]$, where $u \in \overset{\circ}{H}{}^1(\Omega)$, which proves Theorem 2.47.

The reasoning just given illustrates very well the essence of the application of functional-analytic methods in the theory of partial differential equations. Passing to the generalized statement of the problem made it possible to write the problem in a form amenable to the application of the Riesz representation theorem, which immediately gives the existence and uniqueness of the solution. In the Riesz representation theorem the *completeness* of the space under consideration is essential, so that the passage to the Sobolev spaces of Hilbert type played an important role. We note further that the character of the boundary of the region Ω played no role.

The weakness of the proof just given is that it gives no information on the precise smoothness of the solution constructed. This problem can be solved separately, but only for regions with a sufficiently regular boundary. For simplicity we shall assume the boundary $\partial\Omega$ infinitely smooth. It can then be proved that if $f \in H^s(\Omega)$, $s \geq 0$, then $u \in H^{s+2}(\Omega)$. Thus the Laplacian defines an isomorphism

$$\Delta : H^{s+2}(\Omega) \cap \overset{\circ}{H}{}^1(\Omega) \to H^s(\Omega), s \geq 0. \tag{3.20}$$

We now consider briefly the Dirichlet problem for Laplace's equation:

$$\begin{cases} \Delta u(x) = 0, & x \in \Omega, \\ u|_{\partial\Omega} = \varphi. \end{cases} \tag{3.21}$$

In passing to the generalized statement the first question that arises is the interpretation of the boundary condition. If the boundary $\partial\Omega$ is smooth, this condition can be interpreted as we interpreted the trace in Sect. 3.2. In particular, if $\varphi \in H^{3/2}(\partial\Omega)$, then by Theorem 2.46 there exists a function $v \in H^2(\Omega)$ such that $v|_{\partial\Omega} = \varphi$. But then if $u \in H^1(\Omega)$ is a solution of the problem (3.21), we obtain a problem of the form (3.16) for $w = u - v$ with $f = \Delta v \in L^2(\Omega)$, so that we can pass to the generalized statement of (II) or (II'), and in the case of a bounded region Ω we can apply Theorem 2.47, from which it now follows that there exists a unique solution of the problem (3.21). If the boundary $\partial\Omega$ is not smooth, we can first fix the function $v \in H^1(\Omega)$ that gives the boundary condition and pose the problem as follows:

$(\mathcal{D})$: Given a function $v \in H^1(\Omega)$, find a function u such that $u - v \in \overset{\circ}{H}{}^1(\Omega)$ and $\Delta u(x) = 0$ for $x \in \Omega$.

Theorem 2.48. *If Ω is any bounded region in $\mathbb{R}^n$ and $v \in H^1(\Omega)$, then there exists a unique solution u of the problem $(\mathcal{D})$. This solution gives a strict minimum for the Dirichlet integral $\mathcal{D}(u)$ among the functions $u \in H^1(\Omega)$ for which $u - v \in \overset{\circ}{H}{}^1(\Omega)$. Conversely, if u is a stationary point for the Dirichlet integral in the class of functions $u \in H^1(\Omega)$ for which $u - v \in \overset{\circ}{H}{}^1(\Omega)$, then u is a solution of the problem $(\mathcal{D})$ (and therefore the Dirichlet integral has a strict minimum at the function u).*

Thus the Dirichlet problem becomes the problem of minimizing the Dirichlet integral, i.e., minimizing the potential energy. In particular this makes it possible to apply variational methods to this problem, which in turn make it possible to find approximate solutions minimizing the Dirichlet integral over finite-dimensional subspaces of functions which, in a natural sense, approach the whole space $H^1(\Omega)$. For example, one can choose an orthonormal basis $\{\psi_1, \psi_2, \ldots\}$ in the space $\overset{\circ}{H}{}^1(\Omega)$, take the subspace V_N spanned by $\psi_1, \psi_2, \ldots, \psi_N$ and a function $u_N \in V_N$ at which the minimum of $\mathcal{D}(u)$ is attained over the affine subspace $u + V_N$. It can then be shown that if u is a solution of the problem $\mathcal{D}$, then $\|u_N - u\| \to 0$ as $N \to \infty$. This method of solving the Dirichlet problem, which is applicable in many other situations also, was proposed by B. G. Galerkin.

The proof of Theorem 2.48 is just as simple as the proof of Theorem 2.47. The equation $\Delta u = 0$ is equivalent to

$$[u, w] = 0, \quad w \in C_0^\infty(\Omega). \tag{3.22}$$

By continuity one can write $\overset{\circ}{H}{}^1(\Omega)$ here instead of $C_0^\infty(\Omega)$, and then condition (3.22) means that u is the perpendicular (with respect to the inner product $[\cdot, \cdot]$) from the point v to the subspace $\overset{\circ}{H}{}^1(\Omega)$. Hence, in particular, it follows that if another function $u_1 \in H^1(\Omega)$ is given for which $u_1 - v \in \overset{\circ}{H}{}^1(\Omega)$ (i.e., u_1 satisfies the same boundary condition as u), then

$$\mathcal{D}(u_1) = [u_1, u_1] = [(u_1 - u) + u, (u_1 - u) + u] =$$
$$= [u, u] + [u_1 - u, u_1 - u] = \mathcal{D}(u) + \mathcal{D}(u_1 - u),$$

so that $\mathcal{D}(u_1) > \mathcal{D}(u)$ for $u_1 \neq u$. This proves the assertion about the minimality of $\mathcal{D}(u)$ and the uniqueness of the solution.

Unfortunately it is impossible to make direct use of the theorem on the existence of a perpendicular in Hilbert space in this proof, since $[\cdot, \cdot]$ is not an inner product on $H^1(\Omega)$ (taking $u_0 \equiv 1$, we obtain $\mathcal{D}(u_0) = [u_0, u_0] = 0$). But we must try to find $z = u - v \in \overset{\circ}{H}{}^1(\Omega)$, so that (3.22) can be rewritten in the form

$$[z, w] = -[v, w], \quad w \in \overset{\circ}{H}{}^1(\Omega), \tag{3.22'}$$

and then z can also be found as in the proof of Theorem 2.47, i.e., by the Riesz representation theorem, as the vector giving the continuous conjugate-linear functional on the right-hand side of (3.22') in the form of an inner product. The assertion in the statement of Theorem 2.48 – that the condition that the Dirichlet integral be stationary at the function $u \in H^1(\Omega)$ is equivalent to the condition (3.22) – can easily be verified directly.

We note that if u is a solution of the problem $\mathcal{D}$, then $u \in C^\infty(\Omega)$, and moreover, u is analytic in Ω by the results of Sect. 2.5. The question of the precise connection between the smoothness of φ and the smoothness of u is an important one, which we shall discuss immediately for the general problem

$$\begin{cases} \Delta u(x) = f(x), & x \in \partial\Omega, \\ u|_{\partial\Omega} = \varphi. \end{cases} \tag{3.23}$$

We shall assume that the region Ω is bounded and the boundary $\partial\Omega$ is smooth (of class C^∞). Then it is natural to associate the operator

$$H^s(\Omega) \to H^{s-2}(\Omega) \times H^{s-1/2}(\partial\Omega), \quad s > 1/2, \tag{3.24}$$

$$u \mapsto \{\Delta u, u|_{\partial\Omega}\}$$

with the problem (3.23). This operator is a linear topological isomorphism (i.e., a continuous linear operator with a continuous inverse).

B. The Eigenvalue Problem with the Dirichlet Boundary Condition. Consider the simplest *eigenfunction problem*: find functions $\psi \not\equiv 0$ in Ω such that for some $\lambda \in \mathbb{C}$

$$\begin{cases} -\Delta\psi(x) = \lambda\psi(x), & x \in \Omega, \\ \psi|_{\partial\Omega} = 0. \end{cases} \tag{3.25}$$

(The number λ is called an *eigenvalue* and the function ψ an *eigenfunction* for the Dirichlet problem for the operator $-\Delta$.)

In accordance with what was said above, the generalized statement of this problem will have the form

$$(P): \quad -\Delta\psi = \lambda\psi, \quad \psi \in \overset{\circ}{H}{}^1(\Omega).$$

Theorem 2.49. *If Ω is a bounded region in $\mathbb{R}^n$, there exists a system of eigenfunctions of the problem (3.25), interpreted as the problem (P), that is orthonormal in $L^2(\Omega)$ and complete in both $L^2(\Omega)$ and $\overset{\circ}{H}{}^1(\Omega)$.*

To prove this theorem one must consider the unbounded operator A in $L^2(\Omega)$ with domain of definition

$$D(A) = \{u : u \in \overset{\circ}{H}{}^1(\Omega), \quad \Delta u \in L^2(\Omega)\}$$

defined by the formula $Au = -\Delta u$. It follows from Theorem 2.47 that the image $\operatorname{Im} A = A\big(D(A)\big)$ of the operator A coincides with the whole space $L^2(\Omega)$. It is easy to see that the operator A is symmetric, since for $u, v \in D(A)$

$$(Au, v) = [u, v] = (u, Av).$$

It follows from this same relation and Friedrichs' inequality that

$$(Au, u) = [u, u] \geq \varepsilon(u, u), \quad u \in D(A),$$

where $\varepsilon > 0$, from which it is clear that the operator A is invertible and the inverse operator A^{-1} is bounded on $L^2(\Omega)$ and symmetric (hence self-adjoint). Moreover the inverse operator A^{-1} obviously maps $L^2(\Omega)$ into $\overset{\circ}{H}{}^1(\Omega)$. This mapping is continuous, as follows immediately from the closed graph theorem, but can also be obtained using the following chain of inequalities (in which $Au = f$):

$$\|u\|_1'^2 = [u, u] = (Au, u) = (f, u) \leq \|f\|\|u\| \leq C\|f\|\|u\|_1',$$

(here we have used the Cauchy-Bunyakovskij inequality and Friedrichs' inequality). Thus $\|u\|_1' \leq C\|f\| = C\|A^{-1}u\|$, which was to be proved. But now the operator $A^{-1} : L^2(\Omega) \to L^2(\Omega)$ can be represented in the form of the composition of a continuous mapping $L^2(\Omega) \to \overset{\circ}{H}{}^1(\Omega)$ and the imbedding $\overset{\circ}{H}{}^1(\Omega) \to L^2(\Omega)$. By the compactness of this imbedding we find that the operator $A^{-1} : L^2(\Omega) \to L^2(\Omega)$ is compact. Therefore it has a complete orthonormal system of eigenfunctions $\psi_1, \psi_2, \ldots$ with eigenvalues $\mu_1, \mu_2, \ldots$. Moreover $\mu_j \neq 0$ and $\mu_j \to 0$ as $j \to +\infty$. But then these same functions ψ_j will be eigenfunctions of the operator A with eigenvalues $\lambda_j = \mu_j^{-1}$. We note that $\lambda_j > 0$, since if $A\psi_j = \lambda_j \psi_j$, then $\lambda_j(\psi_j, \psi_j) = (A\psi_j, \psi_j) = [\psi_j, \psi_j] > 0$. Therefore $\lambda_j \to +\infty$ as $j \to +\infty$.

The eigenfunctions constructed are also orthogonal in the space $\overset{\circ}{H}{}^1(\Omega)$ with respect to the inner product $[\cdot, \cdot]$, since they satisfy

$$[\psi_i, \psi_j] = (A\psi_i, \psi_j) = \lambda_i(\psi_i, \psi_j) = 0 \quad \text{for } i \neq j.$$

The completeness of the system $\{\psi_j : j = 1, 2, \ldots\}$ in the space $\overset{\circ}{H}{}^1(\Omega)$ follows from the fact that if $u \in \overset{\circ}{H}{}^1(\Omega)$ and $[u, \psi_j] = 0$, then $(u, \psi_j) = 0$, since $[u, \psi_j] = (u, A\psi_j) = \lambda_j(u, \psi_j)$.

We note that all the eigenfunctions ψ_j are analytic in Ω; if the boundary $\partial\Omega$ is smooth, then $\psi_j \in C^\infty(\bar{\Omega})$; and if the boundary $\partial\Omega$ is analytic, then ψ_j are analytic in $\bar{\Omega}$. This follows from the general theory of elliptic boundary-value problems (cf. Sect. 2.2), a particular case of which is the problem (3.25).

C. The Neumann Problem and the Eigenvalue Problem with the Neumann Condition. The homogeneous Neumann problem for Poisson's equation has the form

$$\begin{cases} \Delta u(x) = f(x), & x \in \Omega, \\ \left.\dfrac{\partial u}{\partial n}\right|_{\partial\Omega} = 0. \end{cases} \tag{3.26}$$

In order to pass to the generalized statement of this problem we assume at first that Ω is a bounded region with a smooth boundary and $u \in C^\infty(\bar{\Omega})$. Multiplying both sides of the equation $\Delta u = f$ by the function $\bar{v}$, where $v \in C^\infty(\bar{\Omega})$, and then integrating on Ω, we can use Green's formula

$$\int_\Omega \Delta u(x) \cdot \overline{v(x)} \, dx = -\int_\Omega \nabla u(x) \cdot \nabla \bar{v}(x) \, dx + \int_{\partial\Omega} \bar{v}(x) \frac{\partial u(x)}{\partial n} \, dS_x, \tag{3.27}$$

where dS_x is the element of surface area on the boundary (this formula follows from the general Stokes' theorem). Hence we find by (3.26) that

$$[u, v] = -(f, v). \tag{3.28}$$

By continuity we can take $v \in H^1(\Omega)$ here instead of $v \in C^\infty(\bar{\Omega})$ even when we know only that $u \in H^1(\Omega)$ and $f \in L^2(\Omega)$. This gives us the *generalized statement of the Neumann problem*:

(N): given a function $f \in L^2(\Omega)$, find a function $u \in H^1(\Omega)$ such that (3.28) holds for any function $v \in H^1(\Omega)$.

The solution of this problem is unique up to an arbitrary additive constant: if u_1 is another solution of the problem (N) (with the same function f) and $w = u_1 - u$, then $[w, v] = 0$ for any function $v \in H^1(\Omega)$. Setting $v = w$, we find that $[w, w] = 0$. This means that all the generalized derivatives $\dfrac{\partial w}{\partial x_j}$, $j = 1, 2, \ldots, n$, vanish, from which it is easy to deduce that $w = \text{const}$.

In what follows we shall assume that the region Ω is bounded.

The problem (N) has a solution for those functions $f \in L^2(\Omega)$ and only those that satisfy the condition

$$(f, 1) = \int_\Omega f(x) \, dx = 0, \tag{3.29}$$

i.e., for functions with mean value zero. The necessity of this condition follows immediately from (3.28) with $v \equiv 1$. To prove sufficiency it is simplest to pass to the uniquely solvable problem

$$\begin{cases} -\Delta u(x) + u(x) = g(x), \quad x \in \Omega; \\ \dfrac{\partial u}{\partial n}\bigg|_{\partial\Omega} = 0, \end{cases} \tag{3.30}$$

or, in generalized formulation,

$$[u, v] + (u, v) = (g, v), \quad v \in H^1(\Omega), \tag{3.31}$$

(the unknown function is $u \in H^1(\Omega)$). The unique solvability of this problem for $g \in L^2(\Omega)$ is proved just as for the Dirichlet problem. We now construct an unbounded operator A in $L^2(\Omega)$ (assigning to the function $u \in H^1(\Omega)$ the function $g \in L^2(\Omega)$ connected with u by (3.31)) whose domain of definition consists of the solutions u of problems of the form (3.31) with all possible $g \in L^2(\Omega)$. Reasoning as in Paragraph B above, we find that the operator A is invertible and the inverse operator A^{-1} is compact and self-adjoint in $L^2(\Omega)$. Therefore the operator A has a complete orthonormal system of eigenfunctions in $L^2(\Omega)$ with eigenvalues $\lambda_1, \lambda_2, \ldots$. Moreover $\lambda_j \geq 1$ and $\lambda_j \to +\infty$ as $j \to \infty$. But by using the operator A we can write the problem (N) in the form

$$(A - 1)u = -f,$$

from which it is clear that it has a solution if and only if f is orthogonal to all the eigenfunctions of the operator A with eigenvalue 1. But these eigenfunctions coincide with the solutions of the homogeneous problem (N) (with $f \equiv 0$) and are therefore constant. The solvability of the problem now follows.

We note in passing that along the way we stated and solved an eigenvalue problem with Neumann boundary conditions. The corresponding eigenfunctions are always analytic in Ω, belong to $C^\infty(\bar\Omega)$ when the boundary $\partial\Omega$ is smooth, and are analytic in $\bar\Omega$ when the boundary is analytic. When $s > 3/2$, there is a continuous linear operator

$$\mathfrak{A} : H^s(\Omega) \to H^{s-2}(\Omega) \times H^{s-3/2}(\partial\Omega), \tag{3.32}$$

$$u \mapsto \left\{ \Delta u, \frac{\partial u}{\partial n}\bigg|_{\partial\Omega} \right\}$$

which is no longer a topological isomorphism. This is a Fredholm operator and has a one-dimensional kernel and cokernel. To be specific its kernel consists of the constants and its image consists of the pairs $\{f, \varphi\}$, $f \in H^{s-2}(\Omega)$, $\varphi \in H^{s-3/2}(\Omega)$ such that

$$(f, 1)_{L^2(\Omega)} - (\varphi, 1)_{L^2(\partial\Omega)} = 0. \tag{3.33}$$

The necessity of this condition is easily obtained for smooth $u \in C^\infty(\bar\Omega)$ using Green's formula (3.27) and then extended by passing to the limit. Sufficiency is easily deduced from the properties of Fredholm operators taking account of the fact that the index of the operator $\mathfrak{A}$, given by

$$\operatorname{ind}\mathfrak{A} = \dim\operatorname{Ker}\mathfrak{A} - \dim\operatorname{Coker}\mathfrak{A}, \tag{3.34}$$

is easily found to be 0 (the index is stable, and the operator $\mathfrak{A}$ becomes invertible when Δ is replaced by $\Delta - \varepsilon I$). For this reason it follows from the obvious equality $\dim \operatorname{Ker} \mathfrak{A} = 1$ that $\dim \operatorname{Coker} \mathfrak{A} = 1$, so that the necessary condition (3.33) must also be sufficient.

D. General Elliptic Boundary-Value Problems. Boundary-value problems for general second-order elliptic equations and the corresponding eigenvalue problems can be restated and studied using the devices that we demonstrated above through the example of the Dirichlet and Neumann problems for the Laplacian. We shall now dwell briefly on general elliptic boundary-value problems in the sense of Sect. 2.18. Consider a boundary-value problem such as (2.69) in a bounded region Ω with smooth boundary $\partial\Omega$. We recall that we are assuming $m_j < m$ (and so $m_j \leq m - 1$).

Theorem 2.50. *Suppose the ellipticity conditions hold for the problem* (2.69). *Then*

1) *for any* $s > m - 1/2$ *the operator*

$$\mathfrak{A} : H^s(\Omega) \to H^{s-m}(\Omega) \times \prod_{j=1}^{m/2} H^{s-m_j-1/2}(\partial\Omega),$$

$$(3.35)$$

$$u \mapsto \left\{ Au, B_1 u \big|_{\partial\Omega}, \ldots, B_{m/2} u \big|_{\partial\Omega} \right\}$$

is a Fredholm operator. Moreover its kernel $\operatorname{Ker} \mathfrak{A}$ *lies in* $C^\infty(\bar{\Omega})$ *and the index is independent of* s;

2) *for* $s > m - 1/2$ *the relation* $u \in H^s(\Omega)$ *is equivalent to* $Au = f \in H^{s-m}(\Omega)$ *and* $B_j u \big|_{\partial\Omega} = \varphi_j \in H^{s-m_j-1/2}(\partial\Omega)$, $j = 1, 2, \ldots, m/2$;

3) *for any* $s > m - 1/2$ *the a priori estimate*

$$\|u\|_s \leq C \left(\|Au\|_{s-m} + \sum_{j=1}^{m/2} \|B_j u \big|_{\partial\Omega}\|_{s-m_j-1/2} + \|u\|_0 \right) \qquad (3.36)$$

holds, where the constant $C > 0$ *is independent of* u.

This theorem can be proved by the same method that was used to prove Theorem 2.40. We note that the image of an operator $\mathfrak{A}$ of the form (3.35) can always be defined as the set of finite collections $\{f, \varphi_1, \varphi_2, \ldots, \varphi_{m/2}\}$ satisfying a solvability condition of the form

$$\int_\Omega f g^{(l)} \, dx + \sum_{j=1}^{m/2} \int_{\partial\Omega} \varphi_j \psi_j^{(l)} \, dS_x = 0, \quad l = 1, \ldots, N, \qquad (3.37)$$

where $N = \dim \operatorname{Coker} \mathfrak{A}$. The functions $g^{(l)} \in C^\infty(\bar{\Omega})$, $\psi_1^{(l)}, \psi_2^{(l)}, \ldots, \psi_{m/2}^{(l)} \in C^\infty(\partial\Omega)$ are independent of s. These same solvability conditions define the image of the corresponding operator in Hölder spaces, so that in particular the index of the operator $\mathfrak{A}$ coincides with the index of the boundary-value problem in the sense of Sect. 2.18.

We have, finally, one more theorem.

Theorem 2.51. *If the problem* (2.69) *satisfies the ellipticity condition with parameter* (Ell_Λ) *of Sect. 2.2, then there exists* $R > 0$ *such that for* $|\lambda| > R$ *and* $\lambda \in \Lambda$ *the operator*

$$\mathfrak{A}_\lambda : H^s(\Omega) \to H^{s-m}(\Omega) \times \prod_{j=1}^{m/2} H^{s-m_j-1/2}(\partial\Omega),$$

(3.38)

$$u \mapsto \{(A - \lambda)u, B_1 u\big|_{\partial\Omega}, \ldots, B_{m/2} u\big|_{\partial\Omega}\}$$

is invertible (has a bounded everywhere-defined inverse).

Thus the situation in regard to the Fredholm nature of the operators and solvability is the same in Sobolev spaces as in Hölder spaces. We shall also discuss the eigenvalue and eigenfunction problems.

Consider an elliptic boundary-value problem of the form (2.69) in a bounded region Ω with a smooth boundary $\partial\Omega$. As in Sect. 2.2, we shall assume that the operator A is properly elliptic. In addition we shall assume that the problem itself is *regularly elliptic*. This means, besides the condition of ellipticity already imposed for the boundary-value problem and the condition $m_j < m$ on the orders of the boundary operators B_j, that all the orders m_j of the operators B_j are distinct and that all the operators B_j contain the leading normal derivatives (those of order m_j) with nowhere vanishing coefficients. We shall call such a problem *self-adjoint* if the operator A_0 with domain

$$D(A_0) = \{u : u \in C^\infty(\bar{\Omega}), \ B_j u\big|_{\partial\Omega} = 0, \ j = 1, 2, \ldots, m/2\}, \qquad (3.39)$$

given by the formula $A_0 u = Au$ is symmetric in $L^2(\Omega)$.

Theorem 2.52. *Let the boundary-value problem* (2.69) *be regularly elliptic and self-adjoint. Then the closure in* $L^2(\Omega)$ *of the operator* A_0 *defined above is a self-adjoint operator* $\bar{A}_0$ *with domain*

$$D(\bar{A}_0) = \{u : u \in H^m(\Omega), \ B_j u\big|_{\partial\Omega} = 0\}, \qquad (3.40)$$

and $\bar{A}_0 u = Au$ *for* $u \in D(\bar{A}_0)$. *This operator has a complete orthonormal system of eigenfunctions* $\{\psi_j, \ j = 1, 2, \ldots\}$ *in* $L^2(\Omega)$ *belonging to* $D(A_0)$ *(i.e., they are smooth in* $\bar{\Omega}$ *and satisfy the boundary conditions of the original problem), and if* $\{\lambda_j, \ j = 1, 2, \ldots\}$ *are the corresponding eigenvalues, then* $|\lambda_j| \to +\infty$ *as* $j \to +\infty$.

The operator $\bar{A}_0$ is usually denoted simply A and called the self-adjoint operator in $L^2(\Omega)$ determined by the differential operator and boundary conditions under consideration.

We note that Theorem 2.52 is an example where the Sobolev spaces and the concepts connected with them (for example, the trace on a hypersurface) arise naturally in a problem whose statement contains no Sobolev spaces (in the present instance the problem of describing the closure).

3.4. Generalized Solutions of Parabolic Boundary-Value Problems (cf. Il'in, Kalashnikov, and Olejnik 1962; Agranovich and Vishik 1964; Bers, John, and Schechter 1964; Shilov 1965; Ladyzhenskaya, Solonnikov, and Ural'tseva 1967; Taylor 1972; Ladyzhenskaya 1973; Mikhailov 1983). In this subsection we shall discuss some of the simplest facts about parabolic boundary-value problems, those most closely connected with the elliptic theory and the contents of the present section. For information on other questions of the theory of parabolic equations, see Sect. 5 below.

Consider the mixed (intial-value and boundary-value) problem for a parabolic equation

$$\begin{cases} \dfrac{\partial u(x,t)}{\partial t} = Au(x,t) + f(x,t), & (x,t) \in Q_T, \\ u\big|_{t=0} = \varphi(x), & x \in \Omega, \\ B_j u\big|_{S_T} = \psi_j(x,t), & (x,t) \in S_T,\, j = 1,2,\ldots,b, \end{cases} \qquad (3.41)$$

where Ω is a bounded region in $\mathbf{R}^n$ with smooth boundary $S = \partial\Omega$; $Q_T = \Omega \times (0,T)$ is a cylinder in $\mathbf{R}^{n+1}_{x,t}$; $S_T = S \times (0,T)$ is the lateral boundary of this cylinder; $A = \sum\limits_{|\alpha| \leq 2b} a_\alpha(x,t) D_x^\alpha$ is a differential operator on x of order $2b$ ($b >$ 0 is an integer) with coefficients $a_\alpha \in C^\infty(\bar{Q}_T)$; $B_j = \sum\limits_{|\beta| \leq m_j} b_{j\beta}(x,t) D_x^\beta$ are differential operators of orders $m_j < 2b$ with coefficients $b_{j\beta} \in C^\infty(\bar{S}_T)$; and f, φ, and ψ_j are known functions defined on Q_T, Ω, and S_T respectively. The problem (3.41) is called a *mixed parabolic problem* if the following parabolicity condition holds

(P): for each fixed $t \in [0,T]$ the problem

$$\begin{cases} (A - \lambda)v(x) = g(x), & x \in \Omega, \\ B_j v\big|_{\partial\Omega} = \chi_j, & j = 1,\ldots,b, \end{cases} \qquad (3.42)$$

satisfies the condition for ellipticity with parameter (Ell$_\Lambda$) of Sect. 2.20 on the set $\Lambda = \{\lambda : \operatorname{Re}\lambda \geq 0\}$.

When this condition is satisfied, the problem (3.41) has a unique solution in suitable spaces. However there are two important differences from the elliptic situation here. The first is that Sobolev spaces must be used whose norms contain different numbers of derivatives on x and t, i.e., *anisotropic* Sobolev spaces. The second difference is that if we wish to find a solution u of the problem (3.41) that is sufficiently smooth all the way to the boundary, then the data functions f, φ, and ψ_j must satisfy *consistency conditions*, expressing the consistency of the equation and the initial and boundary conditions at points of the form $(x, 0)$ where $x \in \partial\Omega$, i.e., at the conjunction of the lower base and lateral surface of the cylinder Q_T. The simplest of these conditions are as follows (for simplicity we assume that $f = 0$ and $b_{j\beta}$ is independent of t):

$$\psi_j\big|_{t=0} = B_j\varphi\big|_{t=0, x\in\partial\Omega}, \quad j = 1, 2, \ldots, b. \tag{3.43}$$

Using the equation we find the following consistency conditions:

$$\frac{\partial\psi_j}{\partial t}\bigg|_{t=0} = B_j A\varphi\big|_{t=0, x\in\partial\Omega}, \quad j = 1, 2, \ldots, b. \tag{3.43'}$$

More conditions can be found by differentiating the boundary condition on t and then setting $t = 0$ and by using the equation to replace the derivatives of u on t by derivatives on x, after which one must replace u by φ. If $f \in C^\infty(\bar{Q}_T)$, $\varphi \in C^\infty(\bar{\Omega})$, $\psi_j \in C^\infty(\bar{S}_T)$, and all the consistency conditions are satisfied, then there exists a solution $u \in C^\infty(\bar{Q}_T)$ (we note that $u \in C^\infty(Q_T)$ for any solution u provided $f \in C^\infty(Q_T)$, i.e., at interior points the solution is always smooth when the right-hand side is smooth). When the solutions are of finite smoothness (belong to some Sobolev space), one needs to require only those consistency conditions that make sense by the imbedding theorem.

We omit a detailed description of the spaces and precise statements, as they are somewhat cumbersome, referring the reader to a specialized article devoted to elliptic and parabolic problems and to the literature at the end of the present work. We point out only one of the possible ways of proving that a solution exists, a way that explains why the parabolicity condition is connected with the condition for ellipticity with parameter. We assume at first that the coefficients of the operators A and B_j are independent of t. Then, subtracting from u an arbitrary function w that satisfies the initial and boundary conditions, we reduce the problem to the case when the functions φ and ψ_j are identically zero. When this is done, extending the function f to the infinite cylinder $\Omega \times (0, +\infty)$, we may assume that $f(x, t) = 0$ for $t > T + 1$ We now take the Laplace transform on t, passing from the function $u(x, t)$ to the function

$$\hat{u}(x, \lambda) = \int_0^\infty e^{-t\lambda} u(x, t)\, dt. \tag{3.44}$$

If the integral (3.44) converges sufficiently well for λ in some region of the form $\{\lambda : \operatorname{Re}\lambda > \mu_0\}$ (for example, if the function u and its derivatives u_t and $D_x^\alpha u$ with $|\alpha| \leq 2b$ grow no faster than $e^{\mu_0 t}$ as $t \to +\infty$), then from the equation and the initial condition we find through integration by parts that the equation

$$(A - \lambda)\hat{u}(x, \lambda) = \hat{f}(x, \lambda) \tag{3.45}$$

holds, where $\hat{f}$ is the Laplace transform on t of the function $f = f(x, t)$, obtained as in (3.44). By the parabolicity condition (P) and Theorem 2.51 on the solvability of the problem with a parameter, we can solve the equation (3.45) with zero boundary conditions on $\partial\Omega$ for $\operatorname{Re}\lambda > \mu_0$ if μ_0 is sufficiently large, and we can write

$$\hat{u}(\cdot, \lambda) = (A - \lambda)^{-1}\hat{f}(\cdot, \lambda).$$

It remains only to make use of the inversion formula for the Laplace transform:

$$u(x, t) = \frac{1}{2\pi i} \int\limits_{u-i\infty}^{u+i\infty} e^{t\lambda}\hat{u}(x, \lambda)\, d\lambda, \quad \mu > \mu_0. \tag{3.46}$$

The convergence of this integral and the admissibility of differentiating it on t and x (from which it follows that the desired equation holds along with the initial and boundary conditions on u) follow from estimates for the norm of the operator $(A - \lambda)^{-1}$, usually derived along with the proof of the theorem on solvability of a problem with parameter. One can pass to equations with coefficients depending on t, for example, by means of the abstract theory of evolution equations (cf. Sect. 6)

3.5. Generalized Solutions of Hyperbolic Boundary-Value Problems (cf. Ladyzhenskaya 1953; Bers, John, and Schechter 1964; Ladyzhenskaya 1973; Mizohata 1973). While generalized solutions of elliptic and parabolic equations with smooth coefficients are indeed always smooth in the interior of the region in which they are defined, such is not the case for hyperbolic equations. For example the function $u(x, t) = f(x - at)$ is a generalized solution (in the sense of $\mathcal{D}'(\mathbb{R}^2)$) of the vibrating string equation $u_{tt} = a^2 u_{xx}$ for any locally integrable function f. In particular we can take a function $f = f(\xi)$ having a jump discontinuity at some point ξ_0. Then the solution $u(x, t)$ will have a jump discontinuity along the line $x - at = \xi_0$, which is a characteristic of this equation. Moreover the jump will be constant all along the characteristic. The solutions of general hyperbolic equations and hyperbolic systems may also have discontinuities. In particular these may be jump discontinuities along characteristic hypersurfaces, and if the jump surface is regarded as fixed, then the magnitudes of the jumps of the function and its derivatives are subject to certain conditions (equations) derivable from the validity of the equation in a certain generalized sense (cf. Sect. 4 for more details). Such solutions are especially important in the case of nonlinear hyperbolic equations and systems (for example, the equations of gas dynamics), where they describe shock waves. In the case of nonlinear equations the solution cannot be considered an arbitrary distribution since nonlinear operations (such as multiplication) cannot in general be carried out with distributions of class $\mathcal{D}'$. However, imposing certain a priori restrictions on the solution (for example, requiring that it belong to L^∞ or to some Sobolev space W_p^s), one can interpret the equation as a suitable integral identity. Conditions on the discontinuity in this case usually have a direct physical meaning and are connected with conservation of momentum and energy and with certain laws of thermodynamics (for the equations of gas dynamics these conditions are called Hugoniot conditions).

Thus generalized solutions of hyperbolic equations with a smooth (or zero) right-hand side may have singularities inside the region where the solution is defined. As can be seen from the example given above, these singularities may propagate, and moreover not in an arbitrary fashion, but in a certain correspondence with the equation. The propagation of such interior singularities can be described by means of microlocal analysis, which will be discussed in the next volume of this series and in special articles devoted to microlocal analysis and hyperbolic equations. Roughly speaking the answer is that the singularities of the solutions propagate in accordance with the laws of geometrical optics (along rays associated with a Hamiltonian equal to the principal symbol of the hyperbolic operator under consideration). Serious difficulties arise, however, in describing the singularities near the boundary, which have not yet been completely overcome. It is not difficult to study the case of reflection of a singularity for a wave approaching the boundary transversally (nontangentially), where again the laws of geometrical optics happen to hold: the corresponding rays are reflected from the boundary. But if rays tangent to the boundary arise, the problem becomes incomparably more difficult. The corresponding mathematical theory here must take account of the appearance of the so-called *Rayleigh waves*, which propagate along the boundary (the *whispering gallery effect*) and describe the distribution of energy between the Rayleigh waves and the interior waves. At present this has been done only under certain special hypotheses.

The unique solvability of hyperbolic boundary-value problems, however, can usually be proved by comparatively simple methods, not requiring any detailed study of the singularities of the solutions. Here we should mention first of all the *energy method*, which consists of using the equation to obtain estimates for various norms of the solution (cf. Sect. 1, where such estimates are given in the simplest situations). In addition, in cylindrical regions (of the same form as in Sect. 3.5) and in the case of problems with coefficients independent of t, one can use the method of separation of variables (the Fourier method) to study and solve mixed problems. This technique consists of seeking the solution in the form of a series in eigenfunctions of a self-adjoint elliptic boundary-value problem associated with the given equation and boundary conditions. For example, consider a problem of the form

$$\begin{cases} \dfrac{\partial^2 u}{\partial t^2} = a^2 \Delta u + f(x,t), & (x,t) \in \Omega \times (0,T), \\ u\big|_{x \in \partial \Omega} = 0, \\ u\big|_{t=0} = 0,\ u_t\big|_{t=0} = 0. \end{cases} \tag{3.47}$$

(the case of inhomogeneous boundary and initial conditions can be reduced to this just as was done for a parabolic equation in Sect. 3.5). It is convenient to look for a solution of the form

$$u(x,t) = \sum_{k=1}^{\infty} u_k(t)\psi_k(x), \qquad (3.48)$$

where $\{\psi_k : k = 1, 2, \ldots\}$ is a complete orthonormal system of eigenfunctions of the operator $(-\Delta)$ with the Dirichlet boundary conditions (we shall regard the region Ω as a bounded region in $\mathbb{R}^n$; then the eigenvalue problem can be understood in the sense of Sect. 3.3). Expanding $f(x,t)$ similarly,

$$f(x,t) = \sum_{k=1}^{\infty} f_k(t)\psi_k(x),$$

we find, assuming it is possible to substitute the series (3.48) formally into the equation, that the coefficients $u_k(t)$ must satisfy the following equation and initial conditions:

$$u_k''(t) + \lambda_k a^2 u_k(t) = f_k(t), \quad u_k(0) = u_k'(0) = 0, \qquad (3.49)$$

where λ_k is the eigenvalue of the operator $(-\Delta)$ corresponding to the eigenfunction ψ_k (we recall that $\lambda_k > 0$ for all $k = 1, 2, \ldots$). The coefficients $u_k(t)$ are uniquely determined from (3.49). It is not difficult to show that if, for example, $f \in C([0, T], L^2(\Omega))$, (i.e., $f(\cdot, t)$ is a continuous function of t with values in $L^2(\Omega)$), then the coefficients $f_k(t)$ are continuous in t and then the function $u(x,t)$ found from formula (3.48) will belong to $L^2(\Omega \times (0, T))$ and in a natural sense will be a generalized solution of the problem (3.47). To be specific, it will be the limit of the finite sums $u_N(x,t) = \sum_{k=1}^{N} u_k(t)\psi_k(x)$, which in the case of a region Ω with a smooth boundary will be simply the classical solutions of a problem of type (3.47) with f replaced by the analogous sum f_N; if some integral identity holds for the functions u_N, then by passing to the limit we find that it holds for u also.

For more details on hyperbolic equations and boundary-value problems for them see Sect. 4 below.

§4. Hyperbolic Equations

4.1. Definitions and Examples. We recall the definitions given above (cf. Sects. 1.1.8 and 1.3.3). An operator

$$A = \sum_{|\alpha|+j \leq m} a_{\alpha,j}(t,x)D_x^\alpha D_t^j, \quad t \in \mathbb{R}, \ x \in \mathbb{R}^n, \qquad (4.1)$$

is called *hyperbolic* at the point (t, x) if $a_{0,m}(t, x) \neq 0$ (i.e., the direction of the t-axis is not characteristic) and for any vector $\xi \in \mathbb{R}^n$ all the roots λ of the equation

$$A_m(t, x, \lambda, \xi) \equiv \sum_{|\alpha|+j=m} a_{\alpha,j}(t, x)\xi^\alpha \lambda^j = 0 \qquad (4.2)$$

are real. The operator A is called *strictly hyperbolic* at the point (t, x) if for $\xi \neq 0$ the roots of Eq. (4.2) are real and distinct.

As usual an operator A is called (strictly) hyperbolic in a region $\Omega \subset \mathbf{R}^{n+1}$ if it is (strictly) hyperbolic at each point $(t, x) \in \Omega$.

Similar definitions are made in the case when $a_{\alpha,j}(t, x)$ are square matrices. In this situation Eq. (4.2) is replaced by the characteristic equation

$$\det \sum_{|\alpha|+j=m} a_{\alpha,j}(t, x)\xi^\alpha \lambda^j = 0.$$

The simplest hyperbolic operators are the first-order operators

$$\frac{\partial}{\partial t} + \sum_{j=1}^{n} a_j(t, x)\frac{\partial}{\partial x_j} + b(t, x)$$

with real coefficients $a_j(t, x)$ and the second-order operators

$$\frac{\partial^2}{\partial t^2} - \sum_{i,j=1}^{n} a_{ij}(t, x)\frac{\partial^2}{\partial x_i \partial x_j} + \sum_{j=1}^{n} b_j(t, x)\frac{\partial}{\partial x_j} + c(t, x)$$

with real coefficients, if the matrix $\|a_{ij}\|$ is positive-definite. For example, the wave operator $\Box$ studied in Sect. 1.1.7 is assigned to this class.

4.2. Hyperbolicity and Well-Posedness of the Cauchy Problem. From the physical point of view hyperbolic equations describe processes in which disturbances propagate with finite velocity. Sometimes hyperbolic operators are defined as the operators for which the Cauchy problem is well-posed and the velocity with which a disturbance propagates is finite, i.e., the value of the solution $u(t, x)$ is uniquely determined at the point (t, x) by the values of the initial data in a bounded region $\mathcal{D}_{t,x} \subset \mathbf{R}^n$. In some important special cases it is possible to prove that this definition is equivalent to the definition given in Sect. 4.1, for example, when the coefficients $a_{\alpha,j}$ are constant or when the roots of the characteristic equation (4.2) are distinct. In general, however, the well-posedness of the Cauchy problem may depend on the nonleading terms, i.e., on the coefficients $a_{\alpha,j}(t, x)$ for $|\alpha| + j < m$.

Theorem 2.53 (Gårding 1951). *The operator* $A = \sum\limits_{|\alpha|+j\leq m} a_{\alpha,j}D_x^\alpha D_t^j$ *with constant coefficients* $a_{\alpha,j}$ *is hyperbolic if and only if the equation*

$$Au = f$$

has a unique solution $u \in \mathcal{D}'(\mathbf{R}^{n+1})$ *for any function* $f \in C_0^\infty(H)$ *with support contained in* H. *Here* $H = \{(t, x) : t \geq 0, x \in \mathbf{R}^n\}$.

For equations with variable coefficients the results obtained at present are not so complete. We give two of them

Theorem 2.54 (Mizohata 1973). *Suppose the coefficients of the operator $A(t, x, D_t, D_x)$ belong to C^∞. Assume that the problem*

$$\begin{cases} Au = 0 & \text{for } |t| + |x|^2 < \delta, \\ D_t^j u = \varphi_j & \text{for } t = 0, |x|^2 < \delta, j = 1, 2, \ldots, m-1, \end{cases}$$

has a solution $u \in C^m$ for any $\varphi_j \in C^\infty$ and that this solution is unique and depends continuously on the vector $\Phi = (\varphi_0, \varphi_1, \ldots, \varphi_{m-1})$ (i.e., the mapping $(C^\infty)^m \to C^m$ under which Φ goes to the solution u is continuous). Then the roots λ of the equation

$$A_m(0, 0, \lambda, \xi) = 0$$

are real for all $\xi \in \mathbb{R}^n$.

Theorem 2.55 (Flaschka and Strang, 1971). *If the roots λ of the characteristic equation (4.2) have constant multiplicity, then the Cauchy problem for the equation $Au = 0$ is well-posed in the sense of Hadamard (i.e., the hypotheses of Theorem 2.54 are satisfied) if and only if the following condition is satisfied:*

Condition H: *If $\lambda = \lambda(t, x, \xi)$ is a root of Eq. (4.2) of multiplicity k and $f(t, x)$ is an arbitrary infinitely differentiable function, then*

$$A(fe^{it\psi}) = O(t^{m-k}) \quad \text{as } t \to +\infty$$

for each solution $\psi = \psi(t, x)$ of the equation $\dfrac{\partial \psi}{\partial t} = \lambda\left(t, x, \dfrac{\partial \psi}{\partial x}\right)$.

4.3. Energy Estimates. One of the methods of studying the Cauchy problem for hyperbolic equations is the *method of energy estimates.*

Example 2.56. Let $u \in C^2([0, T] \times \mathbb{R}^n)$ be a solution of the problem

$$\Box u \equiv \frac{\partial^2 u}{\partial t^2} - \Delta u = 0, \quad \text{for } 0 \leq t \leq T; \ u = \varphi_0, \frac{\partial u}{\partial t} = \varphi_1 \quad \text{for } t = 0.$$

If the integral

$$E_0 = \int\limits_{\mathbb{R}^n} \left[\varphi_1^2(x) + \sum_{j=1}^n \left(\frac{\partial \varphi_0(x)}{\partial x_j} \right)^2 \right] dx$$

is finite, then the integrals

$$E_t = \int\limits_{\mathbb{R}^n} \left[\left(\frac{\partial u(t, x)}{\partial t} \right)^2 + \sum_{j=1}^n \left(\frac{\partial u(t, x)}{\partial x_j} \right)^2 \right] dx$$

are finite for $0 \leq t \leq T$, and $E_t = E_0$.

Physically (in the case $n = 3$) this equality is an expression of the law of conservation of energy. We shall give energy estimates for the solution of the Cauchy problem in the case of strictly hyperbolic first-order systems and show how they are used. Analogous results hold also for hyperbolic systems and systems of arbitrary order.

Let

$$P(t, x, D_t, D_x) = D_t \cdot I + \sum_{j=1}^{n} a_j(t, x) D_j + a_0(t, x), \tag{4.3}$$

where a_j are $N \times N$ matrices for $j = 0, 1, \ldots, n$ whose entries are smooth functions and are uniformly bounded for $0 \le t \le T$, $x \in \mathbb{R}^n$, along with their derivatives, and I is the identity matrix. We assume that the roots λ of the equation

$$\det \left(\lambda I + \sum_{j=1}^{n} a_j(t, x) \xi_j \right) = 0 \tag{4.4}$$

are real and distinct for $\xi \in \mathbb{R}^n$, $\xi \ne 0$, and that there exists a number $c_0 > 0$ such that for any distinct roots λ and λ' of this equation we have $|\lambda(t, x, \xi) - \lambda'(t, x, \xi)| \ge c_0 |\xi|$ for all $t \in [0, T]$ and $x \in \mathbb{R}^n$.

Theorem 2.57. *For each real number s there exists a constant $C = C(s, T)$ such that*

$$\|u(t, \cdot)\|_s \le C \int_0^T \|Pu(\tau, \cdot)\|_s \, d\tau, \quad 0 \le t \le T, \tag{4.5}$$

where $u = (u_1, \ldots, u_N)$ is a vector-valued function with smooth components and $u(0, x) = 0$ (cf. Nirenberg 1973).

Here $\|f\|_s$ is the norm of the vector-valued function f in the Sobolev space $\left(H^s(\mathbb{R}^n) \right)^N$ (cf. Sect. 3).

It follows immediately from inequality (4.5) that the solution of the Cauchy problem

$$P(t, x, D_t, D_x) u = f(t, x) \quad \text{for } 0 \le t \le T, \quad u(0, x) = \varphi(x), \tag{4.6}$$

is unique in the class C^1.

To prove the existence of a solution of the problem (4.6) we introduce the Hilbert space $\left(L^2((0, T) \times \mathbb{R}^n) \right)^N$ with the inner product

$$(u, v) = \int_0^T \int_{\mathbb{R}^n} \sum_{j=1}^{n} u_j(t, x) \overline{v_j(t, x)} \, dx \, dt.$$

Let

$$P^*(t, x, D_t, D_x)u = D_t u + \sum_{j=1}^{n} D_j \big(a_j^*(t, x)u\big) + a_0^*(t, x)u,$$

where a_j^* are the matrices that are hermitian-conjugate to a_j, for $j = 0, 1, \ldots, n$.

It is not difficult to see that P^* is a strictly hyperbolic operator of the same type as P. Therefore Theorem 2.57 applies to it, and if v is a smooth vector-valued function with $v(T, x) = 0$, then

$$\|v(t, \cdot)\|_s \leq C \int_t^T \|P^* v(\tau, \cdot)\|_s \, d\tau.$$

Consider the Hilbert space $K_s = L^2\big([0, T], \big(H^s(\mathbf{R}^n)\big)^N\big)$ consisting of the functions of t with values in the space $\big(H^s(\mathbf{R}^n)\big)^N$ and norm

$$\|u\|_{K_s} = \left(\int_0^T \|u(t, \cdot)\|_s^2 \, dt \right)^{1/2}.$$

If $f \in K_s$ and v is a smooth vector-valued function that vanishes for large $|x|$ and is such that $v(T, x) = 0$, then by Theorem 2.57 we have the inequality

$$|(v, f)| \leq C_1 \|f\|_{K_s} \|P^* v\|_{K_s}.$$

By the Riesz representation theorem there exists a vector $u \in K_s$ for which

$$(v, f) = (P^* v, u). \tag{4.7}$$

The vector u is thus a generalized solution of the system of equations $Pu = f$ (cf. Sect. 3). If $s > \frac{n}{2} + 1$, then $u \in C^1$ and the vector-valued function u is a classical solution of this system. In this situation it follows from equality (4.7) that

$$\int u(0, x)\overline{v(0, x)} \, dx = 0$$

for all smooth vector-valued functions $v(t, x)$ satisfying $v(t, x) = 0$. Thus we find that $u(0, x) = 0$, so that u is a solution of the Cauchy problem.

Passing to the limit, we can easily prove by using Theorem 2.57 that there exists a strong generalized solution of the problem (4.6) in K_s, i.e., an element $u \in K_s$ for which there exists a sequence of smooth solutions $\{u_j\}$ such that $u_j(0, x) = 0$ and

$$\|u - u_j\|_{K_s} + \|Pu - Pu_j\|_{K_s} \to 0 \quad \text{as } j \to \infty.$$

It follows from inequality (4.5) that the strong solution is unique.

Using the technique of averaging one can show (cf. Agranovich 1969; Lax and Phillips 1967) that the strong solution coincides with the weak solution, i.e., the solution satisfying (4.7). Thereby it is proved that for each $f \in K_s$ there exists a strong generalized solution $u \in K_s$ and such a solution is unique.

4.4. The Speed of Propagation of Disturbances. It follows from the energy inequalities that disturbances propagate with finite speed. This means that when the values of the initial data are changed in a bounded region $\Omega \subset \mathbb{R}^n$ the value of the solution of the Cauchy problem in a region Ω_1 lying a positive distance d away from Ω does not alter for $0 < t < d/a$, where $a > 0$.

Let $\lambda_1, \ldots, \lambda_N$ be the roots of the characteristic equation (4.4) and $\lambda_i \neq \lambda_j$ for $i \neq j$. Let

$$a = \sup_{\substack{x \in \mathbb{R}^n \\ 0 \leq t \leq T}} \sup_{|\xi|=1} \max_{1 \leq j \leq N} |\lambda_j(t, x, \xi)| < \infty.$$

Consider the "past cone"

$$K = \{(t, x) : |x - x_0| \leq a(t_0 - t), \ 0 \leq t \leq t_0\}$$

with vertex at the point (t_0, x_0), where $t_0 > 0$.

Theorem 2.58 (Mizohata 1973). *If $u \in C^m(K)$, $P(t, x, D_t, D_x)u = 0$ in K, and $\dfrac{\partial^j u(0, x)}{\partial t^j} = 0$ for $x \in K \cap \{t = 0\}$, $j = 0, 1, \ldots, m - 1$, then $u(t, x) = 0$ in K.*

Thus the values of the solution in K depend only on the values of the initial data at the base of the cone. We note further that in the case of the wave operator $\Box = \dfrac{\partial^2}{\partial t^2} - \Delta$ the coefficient a coincides exactly with the value of the velocity of propagation of disturbances.

4.5. Solution of the Cauchy Problem for the Wave Equation. The wave equation $\Box u = 0$ is encountered especially often in applications (cf. Sect. 1.1.7). For that reason formulas that give the solution of the Cauchy problem in explicit form are of great significance. The problem is stated as follows:

$$\Box u = 0 \quad \text{for } t \geq 0; \quad u(0, x) = \varphi_0(x), \quad \frac{\partial u(0, x)}{\partial t} = \varphi_1(x). \tag{4.8}$$

For $n = 1$ the solution can be written using *d'Alembert's formula*

$$u(t, x) = \frac{\varphi_0(x - at) + \varphi_0(x + at)}{2} + \frac{1}{2a} \int_{x-at}^{x+at} \varphi_1(\xi) \, d\xi.$$

If $n = 2$, we have *Poisson's formula*

$$u(t, x, y) = \frac{1}{2\pi} \frac{\partial}{\partial t} \iint\limits_{K_{at}} \frac{\varphi_0(\xi, \eta) \, d\xi \, d\eta}{\sqrt{a^2 t^2 - (x - \xi)^2 - (y - \eta)^2}} +$$

$$+ \frac{1}{2\pi} \iint\limits_{K_{at}} \frac{\varphi_1(\xi, \eta) \, d\xi \, d\eta}{\sqrt{a^2 t^2 - (x - \xi)^2 - (y - \eta)^2}},$$

where $K_{at} = \{(\xi, \eta) : (\xi - x)^2 + (\eta - y)^2 \le a^2 t^2\}$.

If $n = 3$, the solution of the problem (4.8) is given by *Kirchhoff's formula*

$$u(t, x, y, z) = \frac{1}{4\pi a} \frac{\partial}{\partial t} \left(\frac{1}{t} \iint\limits_{S_{at}} \varphi_0(\xi, \eta, \zeta) \, dS \right) + \frac{1}{4\pi a t} \iint\limits_{S_{at}} \varphi_1(\xi, \eta, \zeta) \, dS,$$

where $S_{at} = \{(\xi, \eta, \zeta) : (x - \xi)^2 + (y - \eta)^2 + (z - \zeta)^2 = a^2 t^2\}$.

D'Alembert's formula has an elementary derivation (cf. Sect. 1.1.7). Kirchhoff's formula is easily derived by using the spherical symmetry of the Laplacian. To be specific, for each function $f(x_1, x_2, x_3)$ the spherical average

$$I(x, r) = \frac{1}{4\pi} \iint\limits_{|y|=1} f(x + ry) \, dS_y$$

satisfies the *Euler-Poisson-Darboux equation*

$$r \Delta_x I(x, r) = \frac{\partial^2}{\partial r^2} (r I(x, r)),$$

as can be verified directly. Therefore if we introduce an operator S_t by the relation $S_t f(x) = at I(x, at)$, the function $v(t, x) = S_t f(x)$ satisfies the equation

$$a^2 \Delta v(t, x) = \frac{\partial^2}{\partial t^2} v(t, x),$$

and moreover $v(0, x) = 0$ and $\dfrac{\partial v(0, x)}{\partial t} = a f(x)$. From this it can be seen that the solution of the problem (4.8) with $n = 3$ is given by the formula

$$u(t, x) = \frac{1}{a} S_t \varphi_1(x) + \frac{1}{a} \frac{\partial}{\partial t} S_t \varphi_0(x).$$

Finally, Poisson's formula can be obtained from Kirchhoff's by the *method of descent*, which consists of substituting functions $\varphi_0(x, y)$ and $\varphi_1(x, y)$ independent of z into Kirchhoff's formula and carrying out the elementary transformations to reduce the integral over the sphere to an integral over a disk.

The solution of the Cauchy problem (4.8) for $n \ge 2$ and for initial data φ_0 and φ_1 that are either of compact support or rapidly decreasing can be obtained by the Fourier transform method. If

$$v(t, \xi) = \int u(t, x) e^{-ix \cdot \xi} \, dx,$$

then the function v is a solution of the problem

$$\frac{\partial^2 v}{\partial t^2} + a^2 |\xi|^2 v = 0 \quad \text{for } t > 0; \quad v(0, \xi) = \tilde{\varphi}_0(\xi), \quad \frac{\partial v(0, \xi)}{\partial t} = \tilde{\varphi}_1(\xi),$$

where $\tilde{\varphi}_0$ and $\tilde{\varphi}_1$ are the Fourier transforms of the functions φ_0 and φ_1. From this it can be seen that

$$v(t, \xi) = \tilde{\varphi}_0(\xi) \cos at|\xi| + \tilde{\varphi}_1(\xi) \frac{\sin at|\xi|}{a|\xi|}.$$

Thus the problem reduces to calculating the integral

$$F(t, x) = (2\pi)^{-n} \int \frac{\sin at|\xi|}{a|\xi|} e^{ix \cdot \xi} d\xi,$$

(interpreted as a generalized Fourier transform – cf. Sect. 1). Knowing F, we can find $u(t, x)$ by the formula

$$u(t, x) = \int F(t, x - y)\varphi_1(y) \, dy + \frac{\partial}{\partial t} \int F(t, x - y)\varphi_0(y) \, dy.$$

The problem (4.8) can also be solved without using the Fourier transform. For simplicity we shall assume that $a = 1$. Let $F(t, x)$ be a fundamental solution for the operator $\square$ equal to 0 for $t < 0$. From symmetry considerations it depends only on t and $r = |x|$ and therefore for $t > 0$ satisfies the equation

$$\frac{\partial^2 F}{\partial t^2} - \frac{\partial^2 F}{\partial r^2} - \frac{n-1}{r} \frac{\partial F}{\partial r} = 0.$$

A solution of this equation can be sought in the form $F(t, r) = q\left(\frac{t}{r}\right)$, in which case the function $q(\lambda)$ must satisfy the equation

$$(\lambda^2 - 1)q''(\lambda) - (n - 3)\lambda q'(\lambda) = 0,$$

which is easily integrated. To be specific, for $\lambda > 1$ its solution is given by the formula

$$q(\lambda) = c \int_1^\lambda (\alpha^2 - 1)^{\frac{n-3}{2}} d\alpha + C_1.$$

If we set $q(\lambda) = 0$ for $\lambda \leq 1$, the distribution

$$F(t, x) = \frac{1}{(n - 2)! \sigma_{n-1}} \frac{\partial^{n-1}}{\partial t^{n-1}} q\left(\frac{t}{|x|}\right),$$

where σ_{n-1} is the area of the unit sphere in $\mathbb{R}^n$, is a fundamental solution of the Cauchy problem.

In particular if $n = 3$, then $q(\lambda) = \frac{\lambda}{4\pi}$ for $\lambda > 1$ and

$$F(t, x) = \frac{1}{4\pi} \frac{\partial}{\partial t} \theta\left(\frac{t}{|x|} - 1\right) = \frac{1}{4\pi|x|} \delta(t - |x|).$$

Similarly for each odd $n \geq 3$

$$F(t, x) = \frac{1}{(n-2)! \sigma_{n-1} |x|^{n-2}} \delta(t - |x|).$$

The corresponding formula for even numbers $n \geq 2$ can be obtained from this by the method of descent.

We now give the final formulas (cf. Courant and Hilbert 1962 and John 1955). We recall that the *Legendre polynomials* are the polynomials

$$P_0(t) = 1, \quad P_k(t) = \frac{1}{2^k k!} \frac{d^k}{dt^k} (t^2 - 1)^k, \quad k = 1, 2, \ldots$$

The solution of the Cauchy problem (4.8) is given by the formulas

$$u(t, x) = \frac{\partial}{\partial t} T_{\varphi_0}(t, x) + T_{\varphi_1}(t, x), \tag{4.9}$$

where

$$T_\varphi(t, x) = \frac{1}{2(2\pi)^{k+1}} \sum_{j=0}^{k} \frac{\partial^j}{\partial t^j} \frac{P_k^{(k-j)}(1)}{t^{2k+1-j}} \int_{S_t} \varphi(y) \, dS, \quad n = 2k + 3;$$

$$T_\varphi(t, x) = \frac{1}{(2\pi)^{k+1}} \sum_{j=0}^{k} \frac{\partial^j}{\partial t^j} \frac{P_k^{(k-j)}(1)}{t^{2k-j}} \int_{|y| < t} \frac{\varphi(y)}{\sqrt{t^2 - |y|^2}} \, dy, \quad n = 2k + 2.$$

These formulas are called the *Herglotz-Petrovskij formulas*.

The solution of the Cauchy problem for the inhomogeneous wave equation $\Box u = f(t, x)$ can now be found by Duhamel's principle (cf. Sect. 1.15).

4.6. Huyghens' Principle. Let us consider more closely Kirchhoff's formula, which gives the solution of the Cauchy problem (4.8) when $n = 3$. For simplicity let us suppose $\varphi_0 = 0$. Let $\varphi_1 \in C^2(\mathbb{R}^3)$ and $\varphi_1(x) = 0$ in $\mathbb{R}^3 \setminus \Omega$, where Ω is a bounded region in $\mathbb{R}^3$ and $\varphi_1(x) > 0$ in Ω. Let A be a point in $\mathbb{R}^3$ lying outside Ω. Then

$$u(t, x) = \frac{1}{4\pi a t} \iint_{S_{at}} \varphi_1(\xi, \eta, \zeta) \, dS$$

and we see that $u(t, A) = 0$ for $t \leq t_1 = \dfrac{d}{a}$, where d is the distance from the point A to $\bar{\Omega}$. Moreover $u(t, A) = 0$ for $t \geq t_2 = \dfrac{d_1}{a}$, where d_1 is the distance from A to the point of $\bar{\Omega}$ at maximal distance from A. But if $t_1 < t < t_2$, then $u(t, A) > 0$.

If the region Ω is contracted to the origin, the influence of the initial data will be felt at the point A only at the particular instant $t = \dfrac{d}{a}$, where d is the distance from the point A to the origin. This phenomenon is known as

Huyghens' principle. In essence it says that a sharply localized initial state is observed at each point of the space after a certain time as an equally sharply localized phenomenon. Huyghens' principle is equivalent to the assertion that the values of the solution of the Cauchy problem at the point (t, x) do not depend on all the values of the initial data at the base of the characteristic cone with vertex at the point (t, x), but only on the values of the initial data at the boundary of the base. In particular a fundamental solution of the Cauchy problem for $n = 3$ has the form $\dfrac{1}{4\pi at}\delta(at - |x|)$ and vanishes everywhere outside the surface of the characteristic cone $|x| = at$.

Huyghens' principle can also be described as follows. Each point x at which the initial data are different from zero is the center of a spherical wave having velocity a. Therefore if the initial data vanish outside the region Ω, then at the instant t the solution vanishes outside a region Ω_t with boundary surface S_t, which is the envelope of the family of spheres having radii at and centers at the points of the region Ω. If the region Ω is bounded, then, from some instant t on the surface consists of two connected components. One of them, more distant from Ω, is called the *wave front* and the other the *wave back*.

The Herglotz-Petrovskij formulas (4.9) show that Huyghens' principle holds in the space $\mathbf{R}^n$ if n is odd and $n \geq 3$.

In the case of even n, $n \geq 2$, Huyghens' principle does not hold. If the initial data are nonzero in only a bounded region Ω, the solution vanishes at the instant t at all points at a distance larger than at from Ω, but in general is nonzero at the remaining points. This can be seen from the formulas for T_φ, where the integrals are not taken over the sphere $|x| = at$, as was done in the case of odd n, but over the ball $|x| \leq at$. Thus if n is even, there is a wave front, but not a wave back. This phenomenon is called *diffusion* or the dissipation of the wave back. An approximate model of this phenomenon is the waves arising on the surface of a pond when a stone is thrown into it.

We note further that for even n the denominator $\sqrt{a^2t^2 - |x|^2}$ is present in the integrand, showing that the most important contribution to the value of the integral comes from the values of φ at points near the surface $|x| = at$.

4.7. The Plane Wave Method. We shall exhibit another method of solving the Cauchy problem (4.8) for the wave equation (cf. Courant and Hilbert 1962). This method is based on the expansion of the desired solution as a sum of plane waves, more precisely on a representation of it in the form of an integral

$$\int_{|\alpha|=1} f(t, \alpha \cdot x)\, dS_\alpha,$$

where dS_α is the standard measure on the unit sphere. As with Fourier's method, the plane wave method is applicable to the solution of the Cauchy problem for a general hyperbolic equation $P(D_t, D_x)u = f$ of order m with constant coefficients.

The basis of the method is *Poisson's formula*:

$$\Delta_x \int \frac{f(y)}{|x-y|^{n-2}} \, dy = (2-n)\sigma_{n-1} f(x),$$

which holds for $f \in \mathcal{E}'(\mathbf{R}^n)$ when $n \geq 3$.

Let $f \in C_0^\infty(\mathbf{R}^n)$. For $\alpha \in \mathbf{R}^n$, $|\alpha| = 1$, we define the integral

$$I_\alpha(p) = \int_{\substack{y \cdot \alpha = p \\ y \in \mathbf{R}^n}} f(y) \, d\sigma(y),$$

where $d\sigma(y)$ is Lebesgue measure on the hyperplane $y \cdot \alpha = p$. The function $I_\alpha(p)$ is called the *Radon transform* of the function f. Let

$$V(x) = \int_{\substack{|\alpha|=1 \\ \alpha \in \mathbf{R}^n}} I_\alpha(\alpha \cdot x) \, dS_\alpha.$$

Then

$$V(x) = \int_{|\alpha|=1} dS_\alpha \int_{y \cdot \alpha = x \cdot \alpha} f(y) \, d\sigma(y) = \int f(y) \, dy \int_{\substack{|\alpha|=1 \\ (x-y) \cdot \alpha = 0}} dS'_\alpha,$$

where dS'_α is measure on the intersection of the sphere $|\alpha| = 1$ and a plane passing through its center. This measure depends only on $|x - y|$. It is clear that the integral over this intersection depends only on $|x - y|$, so that

$$V(x) = \int f(y) w(|x-y|) \, dy.$$

The form of the function $w(r)$ is easily found by applying this formula in the case when $f = f_1(r)$, $r = |y - x|$. We have

$$V(x) = \int f_1(r) w(r) \, dy = \sigma_{n-1} \int_0^\infty f_1(r) w(r) r^{n-1} \, dr.$$

On the other hand,

$$V(x) = \int_{|\alpha|=1} dS_\alpha \int_{(y-x) \cdot \alpha = 0} f_1(r) \, d\sigma(y) = \int_{|\alpha|=1} dS_\alpha \int_0^\infty f_1(r) \sigma_{n-2} r^{n-2} \, dr$$

$$= \sigma_{n-1} \sigma_{n-2} \int_0^\infty f_1(r) r^{n-2} \, dr.$$

Comparing the formulas obtained, we find that

$$w(r) = \sigma_{n-2} r^{-1}.$$

Consequently

$$V(x) = \sigma_{n-2} \int \frac{f(y)\,dy}{|y - x|}.$$

Suppose the dimension n is odd. Then

$$\Delta^{\frac{n-3}{2}} V(x) = a_n \int \frac{f(y)\,dy}{|y - x|^{n-2}},$$

where a_n is a constant depending only on n. Applying Poisson's formula, we find that

$$f(x) = \frac{a_n}{(2 - n)\sigma_{n-1}} \Delta^{\frac{n-1}{2}} V(x).$$

The constant a_n can be found by substituting a specific function for $f(x)$, for example $f(x) = e^{-|x|^2}$. Thus we obtain

$$f(x) = \frac{1}{2(2\pi)^{n-1}} \left(-\Delta \right)^{\frac{n-1}{2}} V(x). \tag{4.10}$$

The formula thus obtained also makes sense for even n also provided the operator $\Delta^{1/2}$ is suitably defined (cf. John 1955).

Now let L be a hyperbolic operator of order m with constant coefficients. Consider the Cauchy problem

$$Lu = f(t, x) \quad \text{for } t > 0; \quad \frac{\partial^i u(0, x)}{\partial t^i} = \varphi_i(x), \quad i = 0, 1, \ldots, m - 1,$$

assuming that the functions f and φ_i are smooth functions of compact support. Let

$$I_\alpha(t, p) = \int_{y \cdot \alpha = p} u(t, y)\,d\sigma_y,$$

$$F_\alpha(t, p) = \int_{y \cdot \alpha = p} f(t, y)\,d\sigma_y, \quad \Phi_\alpha^i(p) = \int_{y \cdot \alpha = p} \varphi_i(y)\,d\sigma_y, \quad i = 0, 1, \ldots, m - 1.$$

Then for each $\alpha \in \mathbb{R}^n$ with $|\alpha| = 1$ we obtain

$$L_\alpha\big(I_\alpha(t, p)\big) = F_\alpha(t, p) \quad \text{for } t > 0,$$

$$\frac{\partial^i I_\alpha(0, p)}{\partial t^i} = \Phi_\alpha^i(p), \quad i = 0, 1, \ldots, m - 1,$$

where L_α is a hyperbolic differential operator in the space of the two variables t and p. The solution of this problem will be discussed below. At present we note that if this problem is solved, then by formula (4.10) we have

$$u(t, x) = \frac{1}{2(2\pi)^{n-1}} \left(-\Delta \right)^{\frac{n-1}{2}} \int_{|\alpha|=1} I_\alpha(t, \alpha \cdot x)\,dS_\alpha.$$

In particular if $L = \Box$ and $f = 0$, then

$$I_\alpha(t,p) = \frac{\Phi_\alpha^0(p-at) + \Phi_\alpha^0(p+at)}{2} + \frac{1}{2} \int\limits_{p-at}^{p+at} \Phi_\alpha^1(\xi)\, d\xi.$$

From this one can obtain formula (4.9).

4.8. The Solution of the Cauchy Problem in the Plane. We have shown above how the Cauchy problem for an equation of order m with constant coefficients in the space $\mathbb{R}^n$ can be reduced to the Cauchy problem in the plane. This same method is applicable for systems of equations also.

Consider the operator

$$P(t,x,D_t,D_x) = \sum_{i+j\le m} a_{ij}(t,x)D_t^i D_x^j$$

in the plane of the variables t and x with smooth coefficients a_{ij} and the Cauchy problem for it

$$P(t,x,D_t,D_x)u = f(t,x) \quad \text{for } t > 0,$$

$$u = \varphi_0(x), \quad \frac{\partial u}{\partial t} = \varphi_1(x),\dots, \frac{\partial^{m-1}u}{\partial t^{m-1}} = \varphi_{m-1}(x) \quad \text{for } t = 0.$$

We assume that P is a strictly hyperbolic operator, i.e., $a_{m,0}(t,x) \ne 0$ and m distinct values $\lambda_1(t,x),\dots,\lambda_m(t,x)$ are defined at each point (t,x) satisfying the characteristic equation $P_0(t,x,\lambda_j(t,x),1) = 0$, where $P_0(t,x,\xi,\eta) = \sum_{i+j=m} a_{ij}(t,x)\xi^i\eta^j$. For simplicity we shall assume that $a_{m,0}(t,x) = 1$. Let $Q_j = D_t - \lambda_j(t,x)D_x$. Then the operator

$$P(t,x,D_t,D_x) - Q_1 \circ Q_2 \circ \cdots \circ Q_m$$

has order $m-1$. As it happens (cf. Courant and Hilbert 1962) there exist operators $L_0, L_1,\dots L_m$ for which

$$
\begin{aligned}
L_0 \; &= I \\
L_1 \; &= Q_1 \circ L_0 + R_0, \\
L_2 \; &= Q_2 \circ L_1 + R_1, \\
&\cdots\cdots\cdots\cdots\cdots\cdots\cdots\cdots \\
L_m \; &= Q_m \circ L_{m-1} + R_{m-1} = P,
\end{aligned}
$$

and moreover $R_j = \sum_{i=0}^{j} b_{ij}(t,x)L_i$. If we set $u_j = L_{j-1}u$ for $j = 1,\dots,m$, we obtain a system of m equations

$$Q_j u_j + \sum_{i=1}^{j} b_{i-1,j-1}(t,x)u_i - u_{j+1} = 0, \quad j = 1,\dots,m-1,$$

$$Q_m u_m + \sum_{i=1}^{m} b_{i-1,m-1}(t,x)u_i = f(t,x),$$

whose leading part is diagonal. The functions u_j satisfy the initial conditions

$$u_j(0, x) = \psi_j(x), \quad j = 1, \ldots, m,$$

where the functions $\psi_j(x)$ are easily found using the data functions $\varphi_j(x)$.

The Cauchy problem obtained in this way for the system of first-order equations is equivalent to the system of integral equations

$$u_j(t, x) = \psi_j(x_j) + \int_{l_j} \left(u_{j+1} - \sum_{i=1}^{j} b_{i-1,j-1} u_i \right) dl_j, \quad j = 1, \ldots, m - 1,$$

$$u_m(t, x) = \psi_m(x_m) + \int_{l_m} \left(f - \sum_{i=1}^{m} b_{i-1,m-1} u_i \right) dl_m,$$

where l_j is the segment of the integral curve of the equation $\dot{x}(t) = \lambda_j(t, x)$ starting at the point (t, x) for $t \neq 0$ and ending at the point of intersection $x_j = x_j(t, x)$ of this curve with the x-axis. The system of Volterra integral equations so obtained can be solved by the method of successive approximations.

4.9. Lacunae. In $\mathbb{R}_t \times \mathbb{R}_x^n$, $n \geq 2$, consider a homogeneous strictly hyperbolic operator with constant real coefficients and $a_{m,0} = 1$. The points $(\tau, \xi) \in \mathbb{R}^{n+1}$ that satisfy the equation $P(\tau, \xi) = 0$ form a conical surface Σ_1.

Let Σ be the image of this surface in the real projective space $\mathbb{RP}^n$. The surface Σ consists of several pairwise disjoint sheets (connected components). We denote by T the pencil of lines in $\mathbb{RP}^n$ passing through the point corresponding to the t-axis in $\mathbb{R}_{t,x}^{n+1}$. Strict hyperbolicity means that each line of T intersects the surface Σ in m distinct points different from the origin. If m is even, then Σ consists of $m/2$ separate sheets, each of which is intersected by a line from T in two different points. For m odd there are $(m+1)/2$ separate sheets. Of these $(m-1)/2$ are intersected by every line of T in two distinct points. In addition there is one sheet that intersects every line of T in exactly one point. Consequently this sheet does not separate the space $\mathbb{RP}^n$ into two parts. If m is even, then Σ consists of $m/2$ nested ovals, the innermost of which is convex.

The fundamental solution $E(t, x)$ for the operator P can be chosen in the form of an integral

$$E(t, x) = (2\pi)^{-n-1} \int \frac{e^{it(\tau + i\theta(\tau, \xi)) + ix \cdot \xi}}{P(\tau + i\theta(\tau, \xi), \xi)} \, d\tau \, d\xi, \tag{4.11}$$

where $\theta(\tau, \xi)$ is a real-valued function that is positive-homogeneous of degree 1, chosen so that $P(\tau + i\theta(\tau, \xi), \xi) \neq 0$ for $(\tau, \xi) \in \mathbb{R}^{n+1} \setminus \{0\}$. If $\theta(\tau, \xi) < 0$ when $|\tau| + |\xi| = 1$, then $E(t, x) = 0$ for $t < 0$.

Consider the largest open subset of the half-space $t > 0$ on which E is an infinitely differentiable function. Let L be one of its connected components.

The region L is called a *lacuna* if the function $E(t,x)$ can be extended to an infinitely differentiable function on $\bar{L} \setminus (0,0)$. A lacuna is called *strong* if $E(t,x) = 0$ in L.

The study of lacunae is based on the possibility of choosing the function $\theta(\tau, \xi)$ in formula (4.11) in different ways. By the homogeneity of the functions P and θ, the integral in (4.11) can be replaced by an integral over the surface Σ. The existence of lacunae depends on the topological properties of the surface Σ. A complete study of this question for strong lacunae was carried out by I. G. Petrovskij (1986). Conditions for the existence of lacunae that are not strong are found in a paper of Atiyah, Bott, and Gårding (1970).

We note that the exterior of the cone of propagation is always a strong lacuna and is called the trivial lacuna. If a strong lacuna contains the point $(0,0)$, then there is no wave diffusion.

It can be seen from the Herglotz-Petrovskij formulas that strong nontrivial lacunae for the wave operator exist for odd $n \geq 3$ but not for $n = 1$ or n even. For a strictly hyperbolic first-order operator

$$I\frac{\partial}{\partial t} + \sum_{j=1}^{n} \mathcal{A}_j \frac{\partial}{\partial x_j},$$

where $\mathcal{A}_j$ are constant $N \times N$ matrices, I is the identity matrix, and the matrix $\sum_{j=1}^{n} \xi_j \mathcal{A}_j$ is nonsingular for all $\xi \in \mathbf{R}^n \setminus \{0\}$, the set of (t, x) for which $t > 0$ and the matrix $tI + \sum_{j=1}^{n} x_j \mathcal{A}_j$ is positive definite is a strong lacuna (cf. Courant and Hilbert 1962).

For equations with variable coefficients strong lacunae occur very rarely, and the question of their existence in the general case is not solved.

4.10. The Cauchy Problem for a Strictly Hyperbolic System with Rapidly Oscillating Initial Data. Problems of this kind are encountered rather frequently in mathematical physics. Their solution is of great importance for the foundations of ray approximation in optics. An analogous problem occurs in the construction of quasiclassical approximations in quantum mechanics (for equations of Schrödinger type).

Consider a strictly hyperbolic system of first-order linear equations

$$Lu \equiv \frac{\partial u}{\partial t} + \sum_{j=1}^{n} \mathcal{A}_j(t, x) \frac{\partial u}{\partial x_j} + \mathcal{B}(t, x)u = 0, \tag{4.12}$$

with initial conditions

$$u(0, x) = \varphi(x)e^{i\lambda S(x)}, \tag{4.12'}$$

where $u = (u_1, \ldots, u_N)$, $\varphi = (\varphi_1, \ldots, \varphi_N)$, A_j, and B are $N \times N$ matrices whose entries are smooth real-valued functions, S is a smooth real-valued function for which $\nabla S(x) \neq 0$ for all x and λ is a real parameter, $\lambda \gg 1$.

We shall seek an approximate solution of this problem in the form of a sum

$$u^{(M)}(t, x) = e^{i\lambda S(t,x)} \sum_{j=0}^{M} u_j(t, x) \lambda^{-j}.$$

Here the functions $u_j(t, x)$ must be independent of λ and

$$|D^\alpha(u - u^{(M)})| = O(\lambda^{-M-1})$$

as $\lambda \to \infty$ for any α.

Substituting $u^{(M)}(t, x)$ into Eq. (4.12) and setting the coefficients of λ^k equal to zero for $k = 1, 0, -1, \ldots$, we obtain

$$\left(\frac{\partial S}{\partial t} I + \sum_{j=1}^{n} A_j(t, x) \frac{\partial S}{\partial x_j}\right) u_0 = 0, \tag{4.13}$$

$$i\left(\frac{\partial S}{\partial t} I + \sum_{j=1}^{n} A_j(t, x) \frac{\partial S}{\partial x_j}\right) u_{k+1} + \frac{\partial u_k}{\partial t} + \tag{4.14$_k$}$$

$$+ \sum_{j=1}^{n} A_j(t, x) \frac{\partial u_k}{\partial x_j} + B u_k = 0.$$

In addition we require that for $t = 0$

$$S(0, x) = S(x), \quad u_0(0, x) = \varphi(x), \quad u_k(0, x) = 0 \quad \text{for } k = 1, 2, \ldots. \tag{4.15}$$

It follows from (4.13) that on the support of the vector-valued function u_0

$$\det\left(\frac{\partial S}{\partial t} I + \sum_{j=1}^{n} A_j(t, x) \frac{\partial S}{\partial x_j}\right) = 0,$$

and we require that this condition hold everywhere, i.e. that the surface $S(t, x) = C$ be characteristic. The function $S(t, x)$ is uniquely determined by this condition and the condition $S(0, x) = S(x)$, at least for small t.

Since the operator L is strictly hyperbolic, the rank of the matrix

$$A(t, x) = \frac{\partial S}{\partial t} I + \sum_{j=1}^{n} A_j(t, x) \frac{\partial S}{\partial x_j}$$

is $N-1$. Let l and r be left and right eigenvectors of this matrix corresponding to the eigenvalue zero, so that

$$Ar = 0, \quad lA = 0$$

(here r is a column-vector and l is a row-vector). It is easy to see that the vector-valued functions $r(t, x)$ and $l(t, x)$ can be chosen to be smooth in

t and x and different from 0 everywhere. It then follows from (4.13) that $u_0 = \sigma(t, x)r(t, x)$, where σ is a scalar-valued function. If we multiply the equation on the left by l, we obtain the following differential equation for σ:

$$lL(\sigma r) = 0.$$

Thus the function σ can be found if its value is known for $t = 0$.

It follows from $(4.14)_0$ that

$$iAu_1 = -Lu_0,$$

i.e., $u_1 = \sigma_1 r + h_1$, where σ_1 is a scalar-valued function and h_1 is expressed in terms of Lu_0. To find σ_1 it is necessary to multiply Eq. $(4.14)_1$ on the left by l. We have

$$lL(\sigma_1 r) + lL(h_1) = 0.$$

From this equation σ_1 can be found if the function $\sigma_1(0, x)$ is known. Continuing this process, we obtain

$$u_j = \sigma_j r + h_j, \quad lL(\sigma_j r) + lL(h_j) = 0, \quad j = 1, 2, \ldots,$$

and the quantity h_j is determined if $u_0, u_1, \ldots, u_{j-1}$ are known. We remark that

$$L\big(u^{(M)}(t, x)\big) = e^{iS(t,x)} \sum_{j=0}^{M} \big(iAu_{j+1} + L(u_j)\big)\lambda^{-j} = P_M(t, x, \lambda)\lambda^{-M}$$

and $D^\alpha P_M(t, x, \lambda)$ are bounded as $\lambda \to \infty$ for all α.

The characteristic equation

$$\det\left(\frac{\partial S}{\partial t}I + \sum_{j=1}^{n} A_j(t, x)\frac{\partial S}{\partial x_j}\right) = 0$$

has N solutions $S_1, \ldots, S_N$ satisfying the initial condition $S_j(0, x) = S(x)$. Moreover the left-annihilators $l_1, \ldots, l_N$ form a linearly independent system at each point (t, x), as do the right-annihilators $r_1, \ldots, r_N$. For each function S_j we construct an approximate solution according to this scheme. We obtain

$$u_j^k = \sigma_j^k r_k + h_j^k, \quad l_k L(\sigma_j^k r_k) + l_k L(h_j^k) = 0, \quad j = 1, 2, \ldots.$$

It follows from condition (4.15) that for $t = 0$

$$\sum_{k=1}^{N} \sigma_0^k r_k = \varphi(x), \quad \sum_{k=1}^{N} \sigma_j^k r_k = -\sum_{k=1}^{N} h_j^k, \quad j = 1, 2, \ldots,$$

from which the values of the function σ_j^k are uniquely determined for $t = 0$. Solving the differential equations for σ_j^k, we find these functions for all sufficiently small t. Setting

$$U_M(t,x) = \sum_{j=0}^{M}\sum_{k=1}^{N} u_j^k(t,x)e^{iS_k(t,x)},$$

we find that

$$LU_M(t,x) = Q_M(t,x,\lambda)\lambda^{-M},$$

where Q_M and its derivatives are bounded as $\lambda \to \infty$. The vector-valued function U_M is called an *asymptotic solution* of the Cauchy problem (4.12)–(4.12′) (with precision $O(\lambda^{-M})$). In the present case it differs by little from the actual solution. To be specific, if $V_M(t,x) = u(t,x) - U_M(t,x)$, then

$$LV_M(t,x) = -Q_M(t,x,\lambda)\lambda^{-M}, \quad V_M(0,x) = 0.$$

From this, using the energy estimates, it is easy to obtain the estimate $D^\alpha V_M(t,x) = O(\lambda^{-M})$ for all α.

In the next volume of this series we shall consider the Cauchy problem with rapidly oscillating initial conditions for equations of order $m \geq 2$ and discuss how the asymptotic solutions constructed can be used as the foundation for quasiclassical asymptotics.

4.11. Discontinuous Solutions of Hyperbolic Equations. We begin by considering a solution $u = u(x_0, x_1, \ldots, x_n)$ of a strictly hyperbolic system of first-order equations

$$Lu \equiv \sum_{j=0}^{n} A_j(x)\frac{\partial u}{\partial x_j} + B(x)u = 0, \tag{4.16}$$

where $x = (x_0, x_1, \ldots, x_n)$ and $x_0 = t$, such that u has a *weak discontinuity* on a smooth surface Γ. This means that the vector-valued function $u = (u_1, \ldots, u_N)$ is continuous along with its first derivatives in the directions tangent to Γ while the derivative of u in the direction normal to Γ may have a jump discontinuity. We assume that the matrices $A_j(x)$ and $B(x)$ are continuous. We denote by $[F]$ the jump of the vector-valued function F at the surface Γ.

Let the surface Γ be defined by the equation $S(x) = 0$ $(\nabla S \neq 0)$. The expression

$$\frac{\partial S}{\partial x_j}\frac{\partial u}{\partial x_k} - \frac{\partial S}{\partial x_k}\frac{\partial u}{\partial x_j}$$

is the derivative of u in a direction tangent to Γ and hence continuous. Therefore on Γ

$$\frac{\partial S}{\partial x_j}\Big[\frac{\partial u}{\partial x_k}\Big] = \frac{\partial S}{\partial x_k}\Big[\frac{\partial u}{\partial x_j}\Big].$$

We assume that $\mathrm{grad}\, S(x) \neq 0$ in the region under consideration. It then follows from these relations that

$$\left[\frac{\partial u}{\partial x_j}\right] = \frac{\partial S}{\partial x_j} g(x), \quad j = 0, 1, \ldots, n,$$

where $g(x)$ is a certain vector. It can be seen from the equality $Lu = 0$ that

$$\sum_{j=0}^{n} \mathcal{A}_j(x)\left[\frac{\partial u}{\partial x_j}\right] = 0.$$

Thus $g(x)$ is in the right-hand nullspace of the matrix $\mathcal{A}(x) = \sum_{j=0}^{n} \mathcal{A}_j \frac{\partial S}{\partial x_j}$ and the surface Γ is characteristic.

Since the operator L is strictly hyperbolic, the rank of the matrix $\mathcal{A}(x)$ is $N - 1$. Let r and l be in the right- and left-hand nullspaces of it, so that $l\mathcal{A} = 0$ and $\mathcal{A}r = 0$. As shown above,

$$[\operatorname{grad} u] = \sigma r,$$

where $\sigma(x)$ is a scalar-valued function. Therefore

$$u(t, x) = \frac{1}{2}|S(x)|\sigma r + R(x),$$

and moreover the functions $\sigma(x)$, $r(x)$ and $R(x)$ have continuous derivatives of first order, which in turn have continuous first derivatives in directions tangent to Γ. On both sides of Γ we have

$$Lu = \pm\frac{1}{2}\mathcal{A}(x)\sigma r + \frac{1}{2}|S(x)|L(\sigma r) + L(R) = \frac{1}{2}|S(x)|L(\sigma r) + L(R) = 0.$$

Multiplying this equality on the left by l, we find that

$$\frac{1}{2}|S(x)|l \cdot L(\sigma r) + l \cdot L(R) = 0.$$

But the vector field that is the principal part of the operator lL is tangent to Γ, since $l \cdot \sum_{j=0}^{n} \mathcal{A}_j \frac{\partial S}{\partial x_j} = l\mathcal{A} = 0$ on Γ. Consequently the first-order derivatives of the vector-valued function $lL(R)$ are continuous on Γ. Differentiating the relation obtained in the direction normal to Γ, we find that $lL(\sigma r) = 0$ on Γ, i.e.,

$$l\sum_{j=0}^{n} \mathcal{A}_j(x)r\frac{\partial \sigma}{\partial x_j} + l\left(\sum_{j=0}^{n} \mathcal{A}_j(x)\frac{\partial r}{\partial x_j} + \mathcal{B}(x)r\right)\sigma = 0, \quad x \in \Gamma.$$

This equation can be written in the form

$$\dot{\sigma} + P\sigma = 0, \tag{4.17}$$

where $P = lL(r)$ and the dot denotes differentiation along the bicharacteristic curve defined by the system of differential equations

$$\frac{dx_i}{ds} = l\mathcal{A}_i(x)r, \quad i = 0, 1, \ldots, n,$$

with initial condition $x(0) \in \Gamma$. It is not difficult to see that this curve lies on the surface Γ. In fact

$$\frac{dS(x(s))}{ds} = \sum_{j=0}^{n} \frac{\partial S}{\partial x_j} \frac{dx_j}{ds} = \sum_{j=0}^{n} \frac{\partial S}{\partial x_j} l\mathcal{A}_j(x)r = l\mathcal{A}(x)r = 0,$$

Thus $S(x(s)) \equiv 0$ if $S(x(0)) = 0$. Hence the magnitude of the jump σ cannot be an arbitrary function on Γ, but must satisfy Eq. (4.17).

In mathematical physics solutions with *strong discontinuitiess* are often studied. Such solutions can be regarded as the limits of the usual smooth solutions.

Example 2.59. For $t > 0$ the solution of the equation $\dfrac{\partial^2 u}{\partial t^2} = a^2 \dfrac{\partial^2 u}{\partial x^2}$ with the initial conditions $u(0, x) = \operatorname{sgn} x$, $\dfrac{\partial u}{\partial t}(0, x) = 0$ has the form

$$u(t, x) = \begin{cases} 1 & \text{for } x > at, \\ 0 & \text{for } -at < x < at, \\ -1 & \text{for } x < -at. \end{cases}$$

We now consider once again the system of equations (4.16) and the solution $u(x)$ of this equation for which all the derivatives have a jump discontinuity on the smooth surface Γ. Let Γ be defined by the equation $S(x) = 0$. To study the solution $u(x)$ it is convenient to use the Heaviside function

$$\theta(t) = \begin{cases} 1 & \text{for } t > 0, \\ 0 & \text{for } t \le 0. \end{cases}$$

Consider the function

$$u^{(M)}(x) = u_+(x) + \theta(S(x))u_-(x) + \sum_{j=0}^{M} \Phi_j(S(x))u_j(x),$$

where $\Phi_0(t) = t$ for $t > 0$, $\Phi_0(t) = 0$ for $t \le 0$, and $\Phi'_j(t) = \Phi_{j-1}(t)$ for $j = 1, 2, \ldots, M$. We shall show that smooth functions u_+, u_-, u_j can be chosen for $j = 0, 1, \ldots, M$ such that $u(x) - u^{(M)}(x) \in C^M$.

We remark that

$$\frac{\partial u^{(M)}(x)}{\partial x_k} = \frac{\partial u_+(x)}{\partial x_k} + \theta(S(x)) \frac{\partial u_-(x)}{\partial x_k} + \delta(S(x)) \frac{\partial S(x)}{\partial x_k} u_-(x) +$$

$$+ \theta(S(x)) \frac{\partial S(x)}{\partial x_k} u_0(x) + \sum_{j=1}^{M} \Phi_{j-1}(S(x)) \frac{\partial S(x)}{\partial x_k} u_j(x) +$$

$$+ \sum_{j=0}^{M} \Phi_j(S(x)) \frac{\partial u_j(x)}{\partial x_k}.$$

Let $\mathcal{A}(x) = \sum\limits_{k=0}^{n} \mathcal{A}_k(x) \dfrac{\partial S(x)}{\partial x_k}$. Then

$$Lu = L(u_+) + \theta\big(S(x)\big)\big(L(u_-) + \mathcal{A}(x)u_0(x)\big) + \delta\big(S(x)\big)\mathcal{A}(x)u_-(x) +$$

$$+ \sum_{j=0}^{M} \Phi_j\big(S(x)\big)\big(L(u_j) + \mathcal{A}(x)u_{j+1}(x)\big).$$

Consequently at the points of Γ we have the relations

$$\mathcal{A}(x)u_-(x) = 0,$$
$$L(u_-) + \mathcal{A}(x)u_0(x) = 0,$$
$$L(u_j) + \mathcal{A}(x)u_{j+1}(x) = 0, \quad j = 0, 1, \ldots, M. \qquad (4.18)_j$$

It follows from this that $\det \mathcal{A}(x) = 0$, i.e., the surface Γ is characteristic. As above, we shall assume that the rank of the matrix $\mathcal{A}(x)$ is $N-1$. We denote by $l(x)$ and $r(x)$ vectors satisfying the equalities

$$l(x)\mathcal{A}(x) = 0, \quad \mathcal{A}(x)r(x) = 0,$$

depending smoothly on x and never equal to 0. Then

$$u_-(x) = \sigma(x)r(x),$$

where $\sigma(x)$ is a scalar function. Multiplying $(4.18)_0$ on the left by l, we find that

$$lL(\sigma r) = 0.$$

Similarly it follows from $(4.18)_j$ for $j = 1, \ldots, M$, that

$$lL(u_j) = 0. \qquad (4.19)$$

As above, it follows from this that

$$\dot{\sigma} + lL(r)\sigma = 0, \quad u_j = \sigma_j r + h_j, \quad j = 0, 1, \ldots, M,$$

where σ_j are scalar functions and the vectors h_j are uniquely determined if $h_j \cdot r = 0$ and the functions $L(u_-), L(u_0), \ldots, L(u_{j-1})$ are known. It follows from (4.19) that

$$\dot{\sigma}_j + lL(r)\sigma_j + k_j = 0, \quad j = 0, 1, \ldots, M,$$

and the functions k_j are determined if $L(u_-), L(u_0), \ldots, L(u_{j-1})$ are known. These equations make it possible to find the functions σ_j if their values are known for $t = 0$.

Let $\Gamma_1, \ldots, \Gamma_N$ be characteristic surfaces corresponding to different characteristic roots and coinciding for $x_0 = 0$ with the surface S_0, and let the vector $u(0, x_1, \ldots, x_n)$ be smooth outside S_0 and have a jump discontinuity on S_0. Let

$$[u(0, x_1, \ldots, x_n)]_{S_0} = \varphi(x),$$

and let l_k and r_k be vectors respectively in the left and right nullspaces of the matrix $\sum_{j=0}^{n} A_j(x)\dfrac{\partial S_k}{\partial x_j}$ for $k = 1, \ldots, N$. Let

$$\varphi(x) = \sum_{k=1}^{N} \sigma_0^k r_k.$$

Then $[u]_k$, the jump of the vector u at the surface Γ_k, is determined as $[u]_k = \sigma^k r_k$, where

$$\dot{\sigma}^k + l_k L(r_k)\sigma^k = 0, \quad \sigma^k = \sigma_0^k \quad \text{for } x_0 = 0.$$

If L is not a strictly hyperbolic operator, but the multiplicity of the roots of the characteristic equation is constant, the construction just presented is easily generalized and the magnitude of the jumps of a solution can be found after integrating a system of ordinary differential equations.

If, however, the multiplicity of the roots is not constant, then the picture may become significantly more complicated. In this case the discontinuities of a solution may propagate not along rays, but along submanifolds of Γ having dimension larger than 1. For example the phenomenon of conical refraction discovered by Hamilton occurs when a ray entering a crystal along the direction of an optical axis splits into a collection of rays directed along all the generators of a two-dimensional cone. The discontinuity caused by the entering ray propagates along the surface of the cone and even into the cone, with smaller amplitude.

The constructions just presented are valid also for systems of equations of order m.

4.12. Symmetric Hyperbolic Operators. Consider the operator

$$L = \sum_{j=0}^{n} A_j(x)\frac{\partial}{\partial x_j} + B(x),$$

where $A_j(x)$ are symmetric real-valued matrices of order N. Such an operator L is called a *symmetric hyperbolic operator* if the matrix $\sum_{j=0}^{n} \alpha_j A_j$ is positive definite for some non-zero vector $(\alpha_0, \ldots, \alpha_n) \in \mathbb{R}^{n+1}$ (depending, in general, on x). It is clear that the characteristic roots $\lambda_1, \ldots, \lambda_N$ of the operator L, i.e., the solutions of the equation

$$\left| \lambda \sum_{j=0}^{n} \alpha_j A_j + \sum_{j=0}^{n} \xi_j A_j \right| = 0$$

are real for any real ξ.

Example 2.60. Every second-order hyperbolic equation can be reduced to a symmetric hyperbolic system in the following way. Let

$$M[v] \equiv \frac{\partial^2 v}{\partial t^2} - \sum_{i,j=1}^{n} a_{ij}(t,x)\frac{\partial^2 v}{\partial x_i \partial x_j} + \sum_{j=1}^{n} a_{0j}(t,x)\frac{\partial^2 v}{\partial t \partial x_j} +$$

$$+ b_0(t,x)\frac{\partial v}{\partial t} + \sum_{j=1}^{n} b_j(t,x)\frac{\partial v}{\partial x_j} + c(t,x)v = f(t,x),$$

where $a_{ij}(t,x) = a_{ji}(t,x)$. We set

$$\frac{\partial v}{\partial t} = u_0, \quad \frac{\partial v}{\partial x_j} = u_j, \quad j = 1,\dots,n.$$

The vector $U = (u_1,\dots,u_n,u_0,v)$ satisfies the system of first-order equations

$$\sum_{i=1}^{n} a_{ij}(t,x)\left(\frac{\partial u_i}{\partial t} - \frac{\partial u_0}{\partial x_j}\right) = 0, \quad j = 1,\dots,n$$

$$\frac{\partial u_0}{\partial t} = \sum_{i,j=1}^{n} a_{ij}(t,x)\frac{\partial u_i}{\partial x_j} + \sum_{j=1}^{n} a_{0j}(t,x)\frac{\partial u_0}{\partial x_j} + \sum_{j=0}^{n} b_j(t,x)u_j + c(t,x)v,$$

$$\frac{\partial v}{\partial t} - u_0 = 0,$$

which is a symmetric hyperbolic system. Here $N = n+2$ and

$$\mathcal{A}_0(t,x) = \begin{pmatrix} a_{11} & \cdots & a_{1n} & 0 & 0 \\ \cdots\cdots\cdots\cdots\cdots\cdots\cdots \\ a_{n1} & \cdots & a_{nn} & 0 & 0 \\ 0 & \cdots & 0 & 1 & 0 \\ 0 & \cdots & 0 & 0 & 1 \end{pmatrix},$$

$$\mathcal{A}_j(t,x) = \begin{pmatrix} 0 & \cdots & 0 & -a_{1j} & 0 \\ \cdots\cdots\cdots\cdots\cdots\cdots\cdots\cdots \\ 0 & \cdots & 0 & -a_{nj} & 0 \\ -a_{1j} & \cdots & -a_{nj} & a_{0j} & 0 \\ 0 & \cdots & 0 & 0 & 0 \end{pmatrix}, \quad j = 1,\dots,n.$$

If the values of the functions v and $\frac{\partial v}{\partial t}$ are known for $t = 0$, it is possible to determine the value of the vector $U(0,x)$; moreover the Cauchy problem for the vector U is equivalent to the Cauchy problem for the function v.

Example 2.61. The system of Maxwell equations

$$\varepsilon\frac{\partial \mathbf{E}}{\partial t} - \operatorname{curl}\mathbf{H} = \mathbf{0}, \quad \mu\frac{\partial \mathbf{H}}{\partial t} + \operatorname{curl}\mathbf{E} = \mathbf{0}$$

is a symmetric hyperbolic system.

Definition 2.62. A surface $S(x) = 0$ in the space of the variables $x = (x_0, x_1, \ldots, x_n)$ is a *space-like surface* if the matrix $\sum\limits_{j=0}^{n} A_j(x) \dfrac{\partial S(x)}{\partial x_j}$ is positive or negative definite.

Examples of space-like surfaces are any surface $\sum\limits_{j=0}^{n} \alpha_j x_j = c$ and any surface close to such a surface in the C^1-metric.

The basic method of studying symmetric hyperbolic systems is provided by *energy inequalities*.

Theorem 2.63. *Let* $A_0(x) = I$, $x = (x_0, x')$, $x' = (x_1, \ldots, x_n)$, *and* $u(x) \in C^1([0,t] \times \mathbb{R}^n)$. *Then there exists a constant* $C > 0$ *such that for all* $t \in [0,T]$ *the inequality*

$$\int_{G_t} (u(t,x'), u(t,x'))\, dx' \leq C \left(\int_{G_0} (u(0,x'), u(0,x'))\, dx' + \right.$$

$$\left. + \int_0^t \int_{G_\tau} (Lu(\tau,x'), Lu(\tau,x'))\, dx'\, d\tau \right)$$

holds, where G_t *is the section cut off by the plane* $x_0 = t$ *from the region* Ω *bounded by the plane* $x_0 = 0$ *and a space-like surface* S.

This inequality follows easily upon integrating the equality

$$2(u, Lu) = \sum_{j=0}^{n} \frac{\partial}{\partial x_j} (A_j(x), u, u) + 2(B_0(x)u, u),$$

where $B_0(x) = B(x) - \dfrac{1}{2} \sum\limits_{j=0}^{n} \dfrac{\partial}{\partial x_j} A_j(x)$. We note further that each symmetric hyperbolic system can be reduced to a system in which $A_0(x) = I$ by multiplying on the left by the matrix $\left(\sum\limits_{j=0}^{n} \alpha_j A_j \right)^{-1}$ and carrying out a linear change of variables.

4.13. The Mixed Boundary-Value Problem. The study of the mixed boundary-value problem for hyperbolic differential equations of general type has made significant advances since the 1950's (Kreiss 1958; Kreiss 1970). Up to that time such a problem had been studied only for second-order equations

$$\frac{\partial^2 u}{\partial t^2} = \sum_{i,j=1}^{n} a_{ij}(t,x) \frac{\partial^2 u}{\partial x_i \partial x_j} + \sum_{j=1}^{n} b_j(t,x) \frac{\partial u}{\partial x_j} + c(t,x)u + f(t,x). \quad (4.20)$$

We shall assume that all the functions in this equation are real-valued and
that

$$a_{ij} \in C^1([0,T] \times \bar{\Omega}), \quad b_j, c \in C([0,T] \times \bar{\Omega}), \quad a_{ij} = a_{ji}$$

for $i, j = 1, \ldots, n$, $\sum_{i,j=1}^{n} a_{ij}(t,x)\xi_i\xi_j \geq c_0|\xi|^2$, $c_0 = \text{const} > 0$, and Ω is a
bounded region in $\mathbf{R}^n$ with a piecewise-smooth boundary. The simplest mixed
boundary-value problem is that of finding a solution $u(t,x)$ of Eq. (4.20)
satisfying the initial conditions

$$u(0,x) = \varphi_0(x), \quad \frac{\partial u(0,x)}{\partial t} = \varphi_1(x), \quad x \in \Omega, \tag{4.21}$$

and the boundary condition of first kind (Dirichlet condition)

$$u = g, \quad \text{for } 0 \leq t \leq T, \quad x \in \partial\Omega. \tag{4.22}$$

In addition we consider problems in which this last condition is replaced by
the condition of second kind (Neumann condition)

$$\frac{\partial u}{\partial N} = g \quad \text{for } 0 \leq t \leq T, \quad x \in \partial\Omega, \tag{4.23}$$

or the condition of third kind

$$\frac{\partial u}{\partial N} + \sigma(t,x)u = g \quad \text{for } 0 \leq t \leq T, \quad x \in \partial\Omega. \tag{4.24}$$

Here $\dfrac{\partial u}{\partial N} = \displaystyle\sum_{i,j=1}^{n} a_{ij}(t,x)\dfrac{\partial u}{\partial x_j}\alpha_i$, where $(\alpha_1, \ldots, \alpha_n)$ are the direction co-
sines of the exterior normal to $\partial\Omega$ and $\sigma \geq 0$.

Theorem 2.64. *Let $u \in C^2([0,T] \times \bar{\Omega})$ be a solution of Eq. (4.20) satisfying
the initial conditions (4.21) and one of the conditions (4.22) or (4.23) with
$g = 0$. There exists a constant $C > 0$ independent of u such that*

$$\mathcal{E}(t) \leq C\Big(\mathcal{E}(0) + \iint_Q f^2(t,x)\,dt\,dx\Big), \tag{4.25}$$

where

$$\mathcal{E}(t) = \int_\Omega \Big[u^2 + \Big(\frac{\partial u}{\partial t}\Big)^2 + \sum_{j=1}^{n}\Big(\frac{\partial u}{\partial x_j}\Big)^2\Big]\,dx,$$

$$\mathcal{E}(0) = \int_\Omega \Big[\varphi_0^2(x) + \varphi_1^2(x) + \sum_{j=1}^{n}\Big(\frac{\partial\varphi_0(x)}{\partial x_j}\Big)^2\Big]\,dx,$$

and $Q = [0,T] \times \Omega$.

The integral $\mathcal{E}(t)$ expresses the total energy of a body whose vibrations are described by Eq. (4.20). For that reason such estimates are called *energy estimates*. The proof of the theorem is carried out by multiplying both sides of Eq. (4.20) by $2\dfrac{\partial u}{\partial t}$ and integrating the resulting equality over the region $[0, t) \times \Omega$. After that we use the equalities

$$2\frac{\partial u}{\partial t}\frac{\partial^2 u}{\partial t^2} = \frac{\partial}{\partial t}\left(\frac{\partial u}{\partial t}\right)^2,$$

$$2\frac{\partial u}{\partial t}\frac{\partial}{\partial x_i}\left(a_{ij}(t,x)\frac{\partial u}{\partial x_j}\right) = 2\frac{\partial}{\partial x_i}\left(a_{ij}(t,x)\frac{\partial u}{\partial x_j}\frac{\partial u}{\partial t}\right)$$

$$-\frac{\partial}{\partial t}\left(a_{ij}(t,x)\frac{\partial u}{\partial x_i}\frac{\partial u}{\partial x_j}\right) + \frac{\partial a_{ij}(t,x)}{\partial t}\frac{\partial u}{\partial x_j}\frac{\partial u}{\partial x_j},$$

along with Green's formula and the Cauchy-Bunyakovskij inequality.

In the case of the boundary condition (4.24) the theorem is again true if we add to $\mathcal{E}(t)$ the integral $\displaystyle\int_{\partial\Omega} \sigma(t,x)u^2(t,x)\,dS$.

It follows immediately from inequality (4.25) that the solution of the mixed boundary-value problem is unique.

In addition, this inequality makes it possible to prove the existence of a generalized solution of such a problem, interpreted in the sense of the following definition.

Definition 2.65. A *generalized solution* of the boundary-value problem (4.20), (4.21), (4.22) is a function $u(t,x)$ for which the integral $\mathcal{E}(t)$ is finite for $0 \le t \le T$ and which satisfies the conditions

$$u(0,x) = \varphi_0(x) \quad \text{for } x \in \Omega; \quad u(t,x) = g \quad \text{for } 0 \le t \le T, \quad x \in \partial\Omega,$$

and the equality

$$\int_0^T\!\!\int_\Omega \left[\frac{\partial u}{\partial t}\frac{\partial F}{\partial t} - \sum_{i,j=1}^n a_{ij}(t,x)\frac{\partial u}{\partial x_i}\frac{\partial F}{\partial x_j}+\right.$$

$$\left.+\sum_{j=1}^n\left(b_j(t,x) - \sum_{i=1}^n \frac{\partial a_{ij}(t,x)}{\partial x_i}\right)\frac{\partial u}{\partial x_j}F + (cu - f)F\right]dx\,dt +$$

$$+ \int_\Omega \varphi_1(x)F(0,x)\,dx = 0.$$

for each function F in $C^2(\bar{Q})$ equal to zero for $t = T$, $x \in \Omega$ and for $x \in \partial\Omega$, $0 \le t \le T$.

4.14. The Method of Separation of Variables. Consider the equation

$$\frac{\partial^2 u}{\partial t^2} = Lu, \quad Lu \equiv \sum_{i,j=1}^{n} \frac{\partial}{\partial x_i}\left(a_{ij}(x)\frac{\partial u}{\partial x_j}\right) - q(x)u$$

in a bounded region $\Omega \subset \mathbb{R}^n$ with smooth boundary Γ, assuming that $\sum_{i,j=1}^{n} a_{ij}(x)\xi_i\xi_j \geq c_0|\xi|^2$, where $c_0 = \text{const} > 0$, $q(x) \geq 0$, and the functions $a_{ij}, \frac{\partial a_{ij}}{\partial x_k}$, and q are real-valued and continuous in $\bar{\Omega}$.

We shall seek a solution $u(t,x)$ satisfying the conditions

$$u = \varphi_0, \quad \frac{\partial u}{\partial t} = \varphi_1, \quad \text{for } t = 0, \, x \in \Omega,$$

and

$$u = 0 \quad \text{for } 0 \leq t \leq T, \quad x \in \partial\Omega.$$

The operator L has a complete orthonormal system of eigenfunctions $X_1(x), X_2(x), \ldots$ belonging to $\mathcal{D}(\Omega)$ and corresponding to the eigenvalues $(-\lambda_j^2)$, so that $\lim_{j\to\infty} \lambda_j^2 = \infty$ (cf. Sect. 2).

The function $v_j(t)X_j(x)$ satisfies the equation $\frac{\partial^2 u}{\partial t^2} = Lu$ if and only if

$$\ddot{v}_j + \lambda_j^2 v_j = 0,$$

i.e., $v_j(t) = a_j \cos \lambda_j t + b_j \sin \lambda_j t$. The solution of the boundary-value problem can therefore be sought in the form of the sum of a series

$$u(t,x) = \sum_{j=1}^{\infty} (a_j \cos \lambda_j t + b_j \sin \lambda_j t) X_j(x), \qquad (4.26)$$

whose coefficients a_j and b_j are determined by the initial conditions

$$a_j = \int_{\Omega} \varphi_0(x) X_j(x)\, dx, \quad b_j = \frac{1}{\lambda_j}\int_{\Omega} \varphi_1(x) X_j(x)\, dx, \quad j = 1, 2, \ldots .$$

Theorem 2.66. *If $\varphi_0 \in C^{2+\alpha}(\bar{\Omega})$, $\varphi_1 \in C^{1+\alpha}(\bar{\Omega})$ for $0 < \alpha < 1$ (i.e., $\varphi_0 \in C^2(\bar{\Omega})$, $\varphi_1 \in C^1(\bar{\Omega})$, and the derivatives $D_i D_j \varphi_0$ and $D_i \varphi_1$ satisfy a Hölder condition in $\bar{\Omega}$ with exponent α) and the consistency conditions*

$$\varphi_0(x) = 0, \quad L\varphi_0(x) = 0, \quad \varphi_1(x) = 0 \quad \text{for } x \in \Gamma$$

hold, then the series (4.26) converges in $[0,T] \times \bar{\Omega}$ and defines a function u of class $C^2(\bar{\Omega} \times [0,T])$ for any $T > 0$.

The method just described for solving the mixed boundary-value problem is called the method of *separation of variables* or the *Fourier method*. This method is also applicable to an inhomogeneous equation

$$\frac{\partial^2 u}{\partial t^2} = Lu + f(t, x).$$

The solution of the mixed boundary-value problem is then sought in the form of a series $u(t, x) = \sum_{j=1}^{\infty} v_j(t) X_j(x)$. The functions v_j are determined from the equations $\ddot{v}_j(t) + \lambda_j^2 v_j(t) = f_j(t)$, where $f_j(t) = \int_{\Omega} f(t, x) X_j(x) \, dx$, and the initial conditions

$$v_j(0) = \int_{\Omega} \varphi_0(x) X_j(x) \, dx, \quad \dot{v}_j(0) = \frac{1}{\lambda_j} \int_{\Omega} \varphi_1(x) X_j(x) \, dx.$$

In the case when $\varphi_0 \in W_2^1(\Omega)$, $\varphi_1 \in L^2(\Omega)$, $\varphi_0(x) = 0$ for $x \in \partial\Omega$, and $f \in L^2([0, T] \times \Omega)$, the Fourier method makes it possible to find a generalized solution of the mixed boundary-value problem defined in Sect. 4.13.

§5. Parabolic Equations

5.1. Definitions and Examples. As already pointed out in Sect. 1.3.3 an equation of the form

$$\frac{\partial^m u}{\partial t^m} = \sum_{\substack{|\alpha| + 2b\alpha_0 \leq 2bm \\ \alpha_0 < m}} a_{\alpha, \alpha_0}(t, x) D_x^\alpha \frac{\partial^{\alpha_0} u}{\partial t^{\alpha_0}} + f(t, x) \tag{5.1}$$

is called *parabolic* or *Petrovskij 2b-parabolic* if all the roots $\lambda_j = \lambda_j(t, x, \xi)$ of the characteristic equation

$$\lambda^m = \sum_{|\alpha| + 2b\alpha_0 = 2bm} a_{\alpha, \alpha_0}(t, x) \xi^\alpha \lambda^{\alpha_0}$$

have negative real part for all $\xi \in \mathbb{R}^n \setminus \{0\}$ and for a positive integer b.

Example 2.67. The heat equation $\dfrac{\partial u}{\partial t} = a^2 \Delta u$ is parabolic for $m = 1$ and $b = 1$.

Example 2.68. For $k \geq 3$ the equation $\dfrac{\partial u}{\partial t} = \dfrac{\partial^2 u}{\partial x^2} + i \left(i^{-1} \dfrac{\partial}{\partial x} \right)^k u$ is not Petrovskij-parabolic. It is *Shilov-parabolic*. This means that the condition $\operatorname{Re} \lambda < 0$ is satisfied for large $|\xi|$ by the roots of the equation

$$\lambda^m = \sum_{|\alpha|+2b\alpha_0 \le 2bm} a_{\alpha,\alpha_0} \xi^\alpha \lambda^{\alpha_0}, \quad \xi \in \mathbf{R}^n,$$

in which all the coefficients of Eq. (5.1) are taken into account. The theory of such equations has been worked out in detail in the case of constant coefficients (cf. Gel'fand and Shilov 1958–1959 and Sect. 5.8 below).

It follows from the definition that the order of the derivatives on x occurring on the right-hand side of Eq. (5.1) is larger than m, so that the plane $t = C$ is characteristic for a parabolic equation. This circumstance turns out to be quite significant in the statement of the Cauchy problem. In contrast to hyperbolic equations a perturbation of the initial data, even a localized one, is instantaneously (i.e. for all $t > 0$) propagated to the whole space $\mathbf{R}_x^n$. To get a well-posed Cauchy problem for a parabolic equation it is necessary to impose conditions on the behavior of the solution as $|x| \to \infty$. In addition, a Cauchy problem which is well-posed for $t > 0$ becomes ill-posed when solutions are sought for $t < 0$.

5.2. The Maximum Principle and Its Consequences. Consider a second-order parabolic equation

$$Lu \equiv -\frac{\partial u}{\partial t} + \sum_{i,j=1}^n a_{ij}(t,x)\frac{\partial^2 u}{\partial x_i \partial x_j} + \sum_{i=1}^n b_i(t,x)\frac{\partial u}{\partial x_i} + c(t,x)u = f(t,x) \quad (5.2)$$

assuming that the functions a_{ij}, b_i, and c are continuous on $[0,T] \times \bar{\Omega}$, where Ω is a bounded region in $\mathbf{R}^n$ and

$$\sum_{i,j=1}^n a_{ij}(t,x)\xi_i\xi_j \ge \alpha|\xi|^2, \quad \alpha = \text{const} > 0.$$

Theorem 2.69. *Let $u \in C(\bar{Q})$, where $Q = (0,T] \times \Omega$. Let the derivatives $\dfrac{\partial u}{\partial t}, \dfrac{\partial u}{\partial x_i}, \dfrac{\partial^2 u}{\partial x_i \partial x_j}$ be continuous in Q and $Lu \le 0$ in Q. If $u(t,x) \ge 0$ for $t = 0$, $x \in \Omega$ and for $0 \le t \le T$, $x \in \partial\Omega$, then $u(t,x) \ge 0$ in $\bar{Q}$.*

Proof. Let $c(t,x) < M$ in Q. If u assumes a negative value at some point of Q, then the function $v(t,x) = u(t,x)e^{-Mt}$ assumes a minimal negative value at some point $(t_0, x_0) \in Q$. At this point

$$\frac{\partial v}{\partial t} \le 0, \quad \frac{\partial v}{\partial x_j} = 0, \quad \sum_{i,j=1}^n a_{ij}\frac{\partial^2 v}{\partial x_i \partial x_j} \ge 0, \quad (c-M)v > 0,$$

so that $Lv - Mv > 0$. But the condition $Lu \le 0$ implies that $Lv - Mv \le 0$ in $\bar{Q}$. This contradiction proves the theorem.   □

The following propositions can be deduced from Theorem 2.69 (cf. Il'in, Kalashnikov, and Olejnik 1962).

Corollary 2.70. *If u is a solution of Eq. (5.2) in Q and $u \in C(\bar{Q})$, then*

$$|u(t,x)| \leq K\left(\max_{x \in \bar{\Omega}} |u(0,x)| + \max_{\substack{0 \leq t \leq T \\ x \in \partial\Omega}} |u(t,x)| + t \max_{(t,x) \in \bar{Q}} |f(t,x)| \right).$$

Corollary 2.71. *If u is a solution of the equation $Lu = 0$ in Q, $u \in C(\bar{Q})$, and $c(t,x) = 0$, then for all $(t,x) \in \bar{Q}$ the inequalities*

$$\min_\Gamma u \leq u(t,x) \leq \max_\Gamma u$$

hold, where $\Gamma = \{(t,x) \in \bar{Q} : t = 0 \quad \text{or} \quad x \in \partial\Omega\}$.

Corollary 2.72 (The strong maximum principle). *Let $\mathcal{D}$ be an arbitrary bounded region in $\mathbf{R}_{t,x}^{n+1}$ with smooth boundary. Let $Lu = 0$ in $\mathcal{D}$, $u \in C(\bar{\mathcal{D}})$, and $c(t,x) \leq 0$ in $\mathcal{D}$. If*

$$\max_{\bar{\mathcal{D}}} u(t,x) = u(t_0,x_0) > 0,$$

where $(t_0,x_0) \in \mathcal{D}$, then $u(t,x) = u(t_0,x_0)$ at each point (t,x) of $\mathcal{D}$ for which $t < t_0$ and which can be joined to the point (t_0,x_0) by a curve of the form $x = x(t)$ lying entirely in $\mathcal{D}$.

Corollary 2.73. *Let $\mathcal{D}$ be an arbitrary bounded region in $\mathbf{R}_{t,x}^{n+1}$. Let $Lu = 0$ in $\mathcal{D}$, $u \in C(\bar{\mathcal{D}})$, and $c(t,x) \leq 0$ in $\mathcal{D}$. Let*

$$\max_{\bar{\mathcal{D}}} u = u(t_0,x_0) > 0$$

and $(t_0,x_0) \in \partial\mathcal{D}$. Assume there exists a closed ball K containing the point (t_0,x_0) all of whose points in the region $0 < t < T$ except (t_0,x_0) lie in $\mathcal{D}$. Then either the function $u(t,x)$ is constant in some neighborhood of the point (t_0,x_0) for $t \leq t_0$ or $\dfrac{\partial u(t_0,x_0)}{\partial \gamma} < 0$, where γ is an arbitrary direction forming an acute angle with the radius directed from (t_0,x_0) to the center of the ball K and such that $\dfrac{\partial u(t_0,x_0)}{\partial \gamma}$ exists.

Corollary 2.74. *Let $H = (0,T] \times \mathbf{R}^n$ and let the coefficients of the operator L defined by formula (5.2) be bounded in $\bar{H}$. If $u \in C(\bar{H})$, $Lu \leq 0$, $u(0,x) \geq 0$ and there exist positive constants C and α such that*

$$|u(t,x)| \leq Ce^{\alpha|x|^2} \quad \text{in } H, \tag{5.3}$$

then $u(t,x) \geq 0$ in H.

Corollary 2.75 (Uniqueness of the solution of the Cauchy problem). *Suppose that $H = (0,T] \times \mathbb{R}^n$ and that the coefficients of the operator L defined by the formula (5.2) be bounded in $\bar{H}$. The Cauchy problem for Eq. (5.2) can have only one solution in H in the class of functions of $C(\bar{H})$ satisfying condition (5.3).*

Condition (5.3) can be replaced by the weaker condition

$$|u(t,x)| \leq C e^{\alpha|x|h(|x|)}, \tag{5.4}$$

where $h(r)$ is a nondecreasing positive function and $\displaystyle\int_1^\infty \frac{dr}{h(r)} = \infty$. If the inequality $\displaystyle\int_1^\infty \frac{dr}{h(r)} < \infty$ holds, then there exists a function $u(t,x) \in C(\bar{H})$ satisfying (5.4) and such that $u(0,x) = 0$ and $Lu = 0$ in H (cf. Sect. 5.7 below).

Corollary 2.76 (Bernshtejn's estimates). *Let $Q = \Omega \times (0,T]$, $Q_\delta = \Omega_\delta \times (\delta, T]$, where Ω_δ is the subregion of Ω for which $\mathrm{dist}(\Omega_\delta, \partial\Omega) \geq \delta$, $\delta > 0$. If $u \in C^k(\bar{Q})$, $k \geq 3$, and $Lu = f$ in Q, where $f \in C^{k-1}(\bar{Q})$, then*

$$\max_{\bar{Q}_\delta} |D_x^\alpha D_t^j u(t,x)| \leq M \quad \text{for } |\alpha| + j \leq k,$$

where M depends only on δ, $\max\limits_{\bar{Q}} |u|$, the maximum absolute values of the coefficients of the operator L, and the function f and its derivatives up to order $k - 1$.

The derivation of estimates of this type is based on choosing auxiliary functions and applying the maximum principle.

5.3. Integral Estimates. Let $Q = (0,T] \times \Omega$ and $S = [0,T] \times \partial\Omega$. Let u be a solution of Eq. (5.2) of class $C(\bar{Q}) \cap C_{t,x}^{1,2}(Q)$ (i.e., the derivatives $\dfrac{\partial u}{\partial t}$ and $\dfrac{\partial^2 u}{\partial x_i \partial x_j}$ are continuous in Q) with $u = 0$ on S. Then

$$\int_\Omega u^2(t,x)\,dx + \iint_{Q_t} \sum_{j=1}^n \left(\frac{\partial u(\tau,x)}{\partial x_j}\right)^2 d\tau\,dx \leq$$

$$\leq C\left(\int_\Omega u^2(0,x)\,dx + \iint_{Q_t} f^2(\tau,x)\,d\tau\,dx\right),$$

where $0 < t \leq T$ and $Q_t = (0,t) \times \Omega$.

This estimate is obtained upon multiplying both sides of Eq. (5.2) by u and integrating over Q_t. Similarly, multiplying both sides of this equation by $\frac{\partial u}{\partial t}$ and integrating over Q_t, we obtain

$$
\int_\Omega \sum_{j=1}^n \left(\frac{\partial u(t,x)}{\partial x_j}\right)^2 dx + \iint_{Q_t} \left(\frac{\partial u(\tau,x)}{\partial \tau}\right)^2 d\tau\, dx \le
$$

$$
\le C_1 \left(\iint_{Q_t} f^2(\tau,x)d\tau\, dx + \int_\Omega \left[u^2(0,x) + \sum_{j=1}^n \left(\frac{\partial u(0,x)}{\partial x_j}\right)^2 \right] dx \right).
$$

If we square both sides of Eq. (5.2) and integrate over Q_t, we can obtain the estimate

$$
\iint_{Q_t} \left[u^2 + \left(\frac{\partial u}{\partial \tau}\right)^2 + \sum_{j=1}^n \left(\frac{\partial u}{\partial x_j}\right)^2 + \sum_{i,j=1}^n \left(\frac{\partial^2 u}{\partial x_i x_j}\right)^2 \right] d\tau\, dx \le
$$

$$
\le C_2 \left(\iint_{Q_t} f^2(\tau,x)d\tau\, dx + \int_\Omega \left[u^2(0,x) + \sum_{j=1}^n \left(\frac{\partial u(0,x)}{\partial x_j}\right)^2 \right] dx \right).
$$

Differentiating both sides of Eq. (5.2), we can obtain analogous estimates for the higher-order derivatives of the function u. In doing this, of course, one must assume that the coefficients of the equation are sufficiently smooth.

The analogous estimate for the norms of a solution in the spaces $L^p(Q)$ and $W_p^l(Q)$ with $p > 1$ are also true, but a substantially more complicated technique is required to prove them.

5.4. Estimates in Hölder Spaces. Let

$$
\|u\|_\gamma = \sup_Q |u| + \sup_{P_1, P_2 \in Q} \frac{|u(P_1) - u(P_2)|}{d(P_1, P_2)^\gamma}, \quad 0 < \gamma \le 1,
$$

where $d(P_1, P_2) = \left(|x_1 - x_2|^2 + |t_1 - t_2|\right)^{1/2}$, $P_1 = (t_1, x_1)$, $P_2 = (t_2, x_2)$. Let

$$
\|u\|_{2+\gamma} = \|u\|_\gamma + \left\|\frac{\partial u}{\partial t}\right\|_\gamma + \sum_{j=1}^n \left\|\frac{\partial u}{\partial x_j}\right\|_\gamma + \sum_{i,j=1}^n \left\|\frac{\partial^2 u}{\partial x_i \partial x_j}\right\|_\gamma.
$$

Estimates for these norms of a solution make it possible to study the regularity of the solution. They are especially important in the study of quasilinear parabolic equations. In this situation it is important that the constants occurring in the estimates be independent of the smoothness of the coefficients.

As examples we give two theorems.

Theorem 2.77 (Nash, see Il'in, Kalashnikov, and Olejnik 1962). *Let $u(t,x)$ be a solution of the equation*

$$-\frac{\partial u}{\partial t} + \sum_{i,j=1}^{n} \frac{\partial}{\partial x_i}\left(A_{ij}(t,x)\frac{\partial u}{\partial x_j}\right) + \sum_{j=1}^{n} B_j(t,x)\frac{\partial u}{\partial x_j} + C(t,x)u = F(t,x) \quad (5.5)$$

in the region $Q = (0,T) \times \Omega$. Then the inequality $\|u\|_\gamma \leq K$ holds in the region Q_δ (defined in Corollary 2.76), where $0 < \gamma < \frac{1}{4}$ and the constant K depends only on γ, n, δ, and the quantities

$$M = \max_{\bar{Q}} |u|, \quad C_0 = \inf_{\substack{|\xi|=1 \\ (t,x)\in\bar{Q}}} \sum_{i,j=1}^{n} A_{ij}(t,x)\xi_i\xi_j,$$

$$N = \max_{\bar{Q}}\left(\max_{|\xi|=1}\sum_{i,j=1}^{n} A_{ij}(t,x)\xi_i\xi_j + \sum_{j=1}^{n}|B_j| + |C| + |F|\right).$$

Theorem 2.78 (Il'in, Kalashnikov, and Olejnik 1962). *Let $0 < \gamma < 1$ and let the solution $u(t,x)$ of Eq. (5.5) belong to the class $C^{2+\gamma}(\bar{Q})$. Then*

$$\|u\|_{2+\gamma} \leq C\left(\|F\|_\gamma + \|\varphi\|_{2+\gamma}\right),$$

where $\varphi = u\big|_\Gamma$, $\Gamma = \{(t,x) \in \bar{Q} : t = 0 \text{ or } x \in \partial\Omega\}$. The constant C depends only on n, the region Ω, $\inf\limits_{|\xi|=1,\,(t,x)\in Q} \sum\limits_{i,j=1}^{n} A_{ij}(t,x)\xi_i\xi_j$, and $\sum\limits_{i,j=1}^{n} \|A_{ij}\|_\gamma + \sum\limits_{j=1}^{n} \|B_j\|_\gamma + \|C\|_\gamma$. It is assumed here that $\partial\Omega \in C^{2+\gamma}$.

5.5. The Regularity of Solutions of a Second-Order Parabolic Equation.
Consider Eq. (5.2) with smooth (C^∞) coefficients. To each estimate of a solution in the norms of the spaces W_p^l or $C^{k+\alpha}$ of the same type as the estimates described in Sects. 5.3 and 5.4 there corresponds a theorem on the smoothness of the solution: if the known functions occurring in the problem (for example, F and φ in Theorem 2.78) are such that the norms on the right-hand side of the estimate are finite, then the norm on the left-hand side for the solution of the boundary-value problem is finite.

In addition there are local smoothness theorems: if $\mathcal{D}$ is an arbitrary subregion of Q and $\mathcal{D}' \Subset \mathcal{D}$, i.e., the distance from $\overline{\mathcal{D}'}$ to the boundary of the region $\mathcal{D}$ is positive, then $f \in C^{k+\alpha}(\mathcal{D})$ implies that $u \in C^{k+2+\alpha}(\mathcal{D}')$. Here $C^{k+\alpha}(\mathcal{D})$ is the space of functions defined in $\mathcal{D}$ and having finite norm $\|\cdot\|_{k+\alpha}$. It follows from this in particular that if $\mathcal{D}$ is an arbitrary region in $\mathbb{R}^{n+1}$ and $Lu \in C^\infty(\mathcal{D})$, then $u \in C^\infty(\mathcal{D})$. Thus second-order parabolic equations are hypoelliptic.

If the coefficients of Eq. (5.2) and the function Lu are real-analytic in $\mathcal{D}$ on the variables x, then u is also a real-analytic function with respect to x.

On the variable t this property does not hold: even the very simple equation $\dfrac{\partial u}{\partial t} = \dfrac{\partial^2 u}{\partial x^2}$ has nonzero solutions that vanish for $t < 0$ (cf. Sect. 5.6).

5.6. Poisson's Formula. The formula

$$u(t,x) = \frac{1}{(4\pi a^2 t)^{n/2}} \int_{\mathbf{R}^n} \varphi(y) e^{-\frac{|x-y|^2}{4a^2 t}}\, dy$$

gives a solution of the Cauchy problem

$$\frac{\partial u}{\partial t} = a^2 \Delta u \quad \text{for } t > 0, \quad u(0,x) = \varphi(x).$$

(This formula was presented above in Chapter 1 (cf. (1.40)). It can be obtained for $\varphi \in C_0^\infty(\mathbf{R}^n)$ using the Fourier transform on x. If we set

$$v(t,\xi) = \int u(t,x) e^{-ix\cdot\xi}\, dx,$$

the function v is a solution of the Cauchy problem

$$\frac{\partial v}{\partial t} = -a^2 |\xi|^2 v \quad \text{for } t > 0, \quad v(0,\xi) = \tilde{\varphi}(\xi),$$

where $\tilde{\varphi}(\xi) = \displaystyle\int \varphi(x) e^{-ix\cdot\xi}\, dx$. Therefore $v(t,\xi) = \tilde{\varphi}(\xi) e^{-a^2 t |\xi|^2}$ and

$$u(t,x) = (2\pi)^{-n} \int \tilde{\varphi}(\xi) e^{-a^2 t |\xi|^2} e^{ix\cdot\xi}\, d\xi =$$

$$= (2\pi)^{-n} \iint \varphi(y) e^{i(x-y)\xi - a^2 t |\xi|^2}\, d\xi\, dy.$$

The integral

$$\int e^{iz\xi - a^2 t |\xi|^2}\, d\xi = \prod_{j=1}^{n} \int e^{iz_j \xi_j - a^2 t \xi_j^2}\, d\xi_j$$

is easily computed using a deformation of a contour integral in the complex ξ_j-plane (cf. Vladimirov 1967). This gives the desired formula of Poisson.

Having obtained the formula for $\varphi \in C_0^\infty(\mathbf{R}^n)$, we can verify that the same formula gives the solution of the Cauchy problem for each piecewise-continuous function $\varphi(x)$ satisfying the inequality

$$|\varphi(x)| \le C e^{b|x|^2}.$$

The Poisson formula then defines a solution for $0 < t < (4ab)^{-1}$.

Example 2.79. Let $f \in C^\infty(\mathbf{R})$, with $f(t) > 0$ for $t > 0$, $f(t) = 0$ for $t \le 0$ and suppose the inequalities

$$|f^{(k)}(t)| \leq Ck^{k\gamma} \quad \text{for } k = 1, 2, \ldots, \quad \gamma \in (1, 2),$$

hold. (For the existence of such functions cf. Hörmander 1983–1985, Sect. 1.3.) Then for any $R > 0$ the series

$$u(t, x) = \sum_{k=0}^{\infty} f^{(k)}(t) \frac{x^{2k}}{(2k)!}$$

converges uniformly on x for $|x| \leq R$ and defines an infinitely differentiable function in $\mathbf{R}^2$. This function satisfies the equation $\dfrac{\partial u}{\partial t} = \dfrac{\partial^2 u}{\partial x^2}$ and the condition $u(0, x) = 0$. It can be shown that

$$|u(t, x)| \leq C_1 e^{C_2 |x|^{\frac{2}{2-\gamma}}}.$$

Example 2.80. The function

$$u(t, x) = \frac{1}{\sqrt{1 - 4tA}} e^{\frac{Ax^2}{1 - 4tA}} \quad \text{for } 0 \leq t < \frac{1}{4A},$$

is a solution of the Cauchy problem

$$\frac{\partial u}{\partial t} = \frac{\partial^2 u}{\partial x^2} \quad \text{for } t > 0, \quad u(0, x) = e^{Ax^2}.$$

Here $\lim\limits_{t \to \frac{1}{4A}} u(t, x) = \infty$. This solution is the only solution in the class of functions satisfying the inequality

$$|u(t, x)| \leq C e^{C_1 x^2}.$$

Thus the solution of the Cauchy problem in such a class may fail to exist on a large time interval $[0, T]$.

5.7. A Fundamental Solution of the Cauchy Problem for a Second-Order Equation with Variable Coefficients. Poisson's Formula generalizes easily to an equation of the form

$$\frac{\partial u}{\partial t} = \sum_{i,j=1}^{n} a_{ij} \frac{\partial^2 u}{\partial x_i \partial x_j}$$

with constant coefficients if the matrix a_{ij} is positive-definite. Transforming this matrix into diagonal form, one can show easily that the solution of the Cauchy problem is given by the formula

$$u(t, x) = \frac{1}{(4\pi t)^{n/2} \sqrt{A}} \int \varphi(y) \exp\left(- \frac{\sum\limits_{i,j=1}^{n} a^{ij}(x_i - y_i)(x_j - y_j)}{4t} \right) dy,$$

where a^{ij} is the matrix inverse to (a_{ij}) and $A = \det(a_{ij})$.

Now consider Eq. (5.2) with variable coefficients that are defined and continuous for $t \geq 0$ and $x \in \mathbb{R}^n$. Assume that the coefficients a_{ij}, b_i, c satisfy a Hölder condition on x and that the coefficients a_{ij} satisfy a Hölder condition on t.

Definition 2.81. A *fundamental solution* of the Cauchy problem for Eq. (5.2) is a function $Z(t, \tau, x, y)$ defined for $t > \tau$ and $x, y \in \mathbb{R}^n$ such that for each continuous bounded function φ on $\mathbb{R}^n$ the integral

$$u(t, x) = \int_{\mathbb{R}^n} Z(t, \tau, x, y) \varphi(y) \, dy$$

converges, $Lu(t, x) = 0$ for $t > \tau$, and

$$\lim_{t \to \tau + 0} u(t, x) = \varphi(x).$$

It is natural to seek the solution Z in the form

$$Z(t, \tau, x, y) = W(t, \tau, x, y) + \int_\tau^t \int_{\mathbb{R}^n} W(t, s, x, z) \Phi(s, \tau, z, y) dz \, ds,$$

where

$$W(t, \tau, x, y) =$$

$$= \frac{1}{[4\pi(t-\tau)]^{n/2} A(\tau, y)^{1/2}} \exp \left[\frac{-\sum_{i,j=1}^n a^{ij}(\tau, y)(y_i - x_i)(y_j - x_j)}{4(t-\tau)} \right],$$

and the function Φ satisfies the integral equation

$$\Phi(t, \tau, x, y) = L_{t,x} W(t, \tau, x, y) + \int_\tau^t \int_{\mathbb{R}^n} L_{t,x} W(t, s, x, z) \Phi(s, \tau, z, y) \, dz \, ds.$$

This integral equation can be solved by the method of successive approximations (cf. Il'in, Kalashnikov, and Olejnik 1962; Eidelman 1964; Friedman 1964).

Theorem 2.82. *Let the function φ be continuous on $\mathbb{R}^n$ and let f be continuous on $[0, T] \times \mathbb{R}^n$ and satisfy a Hölder condition on x. Let*

$$|\varphi(x)| + |f(t, x)| \leq Ce^{C_1|x|^{2-\varepsilon}}, \quad \varepsilon > 0, \quad 0 \leq t \leq T.$$

Then for $0 < t \leq T$ the function

$$u(t, x) = \int Z(t, 0, x, y) \varphi(y) \, dy = \int_0^t \int Z(t, \tau, x, y) f(\tau, y) \, dy \, d\tau$$

satisfies Eq.(5.2) *and*

$$\lim_{t\to+0} u(t,x) = \varphi(x).$$

Here $|u(t,x)| \le C_2 e^{C_1|x|^{2-\epsilon}}$ *for* $0 \le t \le T$.

5.8. Shilov-Parabolic Systems. Consider the system of equations

$$\frac{\partial u}{\partial t} = P(D_x)u, \quad u = (u_1, \ldots, u_N) \tag{5.6}$$

with constant coefficients.

Definition 2.83. The system (5.6) is called *Shilov-parabolic* if all the roots $\lambda_j(\xi)$ of the equation

$$\det|\lambda I - P(\xi)| = 0$$

satisfy the inequalities

$$\operatorname{Re}\lambda_j(\xi) \le C_1 - C_2|\xi|^h, \quad C_2 > 0, \quad h > 0.$$

The number h is called the index of parabolicity of the system.

Every system of the form (5.6) that is Petrovskij-parabolic is also Shilov-parabolic. Let

$$\Lambda(\zeta) = \max_j \operatorname{Re}\lambda_j(\zeta).$$

For complex ζ the function $\Lambda(\zeta)$ grows polynomially as $|\zeta| \to \infty$, so that $|\Lambda(\zeta)| \le C(1+|\zeta|)^{p_0}$. If the system (5.6) is Petrovskij-parabolic, then $p_0 = h$.

Let $Q(t,\zeta) = e^{tP(\zeta)}$. There exist K and μ such that $\mu \ge 1 - (p_0 - h)$ and the estimate

$$|Q(t,\zeta)| \le Ce^{-at|\operatorname{Re}\zeta|^h}$$

holds in the region

$$\{\zeta : |\operatorname{Im}\zeta| < K(1 + |\operatorname{Re}\zeta|)^{\mu}\}.$$

The supremum of such numbers μ is called the *genus* of the system.

Theorem 2.84 (cf. Gel'fand and Shilov 1958–1959). *If* $\mu > 0$, *the Cauchy problem for the system* (5.6) *with initial condition* $u(0,x) = \varphi(x)$ *has a solution when* $|\varphi(x)| \le Ce^{b|x|^{p_1}}$, *where* $p_1 = p_0(p_0 - \mu)^{-1}$ *and* $|u(t,x)| \le Ce^{(b+\delta)|x|^{p_1}}$.

For $\mu \le 0$ *a solution exists if* $|\varphi(x)| \le C_\epsilon e^{\epsilon|x|^{p_2}}$ *for all* $\epsilon > 0$, *where* $p_2 = \dfrac{h}{h - \mu}$. *When this happens,* $|u(t,x)| \le C'_\epsilon e^{\epsilon|x|^{p_2}}$.

If the system (5.6) *is Petrovskij-parabolic, we have* $\mu = 1$, $h = p_0 = m$, *and the solution exists when*

$$|\varphi(x)| \leq Ce^{b|x|^{p'}}, \quad \text{where } p' = \frac{p_0}{p_0 - 1}.$$

When this happens, we have for all $\delta > 0$

$$|u(t, x)| \leq C_\delta e^{(b+\delta)|x|^{p'}}, \quad 0 \leq t \leq T.$$

Example 2.85. Let $m > 2$ and

$$\frac{\partial u}{\partial t} = \frac{\partial^2 u}{\partial x^2} + iD_x^m u.$$

Then $\lambda(\zeta) = -\zeta^2 + i\zeta^m$, $\mathrm{Re}\,\lambda(\xi) = -\xi^2$. It can be verified that

$$h = 2, \quad \mu = 3 - m \leq 0, \quad p_0 = m, \quad p_1 = \frac{2}{m-1}.$$

Therefore there exists a solution of the Cauchy problem if

$$|\varphi(x)| \leq C_\varepsilon e^{\varepsilon |x|^{\frac{2}{m-1}}}$$

for any $\varepsilon > 0$. When this happens,

$$|u(t, x)| \leq C_\varepsilon^1 e^{\varepsilon |x|^{\frac{2}{m-1}}}.$$

5.9. Systems with Variable Coefficients. Consider the Cauchy problem for a parabolic system of equations with smooth bounded coefficients:

$$\frac{\partial u}{\partial t} = L(t, x, D_x)u + f(t, x) \quad \text{for } t > 0, \quad u(0, x) = \varphi(x). \tag{5.7}$$

Here $L(t, x, D_x)$ is a matrix of differential operators of order m. Suppose that the system is uniformly parabolic, i.e., the roots λ of the equation

$$|\lambda I - L_0(t, x, \xi)| = 0$$

are such that $\mathrm{Re}\,\lambda(t, x, \xi) \leq -\delta < 0$ for $\xi \in \mathbb{R}^n$, $|\xi| = 1$ and $(t, x) \in H$, where $H = [0, T] \times \mathbb{R}^n$.

Theorem 2.86 (Il'in, Kalashnikov, and Olejnik 1962). *Suppose the coefficients of the operator L are continuous and bounded in H and satisfy a Hölder condition with respect to x in H and that the leading coefficients are uniformly continuous in H with respect to t. Suppose further that $\varphi \in C(\mathbb{R}^n)$, $f \in C(H)$, and f satisfies a Hölder condition with respect to x uniformly on each bounded subset of H. Finally suppose $|f(t, x)| \leq Ae^{a|x|^q}$ in H and $|\varphi(x)| \leq Ae^{a|x|^q}$ in*

$\mathbb{R}^n$, where $q = m(m-1)^{-1}$. Then the Cauchy problem (5.7) has a solution u for $0 \le t \le t_0 = \min\left(T, (c_0/a)^{m-1}\right)$. When this happens,

$$|u(t,x)| \le Ce^{a_0|x|^q} \quad \text{for } 0 \le t \le t_0,$$

where the constants C, a, and c_0 are independent of T.

If the coefficients of the system and their derivatives of orders up to $m+1$ are continuous and bounded in H, then the Cauchy problem (5.7) can have only one solution in the class of functions u satisfying the relation

$$\int_H |u(t,x)|e^{-k|x|^q} dx\, dt < \infty.$$

The proof of the existence of the solution in Theorem 2.86 follows the same scheme as in Sect. 5.8. We first use the Fourier method to study the Cauchy problem for a system of equations with coefficients independent of x. Then a fundamental solution of the Cauchy problem (5.7) is found as the solution of a certain integral equation by the method of successive approximations. In this situation it can be shown that the estimates

$$|D_x^\alpha \Gamma(t,\tau,x,y)| \le C|t-\tau|^{-(n+|\alpha|)/m} e^{-C_1\left(\frac{|x-y|^m}{t-\tau}\right)^{\frac{1}{(m-1)}}}.$$

hold for the fundamental solution $\Gamma(t,\tau,x,y)$ with $0 \le |\alpha| \le m$, where $C_1 = \text{const} > 0$. The uniqueness of the solution is proved using Holmgren's principle.

5.10. The Mixed Boundary-Value Problem. Consider the parabolic equation of order m

$$\frac{\partial u}{\partial t} = L(t,x,D_x)u + f(t,x)$$

with smooth coefficients in the region $Q = (0,T] \times \Omega$. The mixed boundary-value problem is posed as follows: find a solution of this equation satisfying the initial condition

$$u(0,x) = \varphi(x), \quad x \in \Omega,$$

with the boundary condition

$$B_j(t,x,D_x)u = g_j(t,x), \quad 0 < t \le T, \quad x \in \partial\Omega, \quad j = 1,\ldots,k. \tag{5.8}$$

Example 2.87. For the second-order equation (5.2) the boundary-value problem with boundary condition $u = g$ for $0 < t \le T$, $x \in \partial\Omega$, is well-posed. The uniqueness of the solution of this problem and the continuous dependence on the data of the problem follow from the maximum principle or from the integral estimates given in Sect. 5.3.

Condition (5.8) defines a well-posed boundary-value problem if $k = m/2$ and a certain condition is satisfied, which we now state. Let x be an arbitrary point of $\partial\Omega$, $\zeta(x)$ an arbitrary vector tangent to $\partial\Omega$ at the point x, and $\nu(x)$ the unit vector normal to $\partial\Omega$ at the point x. Consider the equation

$$\Phi(t, \tau, x, \rho) \equiv \tau - L_0(t, x, \zeta(x) + \rho\nu(x)) = 0$$

with respect to the variable ρ. It follows from the parabolicity that this equation has no real roots and exactly $m/2$ roots with positive imaginary part for $\operatorname{Re}\tau \geq -\delta|\zeta|^p$ and $|\tau| + |\zeta| \neq 0$. Let

$$\Phi_+(t, x, \tau, \rho) = \prod_{j=1}^{m/2} [\rho - \rho_j(t, x, \tau)],$$

where the product contains the roots ρ_j for which $\operatorname{Im}\rho_j > 0$.

The Complementarity Condition. At each point $(t, x) \in (0, T] \times \partial\Omega$ and for each tangent vector $\zeta(x)$ the functions

$$B_{j0}(t, x, \zeta(x) + \rho\nu(x)), \quad j = 1, \ldots, k,$$

are linearly independent modulo the polynomial $\Phi_+(t, x, \tau, \rho)$ for $\operatorname{Re}\tau \geq -\delta|\zeta|^p$ and for $|\tau| + |\zeta| \neq 0$, $2k = m$.

Here, as usual, $B_{j0}(t, x, \xi)$ is the leading part of the polynomial $B_j(t, x, \xi)$.

Theorem 2.88 (Solonnikov 1965). *If the complementarity condition is met and the functions f, φ, and g_j are infinitely differentiable, then the boundary-value problems with the conditions (5.8) has a unique solution. This solution is infinitely differentiable for $0 < t \leq T$ and $x \in \bar{\Omega}$. Conversely if such a solution exists and is unique for every set of C^∞-functions f, φ, g_j, then the complementarity condition holds.*

In the case when the coefficients of the operators L and B_j are independent of t, $g_j = 0$ for $j = 1, \ldots, m$, and the operator L with boundary conditions $B_j(x, D)u = 0$ for $j = 1, \ldots, m$ is symmetric, the mixed boundary-value problem can be solved by the method of separation of variables. When this is done, the solution is found in the form of a series

$$u(t, x) = \sum_{k=1}^{\infty} v_k(t) X_k(x),$$

where $\{X_k(x)\}$ is an orthonormal system of eigenfunctions of the operator L (i.e., $LX_k = \lambda_k X_k$) satisfying the conditions

$$B_j(x, D)X_k(x) = 0, \quad j = 1, \ldots, m/2; \quad k = 1, 2, \ldots; \quad x \in \partial\Omega.$$

The functions $v_k(t)$ are found from the equations

$$\dot{v}_k(t) = \lambda_k v_k(t) + f_k(t), \quad k = 1, 2, \ldots,$$

where $f_k(t) = \int\limits_\Omega f(t,x)X_k(x)\,dx$, and the initial conditions

$$v_k(0) = \int\limits_\Omega \varphi(x)X_k(x)\,dx, \quad k = 1,2,\ldots .$$

In the study of the mixed boundary-value problem frequent use is made of the Green's function $G(t,\tau,x,y)$, which makes it possible (in the case when $g_j = 0$) to represent the solution of this problem in the form

$$u(t,x) = \int\limits_\Omega G(t,0,x,y)\varphi(y)\,dy + \int_0^t \int\limits_\Omega G(t,\tau,x,y)f(\tau,y)\,dy\,d\tau.$$

The function G has a singularity at $t = \tau$, $x = y$ of the same type as the fundamental solution $Z(t,\tau,x,y)$ of the operator $\frac{\partial}{\partial t} - L$ (cf. Solonnikov 1965; Eidelman 1964; Friedman 1964).

5.11. Stabilization of the Solutions of the Mixed Boundary-Value Problem and the Cauchy Problem. In the applications of the theory of parabolic equations the behavior of their solutions as $t \to +\infty$ is frequently an important question.

The most thoroughly studied question is that of stabilization of the solutions of the second-order equation (5.2).

Theorem 2.89 (Il'in, Kalashnikov, and Olejnik 1962). *Let the coefficients of Eq. (5.2) and the function f be uniformly bounded; let $c(t,x) \le -c_0$, where $c_0 = \text{const} > 0$, and $f(t,x) \to 0$ as $t \to +\infty$ uniformly on x; finally, let the function $u(0,x)$ be bounded. If u is a solution of the Cauchy problem in $[0,\infty) \times \mathbb{R}^n$ or the boundary-value problem in $[0,\infty) \times \Omega$ with boundary condition of the form*

$$u = 0 \quad \text{or} \quad \frac{\partial u}{\partial \nu} + au = 0 \quad \text{for } x \in \partial\Omega,$$

where ν is a direction in the space $(x_1,\ldots,x_n)$ forming an acute angle with the exterior normal to $\partial\Omega$, $a \ge 0$, then

$$u(t,x) \to 0 \quad \text{as } t \to +\infty \quad \text{uniformly on } x.$$

The hypothesis on the function $c(t,x)$ can be dropped if the boundary condition has the form $u = 0$ or $\frac{\partial u}{\partial \nu} + au = 0$, where $a \ge a_0 = \text{const} > 0$. But if the boundary condition has the form $\frac{\partial u}{\partial \nu} = 0$, the hypothesis on c is essential. For example if $c(t,x) \equiv 0$, the function $u \equiv 1$ satisfies Eq. (5.2) and the condition $\frac{\partial u}{\partial \nu} = 0$ on $[0,\infty) \times \Omega$.

The behavior of the coefficients of the equation and the function $u(0, x)$ as $|x| \to \infty$ plays an essential role in the study of the Cauchy problem.

Theorem 2.90 (Il'in, Kalashnikov, and Olejnik 1962). *Let u be a solution of Eq. (5.2) and*

$$\sum_{j=1}^{n} \left(a_{jj}(t, x) + b_j(t, x) \right) \geq c_0 > 0.$$

If $u(0, x) \to 0$ as $|x| \to \infty$, then $u(t, x) \to 0$ as $t \to +\infty$ uniformly on x.

For equations and systems of order $m > 2$ only sufficient conditions of a rather complicated nature for stabilization exist (cf., for example, Eidelman 1964, Sect. 3.8).

Theorem 2.91 (Bers, John, and Schechter 1964). *Let $u(t, x)$ be a solution of the Cauchy problem for the parabolic equation $\dfrac{\partial u}{\partial t} = \sum\limits_{|\alpha|=m} a_\alpha D^\alpha u$ with constant coefficients, and suppose u is representable in the Poisson integral form. A necessary and sufficient condition for $u(t, x)$ to tend to zero uniformly in x as t tends to $+\infty$ is that*

$$\lim_{N \to \infty} N^{-n} \int_{|x-y|<N} u(0, x) \, dx = 0$$

uniformly in $y \in \mathbf{R}^N$.

§6. General Evolution Equations

6.1. The Cauchy Problem. The Hadamard and Petrovskij Conditions. Consider the system of differential equations

$$\frac{\partial u}{\partial t} = \mathcal{A}(t, x, D_x) u + f(t, x), \tag{6.1}$$

where $u = (u_1, \ldots, u_N)$, $f = (f_1, \ldots, f_N)$, and $\mathcal{A}$ is an $N \times N$ matrix whose entries are differential operators of order m with infinitely differentiable coefficients.

Definition 2.92. The Cauchy problem for the system of equations (6.1) with $f = 0$ is *uniformly well-posed* for $0 \leq t \leq T$ if for each vector φ of $\left(H^\infty(\mathbf{R}^n) \right)^N$, where $H^\infty(\mathbf{R}^n) = \bigcap\limits_s H^s(\mathbf{R}^n)$, and for an arbitrary $t_0 \in [0, T)$ the system has a unique solution for $t_0 \le t \le T$ of the class $C^m\left([t_0, T]; \left(H^\infty(\mathbf{R}^n)\right)^N\right)$ satisfying the condition

$$u(t_0, x) = \varphi(x),$$

which depends continuously on φ. This last condition means that for any $\varepsilon > 0$ and $l \in \mathbf{N}$ there exist $\delta > 0$ and $p \in \mathbf{N}$ such that if $\|\varphi\|_{H^p} < \delta$, then $\max\limits_{t_0 \le t \le T} \|u(t, \cdot)\|_{H^l} < \varepsilon$.

Theorem 2.93 (Hadamard, cf. Mizohata 1973). *Let the matrix A be independent of t and x, $A = A(D_x)$. The Cauchy problem for the system (6.1) is uniformly well-posed for $0 \le t \le T$ if and only if the roots $\lambda = \lambda_j(\xi)$ of the equation*

$$\det(\lambda I - A(\xi)) = 0$$

satisfy the condition

$$\operatorname{Re}\lambda_j(\xi) \le p\ln(1 + |\xi|) + C, \quad j = 1, \ldots, N; \quad \xi \in \mathbf{R}^n,$$

where the constants p and C depend only on T.

Later Gårding showed that this condition is equivalent to the seemingly stronger condition

$$\operatorname{Re}\lambda_j(\xi) \le C_1, \quad j = 1, \ldots, N, \quad \xi \in \mathbf{R}^n.$$

Petrovskij obtained a necessary and sufficient condition for the Cauchy problem to be uniformly well-posed in the case when $A = A(t, D_x)$. This condition has the following form:

(A) Let $v_j(t, t_0, \xi)$ be a solution of the Cauchy problem

$$\frac{dv_j}{dt} = A(t, \xi)v_j, \quad v_j(t_0, t_0, \xi) = (0, \ldots, 0, 1, 0, \ldots, 0),$$

where the 1 is in the jth position. Then

$$|v_j(t, t_0, \xi)| \le C(1 + |\xi|)^p$$

for $j = 1, \ldots, N$ and $t_0 \le t \le T$, where p and C depend only on T.

Example 2.94. (Kreiss). Let

$$A(t) = \begin{pmatrix} \sin t & \cos t \\ -\cos t & \sin t \end{pmatrix}, \quad B = \begin{pmatrix} 1 & 1 \\ 0 & 1 \end{pmatrix}, \quad C = \begin{pmatrix} 0 & -1 \\ 1 & 0 \end{pmatrix}.$$

Consider the system of equations

$$\frac{\partial u}{\partial t} = A(t)BA(t)^* \frac{\partial u}{\partial x}.$$

We remark that the roots of the characteristic equation in this case are pure imaginaries, so that the Hadamard condition holds for each t. If we make the change of variable $u = A(t)v$, we obtain a system of equations with constant coefficients

$$\frac{\partial v}{\partial t} = B \frac{\partial v}{\partial x} - Cv,$$

for which the roots of the equation $\det(\lambda I - iB\xi + C) = 0$ are

$$i\xi \pm \sqrt{-1 - i\xi},$$

and, by Hadamard's theorem, the Cauchy problem is not uniformly well-posed. At the same time it is obvious that the Cauchy problem for the latter system is equivalent to the Cauchy problem for the original system. This explains why condition (A) in Petrovskij's theorem cannot be expressed in terms of the roots λ_j, as was done in Hadamard's theorem.

6.2. Application of the Laplace Transform. Let B be a Banach space and let $u : \mathbb{R} \to B$ satisfy $u(t) = 0$ for $t < 0$. Assume that the function $u(t)$ is such that $\|u(t)\| \leq Ce^{\alpha t}$ and that u is of bounded variation in the norm on each bounded closed interval, i.e.,

$$\sum_{i=1}^{k} \|u(t_i) - u(t_{i-1})\| \leq C_1$$

for each partition of the closed interval $[a, b]$:

$$a = t_0 < t_1 < \cdots < t_k = b.$$

The constant C_1 may depend on $[a, b]$.

The *Laplace transform* of the function $u(t)$ is the function

$$v(p) = \int_0^\infty e^{-pt} u(t)\, dt$$

with values in B defined for $\operatorname{Re} p > \alpha$. Here (cf., for example, Mizohata 1973)

$$\frac{u(t+0) + u(t-0)}{2} = \lim_{N \to \infty} \frac{1}{2\pi i} \int_{x-iN}^{x+iN} v(p)e^{pt}\, dp, \quad x > \alpha,$$

where the convergence is understood to be in the norm of the space B. This convergence is uniform on each closed bounded interval contained in the interior of the interval of continuity of the function $u(t)$.

Now consider the differential equation

$$\frac{du}{dt} = Au(t), \tag{6.2}$$

where A is a closed linear operator with domain of definition $\mathcal{D}(A)$ dense in $\mathcal{B}$. Assume that this equation has a solution $u(t)$ for any initial conditions $u(0) = u_0 \in \mathcal{D}(A)$ which is defined for all $t > 0$, and that

$$\|u(t)\| \le Ce^{\alpha t}\|u(0)\| \quad \text{for } t \ge 0.$$

Set $u(t) = 0$ for $t < 0$ and

$$v(p) = \int_0^\infty e^{-pt}u(t)\,dt, \quad \operatorname{Re}p > \alpha.$$

It then follows from (6.2) that

$$Av(p) = pv(p) - u(0).$$

Theorem 2.95. *If the resolvent* $(pI - A)^{-1}$ *exists for* $\operatorname{Re}p > \alpha$, *then*

$$u(t) = \lim_{N \to +\infty} \frac{1}{2\pi i} \int_{x-iN}^{x+iN} e^{pt}(pI - A)^{-1}u(0)\,dp$$

for $t > 0$, $x > \alpha$.

Set $u(t) = T_t u_0$, so that T_t is a bounded linear operator defined on $\mathcal{D}(A)$. Since by hypothesis $\|T_t u_0\| \le Ce^{\alpha t}\|u_0\|$, this operator can be extended to $\mathcal{B}$ and $\|T_t\| \le Ce^{\alpha t}$ for $t > 0$. When this is done, we have $T_{t+s} = T_t T_s$ for $t \ge 0$, $s \ge 0$, and $\lim_{t \to +0} T_t u_0 = u_0$, $u_0 \in \mathcal{B}$, so that the family of operators T_t forms a semigroup. This semigroup is strongly continuous, i.e., the function $T_t u_0$ is continuous in t for $t \ge 0$ and for $u_0 \in \mathcal{D}(A)$

$$\frac{d}{dt}T_t u_0\Big|_{t=0} = \lim_{t \to +0} \frac{T_t - I}{t} u_0 = Au_0.$$

The operator A is called the *infinitesimal generator* of the semigroup T_t.

If T_t is an arbitrary strongly continuous semigroup of bounded operators on $\mathcal{B}$, then the infinitesimal generator operator A is again defined by the equality

$$\lim_{t \to +0} \frac{T_t - I}{t} u = Au$$

and its domain of definition $\mathcal{D}(A)$ consists of those u for which this limit exists. We note certain properties of the operator A:

(1) The operator A is closed and the subspace $\mathcal{D}(A)$ is dense in $\mathcal{B}$.

(2) Let $\|T_t\| \le Ce^{\alpha t}$ and $T_t(e^{-pt})u = \int_0^\infty e^{-pt}T_t u\,dt$ for $\operatorname{Re}p > \alpha$ and $u \in \mathcal{B}$. Then $T_t(e^{-pt})u \in \mathcal{D}(A)$ and $(pI - A)T_t(e^{-pt})u = u$ for all $u \in \mathcal{B}$.

(3) If $u \in \mathcal{D}(A)$, then $T_t(e^{-pt})(pI - A)u = u$.

Thus the resolvent $(pI - A)^{-1}$ of the operator A is defined for $\operatorname{Re}p > \alpha$ and coincides with the operator $T_t(e^{-pt})$. In addition

$$\|(pI - A)^{-m}\| \leq \frac{C}{(\operatorname{Re} p - \alpha)^m}, \quad m = 1, 2, \ldots .$$

Using the inversion formula for the Laplace transform, one can show that

$$T_t u = \lim_{N \to +\infty} \frac{1}{2\pi i} \int_{x-iN}^{x+iN} e^{pt}(pI - A)^{-1} u \, dp, \quad x > \alpha.$$

The existence theorem for a solution of the Cauchy problem for Eq. (6.1) follows immediately from the following *Hille-Yosida theorem* (cf. Yosida 1965).

Theorem 2.96. *Let A be a closed operator in a Banach space B with domain of definition dense in B, and let its resolvent $(\lambda I - A)^{-1}$ be defined for real $\lambda > \alpha$ and satisfy*

$$\|(\lambda I - A)^{-1}\| \leq \frac{1}{\lambda - \alpha} \quad \text{for } \lambda > \alpha.$$

Then there exists a unique semigroup T_t with A as its infinitesimal generator. This semigroup satisfies $\|T_t\| \leq Ce^{\alpha t}$.

By Duhamel's principle this theory guarantees the solvability of the Cauchy problem for the inhomogeneous equation: if the hypotheses of the preceding theorem hold and the functions $f(t)$ and $Af(t)$ are continuous for $0 \leq t \leq T$, then for any $u_0 \in \mathcal{D}(A)$ there exists a unique solution $u \in C^1([0, t]; B)$ such that

$$\frac{du}{dt} = Au(t) + f(t), \quad u(0) = u_0,$$

and $u(t) = T_t u_0 + \int_0^t T_{t-s} f(s) \, ds.$

6.3. Application of the Theory of Semigroups. Let A be a self-adjoint operator on a Hilbert space H and $A = \int\limits_{-\infty}^{\infty} \lambda \, dE_\lambda$ its spectral decomposition. Then the group of unitary operators

$$e^{itA} u = \int_{-\infty}^{\infty} e^{it\lambda} \, dE_\lambda, \quad u \in H,$$

is defined, making it possible to obtain an existence and uniqueness theorem for the Cauchy problem

$$\frac{du}{dt} = iAu + f(t), \quad u(0) = u_0.$$

For example these conditions are satisfied by the operator $A = \Delta$ in $\overset{\circ}{H}_1(\Omega)$, where Ω is a region in $\mathbb{R}^n$, and we obtain a theorem on the solvability of the mixed boundary-value problem for the Schrödinger equation

$$\frac{\partial u}{\partial t} = ia^2 \Delta u + f(t, x), \quad x \in \Omega, \quad 0 < t \leq T;$$
$$u(0, x) = u_0(x), \quad x \in \Omega; \quad u = 0 \quad \text{on } \partial\Omega \times [0, T].$$

In the case under consideration the operator e^{itA} is unitary. The converse assertion holds as well (*Stone's theorem*): every strongly continuous one-parameter group of unitary operators has the form e^{itA}, where A is a self-adjoint operator.

It is remarkable that the theory of semigroups makes it possible to solve the Cauchy problem for an evolution equation with variable coefficients

$$\frac{du(t)}{dt} = A(t)u(t) + f(t), \quad u(0) = u_0. \tag{6.3}$$

Assume that the following conditions hold:

1. $\mathcal{D}(A(t))$ is independent of t for $0 \leq t \leq T$ and dense in $\mathcal{B}$.

2. For $\lambda > 0$ the resolvent $(I - \lambda A(t))^{-1}$ exists and its norm does not exceed 1.

3. The operator $B(t, s) = (I - A(t))(I - A(s))^{-1}$ is uniformly bounded.

4. For some s and every partition $0 = t_0 < t_1 < \cdots < t_k = T$ the inequality

$$\sum_{j=0}^{k-1} \|B(t_{j+1}, s) - B(t_j, s)\| \leq C$$

holds.

5. For some s the weak derivative $\dfrac{\partial B(t, s)}{\partial t}$ exists and is strongly continuous in t.

Theorem 2.97 (Kato, cf. Yosida 1965). *If the conditions 1–5 hold and $u_0 \in \mathcal{D}(A(t))$ and $f(t) \in \mathcal{D}(A(t))$ for $0 \leq t \leq T$, then the problem* (6.3) *has a unique solution*

$$u(t) = U(t, 0)u_0 + \int_0^t U(t, s)f(s)\, ds,$$

where

$$U(t, s)u_0 = \lim_{\max |t_{j+1} - t_j| \to 0} \left[\prod_{j=0}^{k-1} e^{(t_{j+1} - t_j)A(t_j)} \right] u_0,$$

$$s = t_0 < t_1 < \ldots < t_k = t.$$

The hypotheses of this last theorem hold, in particular, in the case when $A(t)$ is a second-order elliptic operator with smooth coefficients, $B = L^2(\Omega)$,

Ω is a region with smooth boundary, and $u_0 \in \mathcal{D}(A(t))$ (the latter by definition means that $u_0 \in H^2(\Omega)$ and $u = 0$ on $\partial\Omega$).

6.4. Some Examples. Consider the Cauchy problem for the system of differential equations with constant coefficients

$$\frac{\partial u}{\partial t} = A(D)u, \quad u(0, x) = \varphi(x).$$

Here $u = (u_1, \ldots, u_N)$ and $A(D)$ is an $N \times N$ matrix whose entries are differential operators of order m. Consider the question of the conditions under which the solution $u(t, x)$ of this problem is an analytic function of t in a sector containing the positive semiaxis $\operatorname{Re} t > 0$, i.e., for $\operatorname{Re} t > 0$ and $|\arg t| \leq \alpha$, where $\alpha > 0$. The methods of the theory of semigroups make it possible to show that a necessary and sufficient condition for this is that the eigenvalues $\lambda_i(p)$ of the matrix $A(p)$ satisfy the inequality

$$\operatorname{Re} \lambda_i(p) \leq -|\operatorname{Im} \lambda_i(p)| \tan \alpha + b, \quad b = \text{const},$$

(cf. Krejn 1967; Eidelman 1964).

Consider the mixed boundary-value problem

$$\begin{cases} \dfrac{\partial u}{\partial t} = \displaystyle\sum_{j=1}^n A_j(x)\dfrac{\partial u}{\partial x_j} + B(x)u, & 0 \leq t \leq T,\ x \in \Omega, \\ u(0, x) = \varphi(x) & \text{for } x \in \Omega, \\ \displaystyle\int_\Gamma u w_j(x)\, dx = 0, & \text{for } j = 1, \ldots, l, \end{cases}$$

where Ω is a region with a smooth boundary Γ and A_j and B are $N \times N$ matrices with smooth entries. We assume that $A_j(x)$ are symmetric real matrices.

In the study of the well-posedness of this problem by the methods of the theory of semigroups the following conditions arise naturally:

I. $B + B^* - \displaystyle\sum_{j=1}^n \frac{\partial A_j}{\partial x_j} \leq 0$.

II. The rank of the matrix $A_\nu(x) = \displaystyle\sum_{j=1}^n A_j(x)\nu_j(x)$, where $(\nu_1, \ldots, \nu_n)$ is the unit exterior normal vector to the boundary Γ, does not change as x traverses the boundary.

III. The space $N(x)$ of vectors of $\mathbf{R}^n$ that are orthogonal to the vectors $w_1(x), \ldots, w_l(x)$ is the maximal space on which the matrix $A_\nu(x)$ is nonpositive.

Theorem 2.98 (Lax and Phillips 1967). *Conditions* I–III *are sufficient for the problem to be uniformly well-posed in the space* $L^2(\Omega)$. (Lax and Phillips, cf. also Krejn 1967).

§7. Exterior Boundary-Value Problems and Scattering Theory [4]

7.1. Radiation Conditions. In this section we shall consider the simplest equations on the entire space or in the exterior of a bounded region. In the study of elliptic equations in $\mathbf{R}^n$ (or in unbounded regions) the point at infinity plays the role of the boundary of the region, and in order to get a well-posed problem it is necessary to impose some conditions on the behavior of the solutions as $r = |x| \to \infty$. These conditions can be given in the form of estimates of solutions or their asymptotic behavior as $r \to \infty$, or the requirement that the solutions belong to certain function spaces that limit the possibilities for the behavior of the solutions in a neighborhood of infinity. The form of these conditions depends essentially on the behavior of the coefficients of the equation in a neighborhood of infinity, including the behavior of the nonleading terms.

Example 2.99. Let $D = \dfrac{1}{i}\dfrac{\partial}{\partial x}$, and let $P(D)$ be an elliptic operator with constant coefficients in $\mathbf{R}^n$ and $P(\xi) \neq 0$ for all $\xi \in \mathbf{R}^n$. Then for any $f \in L^2(\mathbf{R}^n)$ the equation $P(D)u = f$ has a unique solution in the space $L^2(\mathbf{R}^n)$, and the operator $P(D)$ gives an isomorphism of the Sobolev spaces $H^{s+m}(\mathbf{R}^n)$ and $H^s(\mathbf{R}^n)$, where m is the order of the operator P, an isomorphism of the Schwartz space $S(\mathbf{R}^n)$ into itself, and an isomorphism of the space of distributions $S'(\mathbf{R}^n)$ into itself. These assertions are easily verified using the Fourier transform. In particular, they hold for the equation $(\Delta + C)u = f$, if $C < 0$ or $\operatorname{Im} C \neq 0$.

Example 2.100. The equation

$$\Delta u = f, \quad x \in \mathbf{R}^3, \tag{7.1}$$

has a unique solution in the class of functions vanishing at infinity for each $f \in C_0^\infty$. This solution is given by the convolution $u = -\dfrac{1}{4\pi r} * f$. In contrast to the preceding example, the equation (7.1) in $L^2(\mathbf{R}^3)$ can be not solvable for some $f \in C_0^\infty$.

---

[4] This section was written by B. R. Vajnberg.

Consider now *Helmholtz' equation* in $\mathbf{R}^3$

$$(\Delta + k^2)u = f, \quad f \in L_a^2, \quad k > 0. \tag{7.2}$$

Here $a > 0$ is any fixed constant and L_a^2 is the space of functions of $L^2(\mathbf{R}^3)$ that vanish for $r > a$. We recall that Eq. (7.2) is satisfied by the amplitude of steady-state vibrations caused by a periodic force, i.e., if w is a solution of the wave equation

$$\frac{1}{c^2}\frac{\partial^2 w}{\partial t^2} - \Delta w = -f(x)e^{-i\omega t}, \quad \omega > 0, \tag{7.3}$$

and $w(t, x) = u(x)e^{-i\omega t}$, the amplitude of u will be a solution of Eq. (7.2) with $k = \omega/c$. The following two fundamental solutions of Eq. (7.2) are easily found (for example, if they are sought in the form of spherically symmetric functions)

$$E_{\pm} = -\frac{1}{4\pi r}e^{\pm ikr}.$$

The functions $u_{\pm} = E_{\pm} * f$, where $*$ denotes convolution, are solutions of Eq. (7.2) and have order $O(r^{-1})$ as $r \to \infty$. Thus in the class of functions having order $O(r^{-1})$ at infinity the solution of Eq. (7.2) is not unique. In addition, for any $g \in C^{\infty}$ the functions

$$u = \int\limits_{|\xi|=k} g(\xi)e^{i(\xi, x)}\, dS$$

are solutions of the homogeneous equation (7.2) (this can be verified by substituting into the equation) and have order $O(r^{-1})$ as $r \to \infty$ (this is proved using the stationary phase method). Thus Eq. (7.2) has many solutions such that $u = O(r^{-1})$ as $r \to \infty$. On the other hand, in the class of functions that behave like $o(r^{-1})$ as $r \to \infty$ Eq. (7.2) in general has no solutions.

We have arrived at the result that the conditions distinguishing a unique solution of Eq. (7.2) must be more "delicate" than merely prescribing the order of decrease of the function at infinity. Such conditions are the *Sommerfeld radiation conditions*

$$u = O(r^{-1}), \quad \frac{\partial u}{\partial r} - iku = o(r^{-1}), \quad r \to \infty, \tag{7.4}$$

or

$$u = O(r^{-1}), \quad \frac{\partial u}{\partial r} + iku = o(r^{-1}), \quad r \to \infty. \tag{7.5}$$

Equation (7.2) is uniquely solvable both in the class of functions satisfying conditions (7.4) and in the class satisfying (7.5). Indeed, it is easily verified that the solution u_+ satisfies conditions (7.4) and u_- satisfies (7.5) (cf. the derivation of formula (7.10) below). We shall prove that they are unique. Let u be a solution of the homogeneous equation (7.2) satisfying conditions (7.4), x_0 an arbitrary point, and $R > |x_0|$. Obviously

$$u(x_0) = \int\limits_{|x|<r} \left[u(x)(\Delta + k^2)E_+(x - x_0) - E_+(x - x_0)(\Delta + k^2)u(x) \right] dx =$$

$$= \int\limits_{|x|=R} \left(\frac{\partial E_+}{\partial r} u - E_+ \frac{\partial u}{\partial r} \right) dS =$$

$$= \int\limits_{|x|=R} \left[\left(\frac{\partial E_+}{\partial r} - ikE_+ \right) u - E_+ \left(\frac{\partial u}{\partial r} - iku \right) \right] dS = \int\limits_{|x|=R} o(r^{-2})\, dS.$$

This last equality follows because for fixed x_0 the function $E_+(x - x_0)$, like the function u, satisfies conditions (7.4). Passing to the limit as $R \to \infty$, we find that $u(x_0) = 0$.

We now explain the physical meaning of conditions (7.4) and (7.5). The functions

$$v_{\pm} = \frac{1}{r}e^{i(\pm kr - \omega t)}, \quad \omega > 0,$$

describe spherical waves traveling from the origin to infinity (v_+) or from infinity to the origin (v_-). We note that the absolute value of the amplitude of a spherical wave must be proportional to r^{-1} in order for its energy to be conserved as it propagates. Conditions (7.4) hold for outgoing spherical waves but not for incoming spherical waves. Conversely conditions 7.5 distinguish incoming spherical waves. If the time dependence is defined by the factor $e^{i\omega t}$, then conditions (7.4) will distinguish incoming waves and (7.5) outgoing waves.

We now consider the simplest equations of type (7.2) with variable coefficients:

$$[\Delta + k^2 + v(x)]u = f \in L_a^2, \quad x \in \mathbb{R}^3; \quad u \in H^2_{\text{loc}}(\mathbb{R}^3), \tag{7.6}$$

where $k > 0$ and $v \in C_0^\infty(\mathbb{R}^3)$. The number a may be considered to be so large that $v = 0$ for $r > a$. We recall that the inclusion $u \in H^m_{\text{loc}}(\bar{\Omega})$ means that the function u and all its generalized derivatives of order m and less belong to the space L^2 on any compact subset of $\bar{\Omega}$.

We now give several physical problems that lead to Eq. (7.6). In a homogeneous elastic medium whose vibrations are described by Eq. (7.3) suppose that in addition to a periodic force there is a force $F = h(x)w$ proportional to w. In this case the amplitude of the stationary vibrations satisfies Eq. (7.6) with $v(x) = \text{const}\, h(x)$. This same equation is obtained for the amplitude if $F \equiv 0$, but the medium is inhomogeneous and $c = c(x)$. The only difference is that in this situation, v depends on the frequency ω. Eq. (7.6) also arises in quantum mechanics. If $\psi = u(x)e^{\frac{E}{i\hbar}t}$ is the wave function describing the steady state of a quantum-mechanical particle with energy E and mass m in a potential field $v(x)$, then the function u satisfies *Schrödinger's equation*

$$\frac{\hbar^2}{2m}\Delta u + [E - v(x)]u = 0. \tag{7.7}$$

Here $\hbar$ is Planck's constant. Dividing the equation by $\hbar^2/2m$, we arrive at Eq. (7.6).

We shall seek a solution of the problem (7.4), (7.6) in the form

$$u = E_+ * g, \quad g \in L_a^2, \tag{7.8}$$

with an unknown function g. For any g the convolution (7.8) belongs to $H_{\text{loc}}^2(\mathbf{R}^3)$ and satisfies the radiation conditions (7.4). Substituting (7.8) into Eq. (7.6) gives for g the equation

$$(I + T_k)g \equiv g + v(x)(E_+ * g) = f, \quad g, f \in L_a^2. \tag{7.9}$$

Thus formula (7.8) associates with each solution g of Eq. (7.9) a solution of the problem (7.4), (7.6). Conversely if relations (7.4) and (7.6) hold for u, then

$$(\Delta + k^2)u = g, \quad g = f - vu \in L_a^2,$$

and so the function u can be written in the form (7.8), where Eq. (7.9) holds for g. Since Eq. (7.9) is a Fredholm equation, this one-to-one correspondence implies the following proposition.

Theorem 2.101. *There exists an m-dimensional subspace $H \subset L_a^2$ ($m < \infty$) such that Eq. (7.6) has a solution satisfying conditions (7.4) if and only if the function f is orthogonal to the subspace H; the homogeneous problem (7.4), (7.6) has exactly m linearly independent solutions.*

There exist at most a finite number of values of the parameter $k > 0$ for which the problem (7.4), (7.6) fails to have a unique solution (i.e., for which $m \neq 0$). This is proved as follows: Equation (7.9) can be considered for all complex k, and the operator T_k is an entire function of k. It is easy to show that Eq. (7.9) has a unique solution for $k = i\rho$, $\rho \gg 1$. Therefore by the theorem on the inversion of a family of Fredhom operators depending analytically on a parameter (cf. Vajnberg 1982; Gokhberg and Krejn 1967) the operator $(I + T_k)^{-1}$ is a meromorphic function of k. It can be proved separately (cf. Vajnberg 1982) that Eq. (7.9) has a unique solution for $k \gg 1$.

We now exhibit the asymptotic expansion of the solution of the problem (7.4), (7.6) as $r \to \infty$. Since $(\Delta + k^2)u = f - vu$, we have

$$u = E_+ * g = \int\limits_{|y| < a} \frac{e^{ik|x-y|}}{-4\pi|x-y|} g(y)\, dy, \quad g = f - vu.$$

For $r > a > |y|$ and $\omega = x/r$ we have

$$|x - y| = \sum_{j=0}^{\infty} C_j(y, \omega)r^{1-j}, \quad C_0 = 1, \quad C_1 = (\omega, y).$$

Hence for $r > a$

$$u = \frac{e^{ikr}}{r} \sum_{j=0}^{\infty} a_j(\omega) r^{-j}, \quad a_0 = -\frac{1}{4\pi} \int\limits_{|y|<a} e^{ik(\omega,y)} [f(y) - v(y)u(y)] \, dy. \quad (7.10)$$

If the function v is of compact support, as assumed earlier, or tends to zero sufficiently rapidly at infinity, then the operator $-\Delta - v(x)$ has no positive eigenvalues. More precisely, the following theorem holds (cf. Eidus 1969).

Theorem 2.102 (Kato). *If* $|v(x)| < C(1 + |x|)^{-1-\epsilon}$ *and*

$$[\Delta + k^2 + v(x)]u = 0, \quad u \in L^2(\mathbf{R}^n),$$

then $u \equiv 0$.

A particular consequence of Kato's theorem is the following.

Theorem 2.103. *If the function v is real-valued, then the problem* (7.4), (7.6) *has a unique solution for any $k > 0$ and $f \in L_a^2$.*

Indeed, if u is a solution of the homogeneous problem (7.4), (7.6), then

$$0 = \int\limits_{|x|<R} [\bar{u}(\Delta + k^2 + v)u - u(\Delta + k^2 + v)\bar{u}] \, dx = \int\limits_{|x|=R} \left(\bar{u} \frac{\partial u}{\partial r} - u \frac{\partial \bar{u}}{\partial r} \right) dS.$$

We substitute the expansion (7.10) into this and pass to the limit as $R \to \infty$. We find that $a_0(\omega) \equiv 0$ and so $u \in L^2(\mathbf{R}^3)$. By Kato's theorem then $u \equiv 0$, i.e., the problem (7.4), (7.6) can have at most one solution. Then by Theorem 2.101 a solution exists.

For simplicity we have assumed throughout the preceding that $x \in \mathbf{R}^3$. If $x \in \mathbf{R}^n$, then the fundamental solution E_+ of Eq. (7.2) is

$$E_+ = -\frac{i}{4} \left(\frac{k}{4\pi r} \right)^{(n-2)/2} H^1_{(n-2)/2}(kr),$$

where $H^1_{(n-2)/2}$ is the Hankel function of first kind.

In this case the radiation conditions will have the following form:

$$u = O(r^{\frac{1-n}{2}}), \quad \frac{\partial u}{\partial r} \mp iku = o(r^{\frac{1-n}{2}}), \quad r \to \infty, \quad (7.11)$$

and the expansion (7.10) looks somewhat different.

Let $\mathcal{D}$ be a bounded region with smooth boundary Γ, $\Omega = \mathbf{R}^n \setminus \mathcal{D}$ and L a second-order elliptic differential operator in Ω with coefficients infinitely differentiable in $\bar{\Omega}$ which coincides with the Laplacian for $r > a$. In Ω consider the boundary-value problem

$$(L + k^2)u = f \in L_a^2(\Omega), \quad u \in H^2_{\text{loc}}(\bar{\Omega}); \quad Bu|_\Gamma = 0, \quad (7.12)$$

where B is a boundary operator of the Dirichlet problem, the Neumann problem, or the third boundary-value problem.

Theorem 2.101 remains valid for the problem (7.11) (7.12). The assertion that for all $k > 0$ except possibly a certain set Λ of isolated points the problem (7.11), (7.12) has a unique solution also remains valid, and moreover zero is not a limit point of Λ. Under the additional "non-trapping" condition (cf. Sect. 7.4) some neighborhood of infinity will contain no points of Λ and so the set Λ will contain at most a finite number of points. Finally, Theorem 2.103 holds for the problem (7.11), (7.12) and assumes the following form: if the problem (7.12) is formally self-adjoint, then the problem (7.11), (7.12) has a unique solution for all $k > 0$. All the assertions made here remain valid if the coefficients of the operator L are not constant in a neighborhood of infinity, but tend rapidly enough as $r \to \infty$ to constants for which the operator L becomes the Laplacian. The proofs of these statements can be found in (Vajnberg 1982).

7.2. The Principle of Limiting Absorption and Limiting Amplitude. Consider Eq. (7.2) with the complex number z in place of k:

$$\Delta u + z^2 u = f \in L_a^2, \quad x \in \mathbb{R}^n. \tag{7.13}$$

As pointed out at the beginning of this section (Example 2.99), for $\operatorname{Im} z \neq 0$ this equation has a unique solution $u = u_z$ in the space $H^2(\mathbb{R}^n)$. It can be found using the Fourier transform or in the form of a convolution with a fundamental solution that vanishes at infinity, which is easily constructed if one takes account of the spherical symmetry of Eq. (7.13). Having an explicit formula for u_z, one can easily verify that $u_z \to u_{\pm}$ in the space $H^2_{\mathrm{loc}}(\mathbb{R}^n)$, if $z \to k > 0$, so that $\operatorname{Im} z \gtrless 0$. Here $u_{\pm}$ are the solutions of Eq. (7.2) defined above and satisfying the radiation conditions. This constitutes the *principle of limiting absorption* that makes it possible to exhibit (using passage to the limit as $z \to k$) a unique solution of the equation without ascertaining the asymptotic behavior of the solution at infinity. The principle holds in exactly the same form for problem (7.13) if $k \notin \Lambda$.

Equation (7.3) leads to the principle of limiting absorption when a term $\beta w_t'$ is added to the left-hand side to describe resistance proportional to velocity and guarantee the absorption of kinetic energy. Then the amplitude of the stationary vibrations will be a solution of Eq. (7.13) with $z^2 = k^2 + i\beta\omega$, $k = \omega/c$. The passage to the limit as $\beta \to +0$ leads to the solution u_+ for Eq. (7.2).

The *principle of limiting amplitude* for Eq. (7.2) (proposed in Tikhonov and Samarskij 1948) consists of exhibiting a unique solution of Eq. (7.2) using passage to the limit:

$$u_{\pm} = \lim_{t \to \infty} w(t, x) e^{\pm i k t},$$

where w is a solution of the Cauchy problem

$$w_{tt}'' - \Delta w = -f(x) e^{\mp i k t}, \quad w\big|_{t=0} = w_t'\big|_{t=0} = 0.$$

It can be proved that these limits exist (in the space $H^1_{\text{loc}}(\mathbb{R}^3)$) and coincide with the solutions of Eq. (7.2) introduced earlier. Thus the validity of the principle of limiting amplitude means that the amplitude of the stationary vibrations can be obtained by passing to the limit from nonstationary vibrations caused by a periodic force. In a more general situation, for example for problem (7.13), the principle of limiting amplitude is not always valid (cf. Vajnberg 1982).

7.3. Radiation Conditions and the Principle of Limiting Absorption for Higher-Order Equations and Systems. Let

$$P(D)u = f, \quad x \in \mathbb{R}^n, \tag{7.14}$$

be an elliptic (hypoelliptic) equation of arbitrary order with coefficients that are constant in the entire space, and suppose the following conditions are met: 1) the surface $P(\xi) = 0$, $\xi \in \mathbb{R}^n$, decomposes into $\varkappa$ connected smooth surfaces S_j with nonzero curvature; 2) $\operatorname{grad} P(\xi) \neq 0$ for $\xi \in S_j$, $1 \leq j \leq \varkappa$. We specify orientations on S_j, i.e., independently on each surface we choose a normal direction ν. Let $\omega = x/|x|$, let $\xi^j = \xi^j(\omega)$ be a point on S_j at which ν and ω have the same direction, and let $\mu_j(\omega) = (\xi^j(\omega), \omega)$. Then the function u satisfies the radiation conditions if it is representable in the form

$$u = \sum_{j=1}^{\varkappa} u_j; \quad u_j = O(r^{\frac{1-n}{2}}), \quad r \to \infty;$$

$$\frac{\partial u_j}{\partial r} - i\mu_j(\omega)u_j = o(r^{\frac{1-n}{2}}), \quad r \to \infty.$$

These conditions distinguish a unique solution of Eq. (7.14) for any function (or distribution) f with compact support. Depending on the choice of orientation of the surface S_j the wave u_j will either move off to infinity or come in from infinity. Since this choice is carried out independently for each j there are $2^\varkappa$ distinct radiation conditions (determining in general distinct solutions of Eq. (7.14)).

The principle of limiting absorption for Eq. (7.14) consists of the possibility of passing to the limit to obtain solutions of Eq. (7.14) satisfying the radiation conditions from solutions of nearby equations that are uniquely solvable in the space $L^2(\mathbb{R}^n)$ (not contained in the spectrum). Thus if the polynomial $P(\xi)$ has real coefficients, then these solutions are obtained in the limit as $\varepsilon \to +0$ from the uniquely determined solution $u_\varepsilon \in L^2(\mathbb{R}^n)$ of the elliptic equation

$$P(D)u_\varepsilon + i\varepsilon Q(D)u_\varepsilon = f, \quad x \in \mathbb{R}^n,$$

where $Q(\xi)$ has real coefficients and $Q(\xi) \neq 0$ on S_j. Depending on the choice of the set $\operatorname*{sgn}_{\xi \in S_j} Q(\xi)$, $1 \leq j \leq \varkappa$, we obtain in the limit solutions with radiation conditions corresponding to the choice of one orientation of the surfaces S_j or another.

The radiation conditions and the principle of limiting absorption are valid for elliptic equations and systems with variable coefficients that tend sufficiently rapidly to constants as $r \to \infty$ and for exterior boundary-value problems for such equations and systems (cf. Vajnberg 1966; Vajnberg 1982). In this case the radiation conditions are defined by the zeros of the polynomial $P(\xi) = \lim_{r\to\infty} \det A(x,\xi)$, where $A(x,\xi)$ is the characteristic matrix of the system under consideration. The conditions imposed above on $P(\xi)$ can be weakened. In particular problems have been studied in which the surfaces S_j are not convex.

7.4. Decay of the Local Energy. Let w be a solution of the Cauchy problem

$$\begin{cases} w_{tt}'' = a^2(x)\Delta w, & t > 0, \, x \in \mathbb{R}^n, \\ w\big|_{t=0} = \varphi, & w_t'\big|_{t=0} = \psi, \end{cases} \tag{7.15}$$

where $a(x) > 0$ and the functions $1-a$, φ, and ψ have compact support. This last condition means that all the inhomogeneity of the medium whose vibrations are described by Eq. (7.15) and all initial perturbations are concentrated in a finite region of space. The *local energy* is the expression

$$E(\Omega, t) = \frac{1}{2} \int_{\Omega} [a^{-2}(x)|w_t'|^2 + |\Delta_x w|^2]\, dx,$$

where Ω is a bounded region of $\mathbb{R}^n$. The solution of many problems leads to a need to answer the question whether the energy of an initial perturbation will move off to infinity or remain in a bounded region of space, i.e., the question of how $E(\Omega, t)$ behaves as $t \to \infty$.

The problem (7.15) corresponds to the Hamiltonian $H = \frac{1}{2}a^2(x)|\xi|^2$ and the *Hamiltonian system*

$$x_t' = \frac{\partial H}{\partial \xi} = \xi a^2(x), \quad \xi_t' = -\frac{\partial H}{\partial x} = -\frac{1}{2}|\xi|^2 \nabla a^2(x);$$

$$x\big|_{t=0} = x_0, \quad \xi\big|_{t=0} = \xi_0, \quad H(x_0, \xi_0) = H_0.$$

The solutions of this system are called *bicharacteristics* and their projections into $\mathbb{R}_x^n$ are called *(geometric-optical) rays*. We shall say that the *non-trapping condition* holds if all rays move off to infinity as $t \to \infty$.

Let $a \in C^\infty(\mathbb{R}^n)$. Since the singularities (places where infinite differentiability fails) of the solutions of Eq. (7.15) propagate along bicharacteristics (cf. Vajnberg 1975; Hörmander 1983–1985), it follows from the non-trapping condition that for any functions (or distributions) φ and ψ of compact support the singularities of a solution of the problem (7.15) move off to infinity as $t \to \infty$, i.e., for any bounded region Ω the solution of the problem (7.15) is infinitely differentiable for $x \in \Omega$ and sufficiently large t. It turns out that as $t \to \infty$ the solution of the problem (7.15) not only becomes smooth, it also decays.

Theorem 2.104 (cf. Vajnberg 1982). *Suppose the non-trapping condition holds and let Ω be an arbitrary bounded region with* $\operatorname{supp} \varphi \subset \Omega$ *and* $\operatorname{supp} \psi \subset \Omega$. *Then for a solution of the problem (7.15) with any $\alpha = (\alpha_0, \ldots, \alpha_n)$ and some $T = T(\Omega)$, $C = C(\Omega, \alpha)$, we have the estimates*

$$\left| \frac{\partial^{|\alpha|} w}{\partial t^{\alpha_0} \partial x_1^{\alpha_1} \cdots \partial x_n^{\alpha_n}} \right| \leq C \eta(t) \left[\|\varphi\|_{L^2} + \|\psi\|_{L^2} \right], \quad x \in \Omega, \quad t > T.$$

Here $\eta(t) = t^{-n+1-\alpha_0}$ for even n and $\eta(t) = e^{-\varepsilon t}$ for odd n, where ε is some constant depending only on the function $a = a(x)$.

The decay of the local energy follows in particular. The decay of the local energy was first obtained by Morawetz for the Dirichlet problem for the wave equation with constant coefficients in the exterior of a bounded convex body. The convexity of the body in this case guarantees non-trapping since all the geometric-optical rays (they are straight lines), reflecting according to the laws of geometrical optics, move off to infinity. The asymptotic expansion for $|x| \leq b$, $t \to \infty$, of the solutions of the exterior problems for general hyperbolic equations and systems is obtained in (Vajnberg 1975) and (Vajnberg 1982). It is assumed here that the coefficients of the equations are constant in a neighborhood of infinity and that the non-trapping condition holds.

7.5. Scattering of Plane Waves. A function ψ satisfying the equation

$$[a^2(x)\Delta + k^2]\psi = 0, \quad x \in \mathbb{R}^3, \tag{7.16}$$

and having the form

$$\psi(k, \theta, x) = e^{ik(\theta, x)} + u(k, \theta, x), \tag{7.17}$$

where $\theta \in \mathbb{R}^3$, $|\theta| = 1$, and the function u satisfies the radiation conditions (7.4), is called a solution of the *scattering problem* for plane waves in an inhomogeneous medium. Here the function $a^2(x) > 0$ tends rapidly to 1 at infinity. For simplicity we shall assume that $1 - a^2 \in C_0^\infty(\mathbb{R}^3)$. The function $e^{ik(\theta, x)}$ describes a plane wave traveling in the direction of the vector θ and the function u describes a wave scattered at the inhomogeneities of the medium. Since the function $e^{ik(\theta, x)}$ is a solution of Eq. (7.16) for $a \equiv 1$, the function u satisfies the equation

$$[a^2(x)\Delta + k^2]u = k^2[1 - a^2(x)]e^{ik(\theta, x)}. \tag{7.18}$$

By Theorem 2.103 the problem (7.18), (7.4) has a unique solution. In particular $u \equiv 0$ (there is no scattered wave) if $a \equiv 1$ (the medium is homogeneous) and the right-hand side of (7.18) is zero. According to formula (7.10)

$$u = f(k, \theta, x)r^{-1}e^{ikr} + o(r^{-1}), \quad r \to \infty, \, \omega = \frac{x}{r}. \tag{7.19}$$

The function f is called the *scattering amplitude*. Besides the parameter k it also depends on the direction θ of the impacting wave and the direction ω along which the point x tends to infinity.

The solution of the scattering problem and scattering amplitude are defined similarly in the case when there is an obstacle in the space. In this case the function ψ must have the form (7.17), as before and must be a solution of the homogeneous problem (7.12) (where $L \equiv \Delta$ if the medium is homogeneous). Then

$$\begin{cases} (L + k^2)u = (\Delta - L)e^{ik(\theta, x)}, & u \in H^2_{\text{loc}}(\bar{\Omega}), \\ Bu|_\Gamma = -Be^{ik(\theta, x)}|_\Gamma. \end{cases} \tag{7.20}$$

If the problem (7.12) is formally self-adjoint, then the problem (7.20), (7.4) has a unique solution and the asymptotic expansion (7.19) holds for it.

The steady-state scattering problem for quantum-mechanical particles of mass m in a potential field consists of finding a function ψ satisfying the equation

$$\left(-\frac{1}{2m}\Delta - E + v(x)\right)\psi = 0, \quad x \in \mathbb{R}^3, \quad E = \frac{|p|^2}{2m}, \tag{7.21}$$

and representable in the form

$$\psi(p, x) = e^{i(p, x)} + u(p, x), \tag{7.22}$$

for each $p \in \mathbb{R}^3$, where the function u satisfies the radiation conditions with $k = \sqrt{2mE} = |p|$. For simplicity we assume that $v \in C^\infty_0(\mathbb{R}^3)$. Obviously the scattering problem (7.21), (7.22) reduces to (7.16), (7.17) if we set $p = k\theta$, $v(x) = [a^{-2}(x) - 1]\frac{k^2}{2m}$. Accordingly the scattering amplitude $f(k, \theta, \omega)$ in the quantum-mechanical problem is denoted by $f(|p|\frac{x}{|x|}, p)$. In what follows we shall discuss mainly the quantum-mechanical problem.

7.6. Spectral Analysis. The operator $-\frac{1}{2m}\Delta$ in the space $L^2(\mathbb{R}^n)$ has a purely continuous spectrum coinciding with the nonnegative real axis $\bar{\mathbb{R}}_+$. Indeed for $\lambda \notin \bar{\mathbb{R}}_+$ the operator $-\frac{1}{2m}\Delta - \lambda$ on the space $L^2(\mathbb{R}^n)$ has a bounded inverse (this was noted in Example 2.99 at the beginning of Sect. 7) and hence the spectrum of the operator $-\frac{1}{2m}\Delta$ is contained in $\bar{\mathbb{R}}_+$. The operator $-\dfrac{1}{2m}\Delta$ has no eigenvalues: if $L^2(\mathbf{R}^n) \ni u$ is an eigenfunction corresponding to the eigenvalue $\lambda \in \bar{\mathbf{R}}_+$ and $\tilde{u}(\xi)$ is its Fourier transform, then $[\dfrac{1}{2m}\cdot|\xi|^2 - \lambda]\tilde{u}(\xi) = 0$ and $\tilde{u} \in L^2(\mathbf{R}^n)$. Thus $\tilde{u} \equiv 0$ (as an element of $L^2(\mathbf{R}^n)$). It remains to show that the points of the semiaxis $\bar{\mathbf{R}}_+$ belong to the spectrum.

The functions $e^{i(p,x)}$, $p \in \mathbf{R}^n$, satisfy the equation $-\dfrac{1}{2m}\Delta u = \lambda u$ for $\lambda = \dfrac{|p|^2}{2m}$, but they cannot be considered eigenfunctions in the classical sense because they do not belong to $L^2(\mathbf{R}^n)$. They are called the *eigenfunctions of the continuous spectrum*. Let $\varphi \in C_0^\infty(\mathbf{R}^n)$ and

$$u_j(x) = j^{-\frac{n}{2}}\varphi(\frac{x}{j})e^{i(p,x)}.$$

It is easy to verify that for $\lambda = |p|^2/2m$ we have

$$\|u_j\|_{L^2} = \|\varphi\|_{L^2} = \text{const} \quad \text{and} \quad \|(-\frac{1}{2m}\Delta - \lambda)u_j\|_{L^2} \le C/\sqrt{j}.$$

This proves that the operator $-\dfrac{1}{2m}\Delta - \lambda$ does not have a bounded inverse for $\lambda \in \bar{\mathbf{R}}_+$ and so $\bar{\mathbf{R}}_+$ is contained in the (continuous) spectrum of the operator $-\dfrac{1}{2m}\Delta$. The positive spectrum of the operator $-\dfrac{1}{2m}\Delta$ has infinite multiplicity: the eigenfunctions $e^{i(p,x)}$ corresponding to the value $\lambda = \dfrac{|p|^2}{2m}$ depend on the parameter $\theta = \dfrac{p}{|p|}$.

The inversion formula for the Fourier transform:

$$f(x) = \int_{\mathbf{R}^n} \tilde{f}(p)e^{i(p,x)}\,dp, \quad dp = (2\pi)^{-n}dp, \tag{7.23}$$

where

$$\tilde{f}(p) = \int_{\mathbf{R}^n} f(x)\overline{e^{i(p,x)}}\,dx, \tag{7.24}$$

can be regarded as an expansion theorem for a function $f \in L^2(\mathbf{R}^n)$ in eigenfunctions of the continuous spectrum of the operator $-\dfrac{1}{2m}\Delta$. Obviously the eigenfunctions corresponding to distinct values of p are orthogonal and formula (7.23) means that the system of eigenfunctions is complete.

Now let $n = 3$ (this restriction is imposed because the scattering problem was stated for $n = 3$; both here and above one can take any n), and let the function v be a real-valued function in $C_0^\infty(\mathbf{R}^3)$. The spectrum of the operator $-\dfrac{1}{2m}\Delta + v$ in $L^2(\mathbf{R}^3)$ consists of a continuous component coinciding with $\bar{\mathbf{R}}_+$

and a number of negative eigenvalues that is at most finite. The solutions ψ of the scattering problem (7.21) (7.22) are eigenfunctions of the continuous spectrum of the operator $-\dfrac{1}{2m}\Delta + v$ corresponding to the eigenvalue $\lambda = \dfrac{|p|^2}{2m}$.
Eigenfunctions with distinct p are orthogonal, i.e.

$$\int_{\mathbf{R}^3} \psi(p',x)\overline{\psi(p'',x)}\, dx = 0 \quad \text{for} \quad p' \neq p''.$$

There are many ways of assigning a meaning to this integral over the entire space and to those that occur below (just as in formulas (7.23) and (7.24)). For example, they can be understood as the limits in $L^2(\mathbf{R}^3)$ as $\rho \to \infty$ of the corresponding integrals over the balls $|x| \leq \rho$. The eigenfunctions of the continuous spectrum, together with the eigenfunctions $\psi_j \in L^2(\mathbf{R}^3)$, $1 \leq j \leq m$, corresponding to negative eigenvalues, form a complete system in $L^2(\mathbf{R}^3)$, i.e., for any $f \in L^2(\mathbf{R}^3)$ the expansion

$$f(x) = \sum_{j=1}^{m} f_j \psi_j(x) + \int_{\mathbf{R}^3} \tilde{f}(p)\psi(p,x)\, dp, \qquad (7.25)$$

holds, where, if the functions ψ_j are normalized,

$$f_j = \int_{\mathbf{R}^3} f(x)\overline{\psi_j(x)}\, dx, \quad \tilde{f}(p) = \int_{\mathbf{R}^3} f(x)\overline{\psi(p,x)}\, dx,$$

and in addition

$$\|f\|_{L^2}^2 = \sum_{j=1}^{m} |f_j|^2 + \int_{\mathbf{R}^3} |\tilde{f}(p)|^2 dp.$$

If we give up the hypothesis that the function v has compact support and replace it by the assumption that $v = O(|x|^{-N})$ as $|x| \to \infty$ with N arbitrarily large, then the solutions constructed above for the scattering problem are not all of the eigenfunctions of the continuous spectrum. The eigenfunctions $\psi(p,x)$ that were constructed remain orthogonal, but they remain complete (taken together with $\psi_j(x)$) only if the operator $-\dfrac{1}{2m}\Delta + v$ has no singular spectrum (for more details, cf. Berezanskij 1965; Berezin and Shubin 1983).

7.7. The Scattering Operator and the Scattering Matrix. Suppose there are two unitary groups e^{-itH_0} and e^{-itH} in the Hilbert space L. The first describes a "free" dynamics, which is assumed to be well known, and the second describes a perturbed dynamics, which we wish to compare with the first.

As an example we can take

$$L = L^2(\mathbf{R}^3), \quad H_0 = -\frac{1}{2m}\Delta, \quad H = -\frac{1}{2m}\Delta + v, \qquad (7.26)$$

where $v \in C_0^\infty(\mathbb{R}^3)$. Then for $\varphi \in L^2(\mathbb{R}^3)$ the function $u_0 = e^{-itH_0}\varphi$ is a solution at the instant t of the Cauchy problem for the unperturbed Schrödinger equation

$$iu_t' = -\frac{1}{2m}\Delta u, \quad u\big|_{t=0} = \varphi.$$

The function $u = e^{-itH}\varphi$ is a solution at the instant t of the Cauchy problem for the perturbed Schrödinger equation

$$iu_t' = -\frac{1}{2m}\Delta u + v(x)u, \quad u\big|_{t=0} = \varphi.$$

The *wave operators* are the strong limits

$$W^\pm \stackrel{\text{def}}{=} s - \lim_{t\to\pm\infty} e^{itH}e^{-itH_0}. \qquad (7.27)$$

Suppose the wave operator W^- exists and is defined on all of L. Since the operator e^{-itH} is unitary, it follows from (7.27) that

$$\lim_{t\to-\infty}\left(e^{-itH}W^-\varphi - e^{-itH_0}\varphi\right) = 0.$$

Thus the trajectory $l = e^{-itH}W^-\varphi$, $t \in \mathbb{R}$, in the perturbed dynamics is asymptotically near to the trajectory $l_0 = e^{-itH_0}\varphi$, $t \in \mathbb{R}$, in the free dynamics as $t \to -\infty$, and the wave operator W^- (if it is defined on all of L) defines a trajectory l in the perturbed dynamics over each trajectory l_0 in the free dynamics that is asymptotically near to l_0 as $t \to -\infty$ and moves the position of the trajectory l_0 at $t = 0$ into the position of the trajectory l at $t = 0$. The operator W^+ acts similarly (with $t \to -\infty$ replaced by $t \to +\infty$).

Suppose the ranges of values of the operators $W^\pm$ coincide. This property is called *weak asymptotic completeness* of the wave operators. We denote by D the domain of definition of the operators $W^\pm$. When weak asymptotic completeness holds, the operators $(W^+)^{-1} : D \to L$ and $S = (W^+)^{-1}W^- : L \to L$ are defined. It is obvious that for $\varphi \in D$ the trajectory $l = e^{-itH}\varphi$, $t \in \mathbb{R}$, in the perturbed dynamics is asymptotically near to the trajectory $l_0^\pm = e^{-itH_0}\varphi_\pm$, $t \in \mathbb{R}$, in the free dynamics as $t \to \pm\infty$, and the scattering operator takes φ_- into φ_+. Thus the free trajectory l_0^- becomes the free trajectory l_0^+ under the action of the perturbation $H - H_0$ as $t \to \infty$, and the scattering operator assigns the trajectory l_0^+ to the trajectory l_0^-.

The existence of wave operators is usually proved using the following device, which is known as "Cook's method." We denote by V the operator $H - H_0$. Obviously

$$\frac{d}{dt}\left(e^{itH}e^{-itH_0}\varphi\right) = ie^{itH}\left(H - H_0\right)e^{-itH_0}\varphi = ie^{itH}Ve^{-itH_0}\varphi.$$

Hence

$$e^{itH}e^{-itH_0}\varphi = \varphi + i\int_0^t e^{i\tau H}Ve^{-i\tau H_0}\varphi\, d\tau$$

and for the operator W^+ to exist it suffices that for all $\varphi \in L$

$$\int_0^\infty \left\| e^{i\tau H} V e^{-i\tau H_0}\varphi \right\| d\tau < \infty,$$

or, observing that the operator $e^{i\tau H}$ is unitary, that

$$\int_0^\infty \left\| V e^{-i\tau H_0}\varphi \right\| d\tau < \infty, \quad \varphi \in L. \tag{7.28}$$

Since the expressions on the right-hand side of expression (7.27) on which the limit is taken are unitary operators, the limit (7.27) exists on all of the space L if it exists in a dense subset. Thus for the existence of the wave operator W^+ it suffices that relation (7.28) hold for a set of elements φ that is dense in L (*Cook's criterion*).

We shall show how to prove the existence of the wave operators in Eq. (7.26) using Cook's criterion. We denote by L' the set of functions $\varphi \in L^2(\mathbb{R}^3)$ whose Fourier transform $\tilde{\varphi} = \tilde{\varphi}(p)$ belongs to $C_0^\infty(\mathbb{R}^3)$ and vanishes in a neighborhood of the point $p = 0$. The set L' is dense in $L^2(\mathbb{R}^3)$. For $\varphi \in L'$ we have

$$e^{-itH_0}\varphi = \int_{\mathbb{R}^3} e^{it|p|^2 - i(x,p)}\tilde{\varphi}(p)\, dp.$$

Using the stationary phase method one can easily verify that for any N and $x \in \operatorname{supp} v$ this integral does not exceed $C_N t^{-N}$ as $t \to \infty$. Then (7.28) for $\varphi \in L'$ and the existence of the wave operators follow. It is almost as simple to verify the existence of the wave operators in Eq. (7.26) using Cook's criterion if $v \notin C_0^\infty(\mathbb{R}^3)$, but $|v(x)| < C(1 + |x|)^{-1-\varepsilon}$, where $\varepsilon > 0$.

Weak asymptotic completeness of the wave operators is a more delicate property and harder to prove than the existence of the limits in (7.27). In the example (7.26) the wave operators have the property of *asymptotic completeness* (of which weak asymptotic completeness is a consequence), which consists of the following: the range of values of the operators $W^\pm$ coincides with the orthogonal complement to the subspace spanned by the eigenvectors (corresponding to the discrete spectrum) of the operator H.

The wave operators are isometric (being limits of unitary operators) and intertwine the operators H and H_0, i.e.

$$HW^\pm = W^\pm H_0. \tag{7.29}$$

This last property is a consequence of the following chain of equalities

$$e^{i\tau H}W^\pm = e^{i\tau H}\left[\lim_{t\to\pm\infty} e^{itH}e^{-itH_0}\right] = \lim_{t\to\pm\infty}\left[e^{i(t+\tau)H}e^{-itH_0}\right] =$$

$$= \lim_{t\to\pm\infty}\left[e^{i(t+\tau)H}e^{-i(t+\tau)H_0}\right]e^{i\tau H_0} = W^\pm e^{i\tau H_0}.$$

Differentiating on τ and setting $\tau = 0$, we obtain (7.29). It follows from (7.29) that the scattering operator commutes with H_0:

$$SH_0 = H_0 S. \tag{7.30}$$

In quantum-mechanical problems this property expresses the law of conservation of energy under scattering. Indeed, if the state of a particle is defined by
the function $e^{-itH_0}\varphi_-$ as $t \to -\infty$ and by the function $e^{-itH_0}\varphi_+$ as $t \to \infty$
and $\varphi_+ = S\varphi_-$, then the energy of the particle is $E_- = (\varphi_-, H_0\varphi_-)$ as
$t \to -\infty$ and $E_+ = (\varphi_+, H_0\varphi_+) = (S\varphi_-, H_0S\varphi_-) = (\varphi_-, S^*H_0S\varphi_-)$ as
$t \to \infty$. Since the operator S is unitary (it is an isometry, being the product
of two isometries, and defined on all of L), it follows that $S^* = S^{-1}$, and by
(7.30) we have $E_+ = E_-$.

In what follows we shall consider not the arbitrary self-adjoint operators
H_0 and H on the Hilbert space L, but the particular realization given in
(7.26). We shall denote by $F = F_{x \to p}$ the Fourier transform operator (7.24).
We denote by $\hat{S}$ the operator

$$\hat{S} = FSF^{-1} : L^2(\mathbb{R}^3_p) \to L^2(\mathbb{R}^3_p), \tag{7.31}$$

and by $\mathcal{S}(p, p')$ its kernel, i.e., $\hat{S}f = \int_{\mathbb{R}^3} \mathcal{S}(p, p')f(p')\, dp'$. The function $\mathcal{S}(p, p')$

is called the *scattering matrix* (sometimes the operator $\hat{S}$ is also called the
scattering matrix). It follows formally from (7.31) that

$$\mathcal{S}(p, p') = \big(S\varphi^\circ(p', x), \varphi^\circ(p, x)\big), \quad \varphi^\circ = e^{i(p, x)}.$$

Thus the scattering matrix is the matrix notation for the scattering operator
S in the basis $\{\varphi^\circ\}$ of eigenfunctions (of the continuous spectrum) of the
operator H_0.

The scattering matrix $\mathcal{S}(p, p')$ has the form

$$\mathcal{S}(p, p') = \delta(p - p') + \frac{i}{2\pi m}\delta(E_p - E_{p'})f(p, p'). \tag{7.32}$$

Here the constant m is the same as in (7.26), $E_p = \dfrac{|p|^2}{2m}$, the first δ-function
on the right-hand side is concentrated at the point p', and the second on the
sphere $E_p = E_{p'}$. Let $S = I + R$ be the corresponding decomposition of the
scattering operator (the identity operator corresponds to the first δ-function
in (7.32), and the matrix notation $R(p, p')$ of the operator $R = S - I$ in the
basis $\{\varphi^\circ\}$ is given by the second term on the right-hand side of (7.32)). If
$v \equiv 0$, then $S = I$ and $\mathcal{S} = \delta(p - p')$. The fact that the function $R(p, p')$ is
concentrated on the surface $E_p = E_{p'}$ is equivalent to the law of conservation
of energy for the operator S. Indeed, formula (7.30) is equivalent to the
following: $H_0R = RH_0$, from which, after a Fourier transform, we obtain
$(E_p - E_{p'})R(p, p') = 0$. Finally, it suffices to know the function f on the
right-hand side of (7.32) only for $|p| = |p'|$ (it is the coefficient of the δ-
function). As it happens, for $p' = |p|\dfrac{x}{|x|}$ the function $f(p', p)$ coincides with
the scattering amplitude defined in Sect. 7.5 for the steady-state problem
(7.21), (7.22).

The time-dependent theory of acoustical scattering differs very little from the scattering of quantum-mechanical particles. Consider scattering in an inhomogeneous medium whose state is described by a function that is a solution of the problem (7.15), where the function $1 - a^2$ has compact support (the medium is homogeneous in a neighborhood of infinity). Let L be the completion of the space of pairs of functions $f = \begin{pmatrix} \varphi \\ \psi \end{pmatrix}$ with $\varphi, \psi \in C_0^\infty(\mathbf{R}^3)$ with respect to the norm

$$\|f\|_L^2 = \frac{1}{2} \int_{\mathbf{R}^3} \left[a^{-2}(x)|\psi|^2 + |\nabla\varphi|^2 \right] dx.$$

The group $U(t)$ describing the perturbed dynamics, assigns to each element $f = \begin{pmatrix} \varphi \\ \psi \end{pmatrix}$ the vector $\begin{pmatrix} u(t, \cdot) \\ u_t'(t, \cdot) \end{pmatrix}$, where u is a solution of the problem (7.15). This group is unitary and can be written in the form $U(t) = e^{-itH}$, where H is the closure of the matrix operator $i \begin{pmatrix} 0 & I \\ a^2(x)\Delta & 0 \end{pmatrix}$ originally defined on $C_0^\infty(\mathbf{R}^3)$. For $a \equiv 1$ we denote the space L and the group $U(t)$ by L_0 and $U_0(t)$. Although the free group $U_0(t)$ and the perturbed group $U(t)$ act on different spaces, the set of elements is the same in both spaces (only the norms are different) and the natural isomorphism of the spaces L_0 and L makes it possible to extend the results obtained above without any difficulty to this case. Scattering theory is somewhat more difficult to construct when the spaces L_0 and L differ more essentially (for example, in the case of scattering at an obstacle). For more details on the scattering theory see the monographs Lax and Phillips 1960; Newton 1966; Reed and Simon 1972–1979; and Taylor 1972.

§8. Spectral Theory of One-Dimensional Differential Operators

8.1. Outline of the Method of Separation of Variables. We have already encountered the solution of boundary-value problems by the method of separation of variables (or the Fourier method) in Sects. 4 and 5. The general scheme in which separation of variables is encountered can be described as follows. Suppose we wish to solve an equation of the form

$$Lu = f, \tag{8.1}$$

where the linear operator L has the form

$$L = A \otimes 1 + 1 \otimes B. \tag{8.2}$$

This means that the space $\mathcal{H}$ in which the operator L acts is represented in the form of a space of functions $u(x', x'')$ of two groups of variables x' and x'', the operators A and B act on spaces of functions of x' and x'' respectively, and the operator $A \otimes 1$ (resp. $1 \otimes B$) acts on $u(x', x'')$ as the operator A on the variables x' (resp. B on the variables x'') with the other variables fixed. Now consider a system $\{\psi_k : k = 1, 2, \ldots\}$ of eigenfunctions of the operator B:

$$B\psi_k = \lambda_k \psi_k. \tag{8.3}$$

For $\varphi = \varphi(x')$, $\psi = \psi(x'')$, we set $(\varphi \otimes \psi)(x', x'') = \varphi(x')\psi(x'')$. It then follows from (8.2)–(8.3) that

$$L(\varphi \otimes \psi_k) = (A + \lambda_k)\varphi \otimes \psi_k. \tag{8.4}$$

We now assume that the system $\{\psi_k\}$ is *complete* and *linearly independent* in the sense that any function $g = g(x'')$ of a suitable function space admits a unique expansion

$$g = \sum_{k=1}^{\infty} c_k \psi_k, \tag{8.5}$$

where c_k are constants. A frequently encountered case, for example, occurs when the operator B is self-adjoint and has a discrete spectrum in some Hilbert space $\mathcal{H}''$ and ψ_k is a complete orthogonal system of eigenfunctions of the operator B (it was this case that we dealt with in Sects. 3–5). We expand the function $f = f(x', x'')$ of the right-hand side of (8.1) in a series

$$f = \sum_{k=1}^{\infty} f_k \otimes \psi_k, \tag{8.6}$$

i.e., we expand it in a series in the system $\psi_k = \psi_k(x'')$ for each fixed x', so that $f_k = f_k(x')$ are the coefficients of this expansion. The desired solution u is expandable in an analogous series

$$u = \sum_{k=1}^{\infty} u_k \otimes \psi_k. \tag{8.7}$$

Assuming the operator L can be applied termwise to the series (8.7) and keeping in mind (8.4), we find that the coefficients u_k satisfy the equations

$$(A + \lambda_k)u_k = f_k, \tag{8.8}$$

i.e., the problem has been reduced to a system of equations in the variables x' alone.

The operator B often has a whole family of eigenfunctions $\{\psi_k : k \in M\}$, where M is a space with a measure dk, rather than a countable system of eigenfunctions. In this situation any function $g = g(x'')$ of a suitable function space can be written uniquely in the form

$$g = \int \tilde{g}(k)\psi_k \, dk. \tag{8.9}$$

In this situation the eigenfunctions ψ_k themselves may fail to belong to this function space. The simplest example of such a situation arises when B is taken as the operator $i^{-1}d/dx$ on $\mathbb{R}$. Then its eigenfunctions can be taken as the exponentials $\psi_k = e^{ikx}$ and the expansion (8.9) becomes the inversion formula for the Fourier transform ($\tilde{g}(k)$ is the Fourier transform of g, and g can be taken, for example, in $L^2(\mathbb{R})$). Therefore in the general case the function $\tilde{g}(k)$ in (8.9) is an analog of the Fourier transform.

To solve Eq. (8.1) we must expand u and f similarly (for fixed x') and then, denoting the coefficients in (8.9) respectively by $\tilde{u}(x',k)$ and $\tilde{f}(x',k)$, we obtain for $\tilde{u}$ the equation

$$(A + \lambda_k)\tilde{u}(x',k) = \tilde{f}(x',k), \tag{8.10}$$

where the operator A is applied on x' and λ_k is the eigenvalue corresponding to the eigenfunction ψ_k in formula (8.3). Thus an equation analogous to (8.8) is obtained for $\tilde{u}(x',k)$ with the difference, however, that the parameter k in (8.10) may happen to be continuous, while in (8.8) it was discrete.

Naturally the homogeneous equation $Lu = 0$ can also be solved in the manner just described, as was done in Sects. 2.4 and 2.5. In this case there is usually some arbitrariness in the solution of Eqs. (8.8) and (8.10) which can be used to obtain a solution u satisfying various initial or boundary conditions.

Thus the method of separation of variables reduces to two problems: 1) the construction of a system of eigenfunctions $\{\psi_k : k \in M\}$ (and exhibiting the measure dk on M in the case when the parameter k is continuous); 2) the solution of Eq. (8.8) or (8.10). These problems are effectively solvable as a rule only in the case when the variables x' and x'' are one-dimensional (for example, when L is an evolution operator, $x' = t$ is time, and $x'' = x$ is a one-dimensional variable, i.e., when (8.1) is an evolution problem with one spatial variable). In the latter case the equations (8.8) and (8.10) are ordinary differential equations and for them a Cauchy problem is usually obtained. The first problem, which constitutes the spectral theory of one-dimensional differential operators, is more substantial.

8.2. Regular Self-Adjoint Problems. For simplicity we shall consider only self-adjoint eigenvalue problems for linear second-order ordinary differential operators, i.e., problems of the form

$$Ly \equiv -\frac{d}{dx}\left(p(x)\frac{dy}{dx}\right) + q(x)y = \lambda r(x)y, \tag{8.11}$$

where p, q, and r are real-valued functions, $p \in C^1$, q and r are continuous, and $p(x) > 0$ and $r(x) > 0$ for all x in the open or closed interval under consideration. The operator L occurring in (8.11) is called a *Sturm-Liouville operator*. The problem (8.11) can be reduced to a problem of the same form with $p \equiv 1$ and $r \equiv 1$ by changing the independent variable to $z = \int_a^x \left(\frac{r(x)}{p(x)}\right)^{1/2} dx$ and changing the unknown function to $\psi = \big(r(x)p(x)\big)^{1/4}y$.

Therefore instead of the problem (8.11) it suffices to consider the eigenvalue problem for the very simple Sturm-Liouville operator known as the *one-dimensional Schrödinger operator*

$$L = -\frac{d^2}{dx^2} + q(x), \tag{8.12}$$

as we shall do in what follows. Suppose the problem under consideration is the eigenvalue problem

$$Ly = \lambda y \quad \text{for} \quad x \in [a, b] \tag{8.13}$$

with boundary conditions

$$y(a)\cos\alpha + y'(a)\sin\alpha = 0, \quad y(b)\cos\beta + y'(b)\sin\beta = 0. \tag{8.14}$$

We shall assume that the function q is continuous on $[a, b]$. Then this problem belongs to the class of *regular* self-adjoint boundary-value problems (eigenvalue problems for the operator (8.12) are considered singular in the case when either the potential $q(x)$ has singularities as $x \to a$ or $x \to b$ or the interval under consideration is replaced by a ray or the entire real line).

The problem of finding eigenvalues and eigenfunctions, i.e., values λ and functions $y \in C^2([a, b])$ not identically zero satisfying (8.13)–(8.14), is called the *Sturm-Liouville problem*.

To study the problem in the Hilbert space $L^2([a, b])$ it is useful to enlarge the domain of definition of the operator L somewhat, taking it to be the set $D(L)$ consisting of $y \in C^1([a, b])$ for which $y'' \in L^2([a, b])$ and the boundary conditions (8.14) are satisfied. Here y'' must be understood in the distribution sense; the relation $y'' \in L^2([a, b])$ is equivalent to the statement that the function y' is absolutely continuous and $y'' \in L^2([a, b])$. With the domain of definition $D(L)$ the operator L becomes a self-adjoint operator in $L^2([a, b])$. However the eigenvalues and eigenfunctions of the operator L do not change under this extension: it is easy to verify that every generalized solution $y \in \mathcal{D}'((a, b))$ of the equation $Ly = \lambda y$ actually belongs to $C^2([a, b])$.

The study of the Sturm-Liouville problem (8.13)–(8.14) is conveniently carried out using a *source function* or *Green's function* defined as the kernel $G(x, y, \mu)$ (in the sense of Schwartz) of the operator $(L - \mu I)^{-1}$ and $\mu \in \mathbb{C} \setminus \sigma(L)$, where $\sigma(L)$ is the spectrum of the operator L, i.e., the set of its eigenvalues (we recall that the spectrum $\sigma(L)$ of an operator L in Hilbert space is the set of $\lambda \in \mathbb{C}$ for which there does not exist a bounded everywhere-defined inverse operator to $L - \lambda I$; in the regular case the whole spectrum of the self-adjoint operator in $L^2([a, b])$ described above and defined by the expression (8.12) and the boundary conditions (8.14) coincides with the set of eigenvalues of the operator). The Green's function $G(x, y, \mu)$ is uniquely determined by the equation

$$(L_x - \mu I)G(x, s, \mu) = \delta(x - s), \quad x \in [a, b], \tag{8.15}$$

where $s \in (a, b)$ (the notation L_x means that the operator L is applied on x), and the boundary conditions

$$G(a, s, \mu) \cos \alpha + \frac{\partial G}{\partial x}(a, s, \mu) \sin \alpha = 0,$$

$$G(b, s, \mu) \cos \beta + \frac{\partial G}{\partial x}(b, s, \mu) \sin \beta = 0. \tag{8.16}$$

In fact if there existed (for some s) two solutions of the problem (8.15)–(8.16), their difference, regarded as a function of x, would be a regular eigenfunction of the problem (8.13)–(8.14) with eigenvalue μ, contradicting the assumption $\mu \notin \sigma(L)$. Equation (8.15) means that the following conditions hold:

a) the function $x \mapsto G(x, s, \mu)$ is continuous on $[a, b]$;

b) $(L_x - \mu I)G(x, s, \mu) = 0$ for $x \neq s$;

c) $G'_x(s + 0, s, \mu) - G'_x(s - 0, s, \mu) = -1$.

Here $G'_x = \partial G/\partial x$ and $G'_x(s \pm 0, s, \mu) = \lim_{\varepsilon \to +0} G'_x(s \pm \varepsilon, s, \mu)$.

The Green's function is often considered for fixed $\mu = \mu_0$. For simplicity we shall assume that $\mu_0 = 0$ and set $G(x, s) = G(x, s, 0)$. This function, which is defined on $[a, b] \times [a, b]$, is called the Green's function, although it should be kept in mind that it is defined only under the condition that $0 \notin \sigma(L)$. For example a sufficient condition for this is the following.

$$q(x) \geq 0, \quad \sin \alpha \cdot \cos \alpha \leq 0, \quad \sin \beta \cdot \cos \beta \geq 0. \tag{8.17}$$

In particular if $q \equiv 0$ and $\sin \alpha = \sin \beta = 0$, (i.e., the operator in question is $L = -d^2/dx^2$ with zero Dirichlet boundary conditions), it is easy to see using conditions a)–c) and (8.16) that

$$G(x, s) = (b - a)^{-1}\big[\theta(s - x)(b - s)(x - a) + \tag{8.18}$$

$$+ \theta(x - s)(s - a)(b - x)\big] = \begin{cases} \dfrac{b - s}{b - a}(x - a), & x < s, \\ \dfrac{s - a}{b - a}(b - x), & x > s. \end{cases}$$

In the general case under the condition $0 \notin \sigma(L)$ we consider two nontrivial solutions $y_1(x)$ and $y_2(x)$ of the equation $Ly \equiv 0$ satisfying the boundary conditions

$$y_1(a) \cos \alpha + y'_1(a) \sin \alpha = 0, \quad y_2(b) \cos \beta + y'_2(b) \sin \beta = 0. \tag{8.19}$$

Then

$$G(x, s) = -W^{-1}\big[\theta(s - x)y_1(x)y_2(s) + \theta(x - s)y_1(s)y_2(x)\big], \tag{8.20}$$

where $W = y_1(s)y'_2(s) - y'_1(s)y_2(s)$ is the Wronskian of the solutions y_1 and y_2 (by the well-known formula of Liouville it is independent of s). We note that $W \neq 0$ by the linear independence of the solutions y_1 and y_2 (if y_1 and y_2 were linearly dependent, each of these functions would be an eigenfunction with eigenvalue 0). Formula (8.18) is obtained from (8.20) by taking, for example $y_1(x) = x - a$, $y_2(x) = b - x$.

If $0 \notin \sigma(L)$, then the operator L^{-1} is given by the formula

$$L^{-1}f(x) = \int_a^b G(x,s)f(s)\,ds, \qquad (8.21)$$

This, in particular, means that for $f \in C([a,b])$ the equation $Ly = f$ has a unique solution $y \in C^2([a,b])$ satisfying the boundary conditions (8.14), and this solution is given by the right-hand side of formula (8.21). The same is true for $f \in L^2([a,b])$, but then $y \in C^1([a,b])$ and $y'' \in L^2([a,b])$ in the distribution sense. The eigenvalue problem (8.13)–(8.14) can be rewritten in the equivalent form

$$y(x) = \lambda \int_a^b G(x,s)y(s)\,ds, \qquad (8.22)$$

i.e., in the form of an eigenvalue problem for the integral operator L^{-1} with continuous kernel $G(x,s)$. It is obvious that L^{-1} does not have zero as an eigenvalue and that its eigenvalues in $L^2([a,b])$ are precisely the numbers of the form λ^{-1}, where $\lambda \in \sigma(L)$, since it follows from (8.22), the relation $y \in L^2([a,b])$, and the continuity of G that $y \in C([a,b])$, and then $y \in C^2([a,b])$. The kernel G is symmetric, i.e., $G(x,s) = G(s,x)$ for all $x,s \in [a,b]$, since the operator L^{-1} itself is symmetric in $L^2([a,b])$ due to the symmetry of L. Therefore by the Hilbert-Schmidt theorem (cf., for example, Reed and Simon 1972–1979, Theorem VI.16) the operator L^{-1} has a complete orthogonal system of eigenfunctions. Moreover because the operator L^{-1} is compact, its eigenvalues $\{\mu_j : j = 1,2,\ldots\}$ are such that $\mu_j \to 0$ as $j \to \infty$. Thus $\lambda_j = \mu_j^{-1}$ are the eigenvalues of the operator L and $|\lambda_j| \to \infty$. Moreover we have the following theorem.

Theorem 2.105. a) *There exists a complete orthonormal system $\{\psi_j : j = 1,2,\ldots\}$ of eigenfunctions (in $L^2([a,b])$) of the operator L of the form (8.12) with boundary conditions (8.14).*

b) *All the eigenvalues $\{\lambda_j : j = 1,2,\ldots\}$ are simple and $\lambda_j \to +\infty$ as $j \to +\infty$.*

c) *If the function $f \in C^1([a,b])$ has a piecewise-continuous second derivative f'' and satisfies the boundary conditions (8.14), then it can be expanded in an absolutely and uniformly convergent series in the system of functions $\{\psi_j : j = 1,2,\ldots\}$:*

$$f = \sum_{j=1}^{\infty} c_j\psi_j(x), \quad c_j = (f,\psi_j) = \int_a^b f(x)\psi_j(x)\,dx. \qquad (8.23)$$

The proof can be found, for example, in Vladimirov 1967.

The last part of this theorem (part c)) is called *Steklov's theorem*. In the case of the boundary conditions $y(a) = y(b) = 0$ the assumption that the second derivative exists can be dispensed with. For applications it is also an important question when the series (8.23) can be differentiated termwise without losing its uniform convergence. We shall restrict ourselves to the simple remark, which is nevertheless sufficient for the majority of applications, that if both f and Lf satisfy the conditions of part c) in Theorem 2.105, then the series (8.23) will remain absolutely and uniformly convergent when differentiated termwise twice.

We now give some examples of the explicit solution of a Sturm-Liouville problem in the case when $q \equiv 0$. We shall assume also that $a = 0$ and $b = l$ (this can always be achieved by a translation).

Example 2.106. $L = -d^2/dx^2$, with boundary conditions $y(0) = y(l) = 0$. Then the system

$$\left\{ \sqrt{\frac{2}{l}} \sin \frac{k\pi x}{l} : k = 1, 2, \ldots \right\}$$

is a complete orthonormal system of eigenfunctions with eigenvalues $\lambda_k = (k\pi/l)^2$.

Example 2.107. $L = -d^2/dx^2$, with boundary conditions $y'(0) = y'(l) = 0$. There is a complete orthonormal system of eigenfunctions

$$\left\{ 1/\sqrt{l}; \sqrt{\frac{2}{l}} \cos \frac{k\pi x}{l} : k = 1, 2, \ldots \right\}$$

and the eigenvalues are respectively $\lambda_k = (k\pi/l)^2$, $k = 0, 1, 2, \ldots$.

Example 2.108. $L = -d^2/dx^2$, with boundary conditions $y(0) = y'(l) = 0$. The complete orthonormal system of eigenfunctions is

$$\left\{ \sqrt{\frac{2}{l}} \sin \frac{\pi}{2l}(2k + 1)x : k = 0, 1, \ldots \right\},$$

and the eigenvalues are $\lambda_k = \left[\frac{\pi}{2l}(2k + 1) \right]^2$.

Example 2.109. $L = -d^2/dx^2$, with boundary conditions $y(0) = 0$, $y'(l) + \gamma y(l) = 0$. A complete orthonormal system of eigenfunctions can be constructed in the form

$$\left\{ c_k \sin \mu_k x : k = 1, 2, \ldots \right\},$$

where $\{ \mu_k : k = 1, 2, \ldots \}$ are the positive roots of the equation $\tan \mu l = -\mu/\gamma$, each taken once, and c_k are normalized coefficients determined by the conditions

$$c_k^{-2} = \int_0^l \sin^2 \mu_k x \, dx = \frac{l}{2} - \frac{\sin 2\mu_k l}{4\mu_k}.$$

We mention also the oscillation properties of the eigenfunctions. We shall assume that the eigenvalues are arranged in increasing order:

$$\lambda_1 < \lambda_2 < \lambda_3 \cdots \tag{8.24}$$

and the eigenfunctions $\{\psi_j : j = 1, 2, \ldots\}$ of the problem (8.13)–(8.14) are numbered accordingly. The following theorem holds, known as *Sturm's theorem*, (cf., for example, Levitan and Sargsyan 1970, Chapt. 1, Theorem 3.3).

Theorem 2.110. *The function ψ_j has exactly $(j-1)$ zeros on the open interval (a, b).*

All these zeros are simple zeros, since by the uniqueness theorem the conditions $L\psi = \lambda\psi$, $\psi(x_0) = 0$, $\psi'(x_0) = 0$, would imply that $\psi \equiv 0$.

8.3. Periodic and Antiperiodic Boundary Conditions. For convenience we shall assume that $a = 0$ and $b = l$. Instead of the boundary conditions (8.14) we can consider the more general conditions

$$\gamma_1 y(0) + \gamma_2 y'(0) + \gamma_3 y(l) + \gamma_4 y'(l) = 0,$$
$$\delta_1 y(0) + \delta_2 y'(0) + \delta_3 y(l) + \delta_4 y'(l) = 0,$$

where $\gamma = (\gamma_1, \gamma_2, \gamma_3, \gamma_4)$ and $\delta = (\delta_1, \delta_1, \delta_3, \delta_4)$ are linearly independent real vectors. The most important examples are the *periodic* and *antiperiodic boundary conditions* having the respective forms

$$y(0) = y(l), \quad y'(0) = y'(l) \tag{8.25}$$

and

$$y(0) = -y(l), \quad y'(0) = -y'(l). \tag{8.26}$$

Let us assume that the potential q of the operator L of the form (8.12) is periodic with period l. Then if $y \in C^2(\mathbb{R})$, $Ly = \lambda y$ on the entire real axis, and conditions (8.25) (resp. (8.26)) hold, by the uniqueness theorem the function y is periodic with period l (resp. antiperiodic) i.e., $y(x + l) = y(x)$ identically (resp. $y(x + l) = -y(x)$). Of course the converse is true also, i.e., periodic and antiperiodic functions automatically satisfy (8.25) and (8.26) respectively. Thus if the potential q is periodic with period l, then the eigenvalue problem with conditions (8.25) or (8.26) is equivalent to finding nontrivial periodic (resp. antiperiodic) solutions of the equation $Ly = \lambda y$ defined on the entire axis and finding the corresponding values of the parameter λ. The problem itself is called for short the *periodic* (resp. *antiperiodic*) problem for the operator L.

Assertions a) and c) of Theorem 2.105 hold for the periodic and antiperiodic problem; there is a Green's function that is constructed in analogy with the considerations of Sect. 8.1. However assertion b) of Theorem 2.105 is not always true (the multiplicity of an eigenvalue may be 2), as can be seen in even the very simple example that follows.

Example 2.111. Let $q \equiv 0$ and $l = 2\pi$. Then the periodic problem has as a complete orthogonal system of eigenfunctions the standard Fourier system

$$\left\{ \frac{1}{\sqrt{2\pi}}; \frac{1}{\sqrt{\pi}} \cos nx; \frac{1}{\sqrt{\pi}} \sin nx : n = 1, 2, \dots \right\}$$

with the simple eigenvalue 0 and the double eigenvalues n^2, $n = 1, 2, \dots$.

The antiperiodic eigenfunctions in this case have the form

$$\left\{ \frac{1}{\sqrt{\pi}} \cos \left(nx + \frac{x}{2} \right), \frac{1}{\sqrt{\pi}} \sin \left(nx + \frac{x}{2} \right) : n = 0, 1, 2, \dots \right\},$$

and all eigenvalues are double and equal to $\left(n + \frac{1}{2} \right)^2$, $n = 0, 1, 2, \dots$.

Of course all the periodic and antiperiodic eigenfunctions of an operator L of the form (8.12) with l-periodic potential q are periodic with period $2l$. Let us consider all the $2l$-periodic eigenfunctions and denote the corresponding eigenvalues by $\lambda_0, \lambda_1, \lambda_2, \dots$, arranged in increasing order

$$\lambda_0 \leq \lambda_1 \leq \lambda_2 \leq \dots$$

(and taking account of multiplicity in the enumeration). It then turns out that

$$\lambda_0 < \lambda_1 \leq \lambda_2 < \lambda_3 \leq \lambda_4 < \lambda_5 \leq \lambda_6 < \dots \tag{8.27}$$

and the eigenvalues $\lambda_0, \lambda_3, \lambda_4, \lambda_7, \lambda_8, \dots$ correspond to l-periodic eigenfunctions, while the eigenvalues $\lambda_1, \lambda_2, \lambda_5, \lambda_6, \dots$ correspond to l-antiperiodic eigenfunctions (cf., for example, Coddington and Levinson 1955). Thus in the sequence (8.27) after the simple eigenvalue λ_0 corresponding to an l-periodic eigenfunction pairs of eigenvalues corresponding to antiperiodic and periodic eigenfunctions alternate. In the generic situation all the eigenvalues are simple, but potentials all of whose periodic and antiperiodic eigenvalues from some index on are double also play an important role. Such potentials are called *finite-gap* potentials (for reasons that will become apparent below) and can be explicitly described as the solutions of certain nonlinear Novikov ordinary differential equations. They play an important role in the study of solutions of the Korteweg-de Vries equation and its higher analogs that are periodic on the spatial variables (cf., for example, Zakharov, Manakov, Novikov, and Pitaevskij 1980).

8.4. Asymptotics of the Eigenvalues and Eigenfunctions in the Regular Case. The question is one of asymptotics over the spectral parameter λ or the index j of the eigenvalue and eigenfunction as $j \to \infty$ (or as $\lambda = \lambda_j \to \infty$). The asymptotics can be found using perturbation theory. To be specific, suppose first that $\sin \alpha \neq 0$ in (8.14). For simplicity we shall assume that $a = 0$, $b = \pi$. We rewrite the equation $Ly = \lambda y$ in the form $L_0 y - \lambda y = -qy$, where $L_0 = -d^2/dx^2$ and by explicitly solving the equation $L_0 y - \lambda y = f$, with known right-hand side f and initial conditions $y(0) = 1$, $y'(0) = -\cot \alpha$, we then set $f = -qy$. Then for y we obtain an integral equation

$$y(x, \lambda) = \cos sx + \frac{h}{s} \sin sx + \frac{1}{s} \int_0^x \sin[s(x - \tau)]q(\tau)y(\tau, \lambda) \, d\tau, \qquad (8.28)$$

where $h = -\cot \alpha$ and $s = \sqrt{\lambda}$ (cf. Levitan and Sargsyan 1970, Marchenko 1977). In particular we learn from this that the (unnormalized) eigenfunction ψ of the operator L with eigenvalue $\lambda = s^2$ has the form

$$y(x, \lambda) = \cos sx + O(1/s)$$

for large s. This formula can easily be made more precise by substituting the asymptotic relation it gives into the right-hand side of (8.28), yielding

$$y(x, \lambda) = \cos sx + \frac{h}{s} \sin sx + \frac{\sin sx}{2s} \int_0^x q(\tau) \, d\tau + O(1/s^2). \qquad (8.29)$$

Analyzing the s for which the boundary condition at the right-hand endpoint $x = \pi$ holds, it is not difficult to obtain also the asymptotic formulas for the eigenvalues $\lambda_n = s_n^2$ $(n = 0, 1, 2, \ldots)$ as $n \to \infty$:

$$s_n = n + \frac{c}{n} + O(1/n^2), \quad c = \frac{1}{\pi}\left(h + H + \frac{1}{2}\int_0^\pi q(\tau) \, d\tau\right), \qquad (8.30)$$

where $H = \cot \beta$ (we are assuming here that $\sin \beta \neq 0$). For the normalized eigenfunctions $\psi_n(x) = c_n y(x, \lambda_n)$ we have the asymptotics

$$\psi_n(x) = \sqrt{\frac{2}{\pi}}\left[\cos nx + \frac{\beta(x)}{n} \sin nx\right] + O(1/n^2), \qquad (8.31)$$

where

$$\beta(x) = -cx + h + \frac{1}{2}\int_0^x q(\tau) \, d\tau,$$

and c is the same as in (8.30).

In the case $\sin \alpha = 0$ (i.e., for boundary condition $y(0) = 0$) the integral equation (8.28) must be replaced by the equation

$$y(x, \lambda) = \frac{\sin sx}{s} + \frac{1}{s}\int_0^x \sin[s(x - \tau)]q(\tau)y(\tau, \lambda) \, d\tau \qquad (8.32)$$

and for $\sin \beta \neq 0$ for the eigenvalues $\lambda_n = s_n^2$, $n = 0, 1, 2, \ldots$ and the normalized eigenfunctions $\psi_n(x) = c_n y(x, \lambda_n)$ we obtain asymptotic formulas as $n \to +\infty$:

$$s_n = n + \frac{1}{2} + \frac{H_1}{\pi(n + 1/2)} + O(1/n^2), \quad H_1 = H + \frac{1}{2}\int_0^\pi q(\tau) \, d\tau, \qquad (8.33)$$

where $H = \cot \beta$; and

$$\psi_n(x) = \sqrt{\frac{2}{\pi}} \sin \left(n + \frac{1}{2} \right) x + O(1/n). \qquad (8.34)$$

Finally, in the case of the boundary conditions $y(0) = y(\pi) = 0$ the use of the same integral equation (8.32) leads to the following asymptotic relations for $\lambda_n = s_n^2$, $(n = 1, 2, \ldots)$ and ψ_n as $n \to +\infty$:

$$s_n = n + \frac{\alpha_1}{n} + O(1/n^2), \quad \alpha_1 = \frac{1}{2\pi} \int_0^\pi q(\tau) \, d\tau, \qquad (8.35)$$

$$\psi_n(x) = \sqrt{\frac{2}{\pi}} \sin nx + O(1/n). \qquad (8.36)$$

These asymptotic relations can be sharpened further by making additional assumptions on the smoothness of the potential q. In particular if $q \in C^\infty([0, l])$, there are asymptotic series for s_n and ψ_n (cf. Levitan and Sargsyan 1970 and Marchenko 1977).

The question of the asymptotics of the eigenvalues of the periodic and antiperiodic problem is answered somewhat differently, since the neighbouring periodic or antiperiodic eigenvalues in (8.27) are quite close together in the case of a smooth potential and for that reason it is difficult to distinguish them and write out their asymptotics. Nevertheless it is not difficult to find the asymptotics of the eigenvalues. To be specific, we introduce the following notation:

$$\mu_0^\pm = \lambda_0, \ \mu_1^- = \lambda_1, \ \mu_1^+ = \lambda_2, \ \mu_2^- = \lambda_3, \ \mu_2^+ = \lambda_4, \ \mu_3^- = \lambda_5, \ \mu_3^+ = \lambda_6, \ldots, \tag{8.37}$$

where the λ_j are taken from (8.27), so that $\mu_{2k}^\pm$ are the eigenvalues of the periodic problem and $\mu_{2k+1}^\pm$ are the eigenvalues of the antiperiodic problem. Let the π-periodic extension $\hat{q}$ of the potential q to the whole real axis belong to the Sobolev space $H_{\mathrm{loc}}^n(\mathbf{R})$, i.e., $\hat{q}^{(p)} \in L^2$ on each finite interval wth $p = 0, 1, \ldots, n$. Then as $k \to +\infty$

$$\sqrt{\mu_k^\pm} = k + \sum_{1 \le 2j+1 \le n+2} a_{2j+1}(2k)^{-2j-1} \pm |e_n(2k)|(2k)^{-n-1} + \gamma_k^\pm k^{-n-2},$$

$$(8.38)$$

where k can be either even or odd and the numbers a_{2j+1} are independent of k,

$$a_1 = \frac{1}{\pi} \int_0^\pi q(\tau) \, d\tau, \quad e_n(\xi) = \frac{1}{\pi} \int_0^\pi q^{(n)}(x) e^{-i\xi x} \, dx$$

and

$$\sum_{k=0}^\infty |\gamma_k^\pm|^2 < \infty$$

(cf. Marchenko 1977, Theorem 1.5.2). In particular it can be seen from (8.38) that the numbers $\mu_k^\pm$ are quite close together for smooth q; if $\hat{q} \in C^\infty(\mathbf{R})$, then $\mu_k^+ - \mu_k^- = O(k^{-N})$ as $k \to +\infty$ for any fixed N.

8.5. The Schrödinger Operator on a Half-Line. On the half-line $[0, +\infty)$ consider the one-dimensional Schrödinger operator L of the form (8.12) with a real potential $q = q(x)$ that is continuous on $[0, \infty)$. We assume that q is continuous here and below only for simplicity; it would suffice that q be measurable and locally bounded, and in the majority of cases mere local integrability of q would suffice. Our purpose is to find a way to construct a complete orthogonal system of eigenfunctions of the operator L in the space $L^2([0, +\infty))$ (not necessarily a discrete system). This means that we must find a space M with measure dk and a family of eigenfunctions $\psi_k \in \mathcal{D}'((0, +\infty))$ of the operator L that is measurable with respect to k and has real eigenvalues λ_k such that the "Fourier transform"

$$u \mapsto \tilde{u}(k) = \int u(x)\overline{\psi_k(x)}\, dx, \quad u \in C_0^\infty((0, +\infty)), \qquad (8.39)$$

can be extended to a unitary isomorphism

$$U : L^2([0, +\infty)) \to L^2(M, dk).$$

The unitary character of the operator U means first of all that it is an isometry, i.e., that Plancherel's formula

$$\int_0^\infty |u(x)|^2\, dx = \int_M |\tilde{u}(k)|^2\, dk, \qquad (8.40)$$

holds, and second that the operator U is an epimorphism, a property which can be expressed by saying that the adjoint of the operator U:

$$U^* : L^2(M, dk) \to L^2([0, +\infty)),$$

having the form

$$(U^* f)(x) = \int f(k)\psi_k(x)\, dk, \qquad (8.41)$$

is also an isometry. The isometric character of the operator U^* can be written (neglecting a precise description of the integrals and the justification for reversing the order of integration) in the form of the *orthogonality relations*

$$\int_0^\infty \psi_k(x)\overline{\psi_{k'}(x)}\, dx = \delta(k, k'), \qquad (8.42)$$

where $\delta(k, k')$ is defined by the relation

$$\int \delta(k, k')\varphi(k')\, dk' = \varphi(k)$$

on suitable functions φ on M. The isometric character of U itself (i.e., equality (8.40)), upon substituting the expression $\tilde{u}(k)$ for u and formally changing the order of integration, becomes the relation

$$\int \psi_k(x)\overline{\psi_k(x')}\,dk = \delta(x - x'), \tag{8.43}$$

called the *completeness relation*. For that reason the usual method of proving that the generalized eigenfunctions $\{\psi_k : k \in M\}$ form a complete orthonormal system consists of verifying the orthogonality and completeness relations (8.42) and (8.43) interpreted in one sense or another. We note that the validity of each of these relations depends essentially on the choice of the measure dk on M. This choice of measure, as a rule, is the main nontrivial element in the construction of a complete orthogonal system of eigenfunctions (the space M itself usually consists simply of all the tempered eigenfunctions or, what is more convenient, some labels for them).

We remark that if $\psi_k \in \mathcal{D}'\big((0, +\infty)\big)$ is an eigenfunction of the operator L, i.e. $L\psi_k = \lambda_k \psi_k$, then $\psi_k \in C^2\big((0, +\infty)\big)$ so that the integral in (8.39) can be understood in the ordinary sense. The function $k \mapsto \lambda_k$ is measurable on M, since we assumed that the vector-valued function $k \mapsto \psi_k$ is measurable. The operator L, which is defined a priori on functions $u \in C_0^\infty\big((0, \infty)\big)$, becomes the operator L of multiplication by the function λ_k in the space $L^2(M, dk)$ when it is acted on by the isomorphism U, i.e.,

$$ULu = \Lambda U u, \quad u \in C_0^\infty\big((0, \infty)\big). \tag{8.44}$$

We enlarge the domain of definition of the operator Λ, including in it all functions $v \in L^2(M, dk)$ for which the function $k \mapsto \lambda_k v(k)$ again belongs to $L^2(M, dk)$. We obtain a self-adjoint operator, which we shall again denote by Λ (the identity (8.44), of course, is not lost in this enlargement). But then the operator $A = U^{-1}\Lambda U$ is a self-adjoint operator in $L^2\big((0, +\infty)\big)$ which is an extension of the operator L defined on $C_0^\infty\big((0, +\infty)\big)$.

Thus the possibility of applying the scheme described in Sect. 8.1 is guaranteed in advance if we choose some self-adjoint extension of the operator L. Naturally this choice is not arbitrary as a rule, but dictated by the original problem of mathematical physics that we are trying to solve.

8.6. Essential Self-Adjointness and Self-Adjoint Extensions. The Weyl Circle and the Weyl Point. We now recall certain facts of the abstract theory of unbounded operators in Hilbert space (for details and proofs cf., for example, Najmark 1969; Reed and Simon 1972–1979, Chapt. 8; Birman and Solomyak 1980; or Berezin and Shubin 1983, Appendix I). A self-adjoint operator A in a Hilbert space $\mathcal{H}$ is always *closed*, i.e., if $u_n \in D(A)$, $u_n \to u$, and $Au_n \to v$ in $\mathcal{H}$, then $u \in D(A)$ and $Au = v$. Let us begin our study with a symmetric operator A_0 with domain of definition $D(A_0)$ dense in $\mathcal{H}$. Then we can first consider the closure of the operator A_0, namely the operator $\bar{A}_0$, whose graph is the closure of the graph of the operator A_0. This operator will again be symmetric, but by no means must it be self-adjoint. If the operator $\bar{A}_0$ is self-adjoint, then A_0 is called *essentially self-adjoint*. In this case $\bar{A}_0 = A_0^*$.

The operator A_0^* is always closed and $\bar{A}_0 \subset A_0^*$, in view of the symmetry of A_0. If the self-adjoint operator A is an extension of A_0, then it is contained between $\bar{A}_0$ and A_0^*, i.e.,

$$\bar{A}_0 \subset A \subset A_0^*$$

(this means that $D(\bar{A}_0) \subset D(A) \subset D(A_0^*)$ and A is the restriction of the operator A_0^* to $D(A)$ and A is an extension of the operator $\bar{A}_0$). In particular, if the operator A_0 is essentially self-adjoint, then its unique self-adjoint extension is its closure $\bar{A}_0$. In the general case it is necessary to consider the so-called *defect subspaces*: the closed subspaces in $\mathcal{H}$ of the form

$$N_i = \mathrm{Ker}\,(A_0^* - iI), \quad N_{-i} = \mathrm{Ker}\,(A_0^* + iI)$$

(i.e., the eigenspaces of the operator A_0^* with eigenvalues $\pm i$), whose dimensions n_+ and n_- are called the *defect numbers* or *defect indices* of the operator A_0. There exists a direct sum decomposition

$$D(A_0^*) = D(\bar{A}_0) \dotplus N_i \dotplus N_{-i},$$

from which it follows in particular that the operator A_0 is essentially self-adjoint if and only if both defect numbers $n_\pm$ are 0. A symmetric operator A_0 has a self-adjoint extension if and only if $n_+ = n_-$ and then all its self-adjoint extensions A are obtained by taking some unitary operator $V : N_i \to N_{-i}$ and setting

$$D(A) = \{x_0 + z + Vz : x_0 \in D(\bar{A}_0), \quad z \in N_i\}.$$

When this is done, naturally, A is the restriction of A_0^* to $D(A)$, so that for the vectors in brackets

$$A(x_0 + z + Vz) = \bar{A}_0 x_0 + iz - iVz.$$

We now return to the study of the Schrödinger operator on the half-line $[0, \infty)$. We begin with an operator A_0 that is defined by an expression of the form (8.12) with a real potential $q \in C([0, +\infty))$ having domain of definition $C_0^\infty((0, +\infty))$. Thus the functions of $D(A_0)$ vanish in a neighborhood of 0 and $+\infty$. The operator A_0^* is also defined by the expression for L, but for all those $u \in L^2([0, +\infty))$ such that $Lu \in L^2([0, +\infty))$ if applying L is taken in the sense of distribution theory. The operator $\bar{A}_0$ is called the *minimal operator* defined by the expression L in this case and is denoted $L_{\min}$ while the operator A_0^* is called the *maximal operator* and denoted $L_{\max}$. The defect subspaces $N_{\pm i}$ consist of solutions u of the equations $(L \mp iI)u = 0$ such that $u \in L^2([0, +\infty))$. In particular it is clear from this that $n_\pm \le 2$. Complex conjugation gives an isomorphism between N_i and N_{-i}, so that $n_+ = n_-$. Therefore the operator $L_{\min}$ always has self-adjoint extensions.

The number n_+ $(= n_-)$ may assume one of the values $0, 1, 2$. However the case $n_+ = n_- = 0$ is impossible in the present situation. In fact for $x = 0$ we introduce the boundary conditions

$$y(0)\cos\alpha + y'(0)\sin\alpha = 0 \qquad (8.45)$$

and consider the operator $A_{0\alpha}$ defined by the expression L on functions $y \in C^\infty([0,\infty))$ that equal 0 for large x and satisfy this boundary condition. The operator $A_{0\alpha}$ is symmetric and its closure $L_{\alpha,\min} = \bar{A}_{0\alpha}$ is a symmetric operator that is an extension of the minimal operator $L_{\min}$, but does not coincide with it. This means in particular that $L_{\min} \neq L_{\max}$, so that the defect indices of the operator $L_{\min}$ are nonzero.

Boundary conditions of the form (8.45) arise naturally in problems of mathematical physics, and the question arises whether such a condition suffices for obtaining a self-adjoint operator. In other words, is the operator $A_{0\alpha}$ described above essentially self-adjoint (and the operator $L_{\alpha,\min}$ self-adjoint)? The answer to this question depends on the behavior of the potential $q(x)$ as $|x| \to \infty$. We first point out a sufficient condition for the operator $L_{\alpha,\min}$ to be self-adjoint (cf. Berezin and Shubin 1983, Sect. 1.1).

Theorem 2.112 (Sears). *Let $q(x) \geq -Q(x)$, where Q is a continuous positive nondecreasing function on $[0, +\infty)$ for which*

$$\int_0^\infty Q^{-1/2}(x)\,dx = \infty.$$

Then the operator $L_{\alpha,\min}$ is self-adjoint.

In particular if, for example, $q(x) \geq -Ax^2 - B$, then the hypothesis of the theorem is satisfied and the operator $L_{\alpha,\min}$ is self-adjoint for any α. However, even with $q(x) = -|x|^{2+\varepsilon}$ where $\varepsilon > 0$, the operator $L_{\alpha,\min}$ is no longer self-adjoint (cf. Berezin and Shubin 1983). We note that the defect indices $n_+ = n_-$ of the operator $L_{\alpha,\min}$ may now be only 0 or 1, since the space of solutions of the equation $Ly = iy$ satisfying the boundary condition (8.45) is at most one-dimensional. The defect index here is 1 if and only if a nontrivial solution of the equation $Ly = iy$ satisfying condition (8.45) belongs to $L^2([0, +\infty))$.

The self-adjointness of the operator $L_{\alpha,\min}$ is equivalent to $L_{\alpha,\min} = L_{\alpha,\max}$ and under this condition we can denote the operator $L_{\alpha,\min}$ (= $L_{\alpha,\max}$) simply by L_α.

A necessary and sufficient condition for essential self-adjointness can be given in terms of a certain procedure for passing to the limit as $l \to \infty$ in problems on the closed interval $[0, l]$. To be specific, we choose $\lambda \in \mathbb{C}$ and let $\varphi(x) = \varphi(x, \lambda)$ and $\theta(x) = \theta(x, \lambda)$ be two (complex) solutions of the equation $Ly = \lambda y$ satisfying the initial conditions

$$\varphi(0) = \sin\alpha, \quad \varphi'(0) = -\cos\alpha; \quad \theta(0) = \cos\alpha, \quad \theta'(0) = \sin\alpha.$$

We shall suppose that $\lambda \notin \mathbb{R}$. We set

$$w = -\frac{\theta(l)z + \theta'(l)}{\varphi(l)z + \varphi'(l)}.$$

Then as z ranges over the real axis (including ∞) w ranges over the circle C_l. We denote by K_l the disk of which it is the boundary. Then if $l' > l$, we have $K_{l'} \subset K_l$, i.e., as l increases the disks K_l contract, and then in the limit as $l \to \infty$ the circles C_l tend either to a limiting circle or to a limiting point, called the *Weyl limit-circle* or the *Weyl limit-point* respectively. It can be shown that the operator $L_{\alpha,\min}$ is self-adjoint if and only if for some (or any) nonreal λ the case of a Weyl limit-point holds for the problem $Ly = \lambda y$.

8.7. The Case of an Increasing Potential. Consider the Schrödinger operator (8.12) in $L^2([0, +\infty))$ with the boundary condition (8.45) and a potential $q \in C([0, +\infty))$ such that $q(x) \to +\infty$ as $x \to +\infty$. This case is the closest to the regular case: under this assumption the operator L_α has a discrete spectrum, i.e., there exists a complete orthonormal system of eigenfunctions $\{\psi_j : j = 1, 2, \ldots\}$ of the operator L_α in $L^2([0, +\infty))$; and if $\{\lambda_j : j = 1, 2, \ldots\}$ are the eigenvalues, then $\lambda_j \to +\infty$ as $j \to \infty$. It can be shown in addition that all the eigenvalues are simple and the eigenfunctions $\psi_j = \psi_j(x)$ decay faster than any exponential of the form e^{-ax}, with $a > 0$ as $x \to +\infty$ (cf. Berezin and Shubin 1983). As in the regular case, ψ_j has exactly $j-1$ zeros on $(0, +\infty)$ (cf. Levitan and Sargsyan 1970), if the eigenvalues are arranged in increasing order.

We also point out that the operator L_α has a discrete spectrum under weaker hypotheses on the potential. To be specific, *Molchanov's theorem* (cf., for example, Najmark 1969, Sect. 24, Theorem 13) asserts that if the potential q is bounded below, then the spectrum is discrete if and only if

$$\lim_{x \to +\infty} \int_x^{x+a} q(t)\, dt = +\infty$$

for any $a > 0$.

Under certain regularity assumptions on the behavior of the potential $q(x)$ as $x \to \infty$ one can prove an asymptotic formula for the distribution function of the eigenvalues as $\lambda \to +\infty$:

$$N(\lambda) = \sum_{\lambda_j < \lambda} 1.$$

This formula has the form

$$N(\lambda) \sim \frac{1}{2\pi}\mathrm{mes}\,\{(x, \xi) : \xi^2 + q(x) < \lambda\} = \frac{1}{\pi} \int\limits_{q(x) < \lambda} \sqrt{\lambda - q(x)}\, dx \quad (8.46)$$

(for a precise statement and proof cf. Kostyuchenko and Sargsyan 1979).

We note that in this case, as in the regular case, the space M and the measure dk discussed in Sect. 8.5 can be taken to be the spectrum of the operator L_α (i.e., the set of its eigenvalues) and the discrete measure on it defined by setting the measure of each point equal to 1.

8.8. The Case of a Rapidly Decaying Potential. In the case when the potential decreases sufficiently rapidly one can also exhibit the space M and the measure dk explicitly, and a complete orthogonal system of eigenfunctions can be constructed using asymptotic considerations for the solutions of the equation $Ly = \lambda y$ as $x \to +\infty$.

We assume first for simplicity that the potential q has compact support, i.e., $q(x) = 0$ for large x. Consider all the complex solutions of the equation $Ly = k^2 y$ (where $k > 0$) satisfying the boundary condition (8.45). They are all proportional to any fixed nontrivial solution and for large x have the form $c_1 e^{ikx} + c_2 e^{-ikx}$ and $c_1 c_2 \neq 0$ when the solution under consideration is nontrivial, since there is a nontrivial real-valued solution in the set under consideration. We choose the solution for which $c_1 = 1$ and denote it by $\varphi(x, k)$, so that

$$\varphi(x, k) = e^{ikx} + S(k)e^{-ikx} \tag{8.47}$$

for large x. It is easy to verify that $|S(k)| = 1$. The coefficient $S(k)$ plays an important role in scattering theory in the theory of the inverse scattering problem (cf. Marchenko 1977). We further set $\psi(x, k) = (2\pi)^{-1/2}\varphi(x, k)$. The half-line $[0, +\infty)$ with Lebesgue measure dk is part of the space M with measure dk and the system $\psi(x, k)$ is part of a complete orthogonal system of eigenfunctions of the operator. It is not difficult to prove that the spectrum of the operator L_α in this case consists of the half-line $[0, +\infty)$ together with a finite collection of simple negative eigenvalues $E_j < 0$, $j = 1, 2, \ldots, N$, to which correspond square-integrable (in fact even exponentially decreasing) eigenfunctions satisfying the boundary condition (8.45). We denote by $\psi_j(x)$ the corresponding eigenfunctions normalized in $L^2([0, +\infty))$. We now set $k_j = i\sqrt{-E_j}$, $j = 1, \ldots, N$, and introduce the space $M = [0, \infty) \cup \{k_1, \ldots, k_N\}$ with measure dk equal to Lebesgue measure on $[0, \infty)$ and equal to the standard discrete measure on the set $\{k_1, \ldots, k_N\}$, under which each point k_j has measure 1. Setting $\psi_k(x) = \psi(x, k)$ and $\psi_{k_j}(x) = \psi_j(x)$, we obtain a complete orthogonal system of eigenfunctions of the operator L_α in a sense analogous to that of Sect. 8.5. Moreover in the spectral representation (8.44) (for the operator L_α) the operator Λ to which L_α maps under the action of the similarity transformation given by the unitary operator U will be the operator of multiplication by k^2. In this situation a precise meaning can easily be given to the orthogonality and completeness relations (8.42) and (8.43) using the usual methods of distribution theory. Another way of proving that we obtain a complete orthogonal system of eigenfunctions is to integrate the resolvent and the corresponding Green's function (the kernel of the operator $(L_\alpha - \lambda I)^{-1}$ for $\lambda \notin \sigma(L_\alpha)$) over a contour enclosing the spectrum in $\mathbb{C}$ and subsequently passing to the limit as the contour is contracted to the positive real axis and the collection of negative eigenvalues (the appearance of eigenfunctions in the limit here is connected with a formula analogous to (8.20)).

All the results on the spectrum and the construction of a complete orthogonal system of eigenfunctions of the operator L_α are preserved if instead of assuming that the potential q has compact support we require only that

$$\int_0^\infty x|q(x)|\,dx < \infty. \tag{8.48}$$

Under this hypothesis, passing to an integral equation analogous to (8.28) only with 0 replaced by ∞, one can prove that for $k > 0$ there exists a unique solution $y(x,k)$ of the equation $Ly = k^2 y$ asymptotic to e^{ikx} as $x \to +\infty$, i.e., such that $y(x,k) = e^{ikx}(1 + o(1))$ as $x \to +\infty$. It is now necessary to write the same asymptotic condition instead of (8.47), i.e., to take $\varphi(x,k) = y(x,k) + S(k)y(x,-k)$, where $S(k)$ is chosen so that the boundary condition (8.45) holds for φ. After this everything that was said about operators with potentials of compact support carries over to the case of potentials satisfying (8.48) (cf. Marchenko 1977, Chapter 3 or Berezin and Shubin 1983, Chapter 2).

8.9. The Schrödinger Operator on the Entire Line. We now consider a Schrödinger operator L of the form (8.12) on $L^2(\mathbb{R})$ with a real-valued potential $q \in C(\mathbb{R})$. For the most part the results described above for an operator on the half-line carry over to this case. One need only keep in mind that an operator on the entire line has properties very close to those of a direct sum of two Schrödinger operators with potential q on $[0,\infty)$ and on $(-\infty, 0]$ and with any boundary conditions of the form (8.45) at the point 0 for each of these operators.

To clarify this we shall show how the operator L on $L^2(\mathbb{R})$ can be regarded on the natural decomposition of $L^2(\mathbb{R})$ into the direct sum $L^2(\mathbb{R}) = L^2((-\infty, 0]) \oplus L^2([0, +\infty))$ as an operator acting on the direct sum on the right-hand side of this decomposition and defined by the boundary conditions $y_1(0) = y_2(0)$, $y_1'(0) = y_2'(0)$ (here y_1 and y_2 are functions on $(-\infty, 0]$ and $[0, +\infty)$ respectively). If we now pass to separate boundary conditions for y_1 and y_2, we obtain the direct sum of two operators on the half-lines. But we already know from our study of the operator on a half-line that a change in the boundary conditions at the endpoint (i.e, the number α of (8.45)) has no influence on the qualitative properties of the spectrum, which depend only on the behavior of the potential at infinity. In accordance with this remark (which can be given a precise meaning using the technique of operator splitting, cf. Glazman 1963, Sect. 2) the following assertions about the Schrödinger operator L on $\mathbb{R}$ sound very natural.

The defect indices of the minimal operator $L_{\min}$ (i.e., of the operator obtained by closure in $C_0^\infty(\mathbb{R})$) can be $(0,0)$, $(1,1)$, or $(2,2)$. A sufficient condition for essential self-adjointness (i.e., self-adjointness of $L_{\min}$ or the vanishing of the defect indices) is that the hypotheses of Sears' theorem hold both as $x \to +\infty$ and as $x \to -\infty$, i.e., that they hold on $[0, +\infty)$ for both $q(x)$ and $q(-x)$. In particular if $q(x) \geq -Ax^2 - B$ for all $x \in \mathbb{R}$, then the operator L_{min} is self-adjoint (and then we shall write L instead of $L_{min} = L_{max}$). A necessary and sufficient condition for L_{min} to be self-adjoint is that the case of a Weyl limit point hold for both $x \to +\infty$ and $x \to -\infty$.

If $q(x) \to +\infty$ as $|x| \to \infty$, then the spectrum of the operator L is discrete and the eigenfunctions ψ_j, $j = 1, 2, \ldots$, decrease faster than any exponential of the form $e^{-a|x|}$ as $|x| \to \infty$; moreover the eigenvalues λ_j are simple and tend to $+\infty$ as $j \to +\infty$. If they are arranged in increasing order, then ψ_j has exactly $j - 1$ zeros. The obvious analog of Molchanov's theorem described in Sect. 8.7 also holds.

Formula (8.46) carries over without change (under suitable assumptions, cf. Kostyuchenko and Sargsyan 1979), the only difference being that in the case of the half-line $[0, \infty)$ in (8.46) only the points $x > 0$ had to be considered, while in the case of the entire line arbitrary points x must be considered.

The case of a potential vanishing at infinity becomes somewhat more complicated. Let (8.48) hold for $q(x)$ or $q(-x)$, i.e.

$$\int_{-\infty}^{+\infty} |xq(x)|\, dx < \infty.$$

Then the spectrum of the operator L again consists of the half-line $[0, +\infty)$ and a finite number of simple negative eigenvalues. In this case, however, the half-line $[0, +\infty)$ is a double spectrum. This means that in the spectral representation (8.44) the operator Λ, which is unitarily equivalent to the operator L, will contain a repeated operation of multiplication by k^2 in $L^2([0, \infty))$. We shall describe a complete orthogonal system of eigenfunctions of the operator L. To do this we introduce the solutions $y_1(x, k)$ and $y_2(x, k)$ of the equation $Ly = k^2 y$ with the asymptotic relations

$$y_1(x, k) = e^{ikx}(1 + o(1)), \quad x \to +\infty,$$

$$y_2(x, k) = e^{-ikx}(1 + o(1)), \quad x \to -\infty.$$

We note that $\overline{y_1(x, k)}$ is again a solution of the same equation satisfying the asymptotic relation $e^{-ikx}(1 + o(1))$ as $x \to +\infty$ and $\overline{y_2(x, k)}$ is a solution satisfying the asymptotic relation $e^{ikx}(1 + o(1))$ as $x \to -\infty$, so that $\overline{y_1(x, k)} = y_2(x, -k)$ and $\overline{y_2(x, k)} = y_1(x, -k)$. For $k > 0$ we can construct two fundamental systems of solutions of the equation $Ly = k^2 y$:

$$\{y_1(x, k), y_1(x, -k)\} \quad \text{and} \quad \{y_2(x, k), y_2(x, -k)\}.$$

In particular the solution $y_2(x, k)$ can be written in the form

$$y_2(x, k) = a(k)y_1(x, -k) + b(k)y_1(x, k).$$

We remark that if $q(x) \equiv 0$, then $y_2(x, k) = y_1(x, -k)$, so that $a(k) \equiv 1$ and $b(k) \equiv 0$. From the physical point of view the solution $y_2(x, k)$ corresponds to the wave function of a quantum particle that becomes a free particle with momentum k as $x \to -\infty$. When the particle passes through a potential barrier (after interacting with the field given by the potential $q(x)$) as $x \to +\infty$, we obtain a superposition of two wave functions of particles with momenta k

(the function $y_1(x, -k)$) and $-k$ (the function $y_1(x, k)$). For that reason $a(k)$ is called the *transmission coefficient* and $b(k)$ the *reflection coefficient*.

Now let M consist of two copies of the half-line $[0, \infty)$ denoted $\mathbf{R}_1^+$ and $\mathbf{R}_2^+$, together with a finite collection of points $k_j = i\sqrt{-E_j}$, $j = 1, \ldots, N$, where $\{E_1, \ldots, E_N\}$ is the set of negative eigenvalues of the operator L. The measure dk on M consists of Lebesgue measure dk on $\mathbf{R}_1^+$ and $\mathbf{R}_2^+$ together with the standard discrete measure on the finite set $\{k_1, \ldots, k_N\}$ equal to 1 at each point. For $k \in M$ we set

$$
\psi_k(x) = \begin{cases}
[\sqrt{2\pi}a(k)]^{-1}y_2(x, k), & k \in \mathbf{R}_1^+, \\
[\sqrt{2\pi}a(k)]^{-1}y_1(x, k), & k \in \mathbf{R}_2^+, \\
\psi_j(x), & k = k_j,
\end{cases}
\tag{8.49}
$$

where ψ_j is a normalized (in $L^2(\mathbf{R})$) eigenfunction of the operator L with eigenvalue E_j. Then $\{\psi_k : k \in M\}$ is a complete orthogonal system of eigenfunctions of the operator L in the sense of Sect. 8.5 (cf. Faddeev 1959). Another such system can be obtained from (8.49) by complex conjugation.

The coefficients $a(k)$ and $b(k)$ play an important role in one-dimensional scattering theory, including the inverse scattering problem, which has played an essential role in the development of modern methods of integrating nonlinear equations (cf. Zakharov, Manakov, Novikov, and Pitaevskij 1980; Faddeev 1959).

8.10. The Hill Operator. A *Hill operator* is a Schrödinger operator L of the form (8.12) on $\mathbf{R}$ whose potential q is periodic , i.e. $q(x + l) = q(x)$ for some $l > 0$. The spectrum of the operator L can be described using the eigenvalues (8.27) or (8.37) corresponding to periodic and antiperiodic boundary values on $[0, l]$. To be specific the spectrum $\sigma(L)$ is the union of closed intervals $[\lambda_0, \lambda_1], [\lambda_2, \lambda_3], [\lambda_4, \lambda_5], \ldots$ which may have common endpoints and are called *spectral zones* or *permitted zones* (this last terminology is motivated by the fact that the energy of the quantum particle described by the Hamiltonian L can assume only values in $\sigma(L)$, i.e., in the permitted zones). The intervals $(-\infty, \lambda_0), (\lambda_1, \lambda_2), (\lambda_3, \lambda_4), \ldots$ complementary to $\sigma(L)$ are called *forbidden zones, gaps* or *lacunae* (there may be only a finite number of lacunae, as already mentioned in Sect. 8.3, and then the potential is said to be finite-gap potential). As can be seen from the asymptotic relation (8.38), in the case of smooth q the lacunae become rapidly smaller at infinity.

A complete orthogonal system of eigenfunctions can be constructed using Floquet theory. On the two-dimensional space of solutions of the equation $Ly = \lambda y$ we consider the shift operator $(Ty)(x) = y(x + l)$ called the *monodromy operator*. Its eigenvectors are called the *Bloch eigenfunctions* of the operator L with eigenvalue λ. Thus if ψ is a Bloch eigenfunction, then $L\psi = \lambda\psi$ and $T\psi = \mu\psi$. Calculating the matrix of the operator T in the basis consisting of the two solutions y_1 and y_2 of the equation $Ly = \lambda y$ satisfying the initial conditions

$$y_1(0) = 1, \ y_1'(0) = 0; \quad y_2(0) = 0, \ y_2'(0) = 1,$$

we verify easily that $\det T$ is equal to the Wronskian of the solutions y_1 and y_2 at the point l, so that by Liouville's formula $\det T = 1$. Therefore if μ_1 and μ_2 are the eigenvalues of the operator T (called *Floquet multipliers*), then $\mu_1 \mu_2 = 1$. Let $D(\lambda) = \mu_1 + \mu_2$ be the trace of the monodromy operator. If $|D(\lambda)| > 2$, then μ_1 and μ_2 are real and if $|\mu_1| > |\mu_2|$, then $|\mu_1| > 1 > |\mu_2|$, from which it follows that each Bloch eigenfunction grows exponentially either as $x \to +\infty$ (for the multiplier μ_1) or as $x \to -\infty$ (for the multiplier μ_2). And if $|D(\lambda)| \le 1$, then $\mu_1 = \bar{\mu}_2$ and $|\mu_1| = |\mu_2| = 1$, so that the Bloch eigenfunction is bounded. Here the equality $\mu_1 = \mu_2 = 1$ (or equivalently $D(\lambda) = 2$) means precisely that λ is a periodic eigenvalue, and the equality $\mu_1 = \mu_2 = -1$ (or $D(\lambda) = -2$) means that λ is an antiperiodic eigenvalue. The spectrum $\sigma(L)$ coincides with the set of λ for which the equation $Ly = \lambda y$ has a bounded solution, i.e., $\sigma(L) = \{\lambda : |D(\lambda)| \le 2\}$.

In what follows we shall assume that $\lambda \in \sigma(L)$, so that $|\mu| = 1$ for each multiplier μ.

A Bloch eigenfunction ψ with multiplier μ can be written in the form

$$\psi(x) = e^{ipx}\varphi(x), \tag{8.50}$$

where the function φ is l-periodic and the number $p \in \mathbf{R}$ called the *quasi-momentum* is chosen so that $\mu = e^{ipl}$. The quasimomentum p is determined only up to a term $2\pi k/l$, where $k \in \mathbf{Z}$. Therefore it can always be chosen to be in the interval $B = (-\pi/l, \pi/l]$, called the *Brillouin zone*. We shall do this in what follows.

The function φ of (8.50) satisfies the equation $L_{(p)}\varphi = \lambda\varphi$, where

$$L_{(p)} = -\frac{d^2}{dx^2} - 2ip\frac{d}{dx} + p^2 + q(x),$$

so that it is a periodic eigenfunction of the operator $L_{(p)}$. This operator has a sequence of periodic eigenvalues

$$\varepsilon_1(p) \le \varepsilon_2(p) \le \dots$$

(counted according to multiplicity) depending continuously on p. The functions $\varepsilon_j(p)$ are called *band functions*. We remark that $\varepsilon_j(-p) = \varepsilon_j(p)$ and for $p \in [0, \pi/l]$ the values of the function $\varepsilon_j(p)$ range over the jth zone of the spectrum (it can be proved that the function $\varepsilon_j(p)$ is strictly monotonic on $[0, \pi/l]$, i.e., on half of the Brillouin zone).

Let $\varphi_{j,p}$ be a periodic eigenfunction of the operator $L_{(p)}$ with eigenvalue $\varepsilon_j(p)$ normalized in $L^2([0, l])$. We may assume that the vector-valued function $p \mapsto \varphi_{j,p}$ is chosen so as to be measurable (in p). Setting $\psi_{j,p}(x) = e^{ipx}\varphi_{j,p}(x)$, we obtain a Bloch eigenfunction of the operator L with eigenvalue $\varepsilon_j(p)$.

We now introduce the space $M = \bigcup_{j=1}^{\infty} B_j$, where each B_j is a copy of the Brillouin zone B, and we assume that $B_i \cap B_j = \emptyset$ for $i \ne j$. We construct a measure dk on M equal to $(2\pi)^{-1}l\,dp$ on each B_j. Now if a point of the space M is given by the pair (j,p), where $j = 1, 2, \ldots$ and $p \in B$, then $\{\psi_{j,p} : (j,p) \in M\}$ will be a complete orthogonal system of eigenfunctions of the Hill operator L.

This construction of a complete orthogonal system of Bloch eigenfunctions was first carried out by I. M. Gel'fand. Details and proofs can be found in Titchmarsh 1946, Chapter 21 and Appendix 8.

The examples given above in Sects. 8.7–8.10 are essentially all the situations where the spectral decomposition of the Schrödinger operator in the singular case can be described more or less explicitly. The spectral decomposition can also be constructed using a limiting passage from the decompositions on finite intervals (cf. Levitan and Sargsyan 1970), but the cases when one can carry out this procedure explicitly are extremely rare. We refer the reader to (Najmark 1969) for information on the spectral decompositions of one-dimensional differential operators of higher order, including the nonself-adjoint case, where the situation becomes much more complicated.

§9. Special Functions

In solving the equations of mathematical physics and boundary-value problems for them it is not always possible to get by with the stock of standard elementary functions. Each equation generates a class of solutions that are not always elementary functions. However among the nonelementary functions encountered in solving the simplest and most important equations there are functions that appear repeatedly and therefore have beeen well studied and given various names. Such functions are customarily called *special functions*. These are most often functions of one variable that arise in separation of variables, such as, for example, the eigenfunctions of a Sturm-Liouville operator L of the form

$$Ly = -(k(x)y')' + q(x)y$$

on some finite or infinite interval. A particularly frequent case occurs when the function $k(x)$ vanishes at one of the endpoints of the interval. In this section we shall study some of the more important special functions (we note, however, that we are omitting certain common functions, for example the gamma and beta functions of Euler, on the grounds that they have no direct bearing on the boundary-value problems of mathematical physics).

9.1. Spherical Functions. A *spherical harmonic* of degree $k = 0, 1, \ldots$ is the restriction to the unit sphere $\mathbb{S}^{n-1} \subset \mathbb{R}^n$ of a homogeneous harmonic polynomial of degree k in $\mathbb{R}^n$.

Example 2.113. Consider the Dirichlet problem for Laplace's equation in the unit ball of the space $\mathbb{R}^3$:

$$\Delta u = 0 \quad \text{for } r < 1, \quad u(r, \theta, \varphi) = f(\theta, \varphi) \quad \text{for } r = 1.$$

Here (r, θ, φ) are spherical coordinates on $\mathbb{R}^3$

As usual we first consider solutions u of Laplace's equation of the form $u(r, \theta, \varphi) = R(r)Y(\theta, \varphi)$. To determine the function R we have *Euler's equation*

$$r^2 R'' + 2r R' - \lambda R = 0,$$

and to determine $Y(\theta, \varphi)$ we have the equation

$$\frac{1}{\sin \theta} \frac{\partial}{\partial \theta} \left(\sin \theta \frac{\partial Y}{\partial \theta} \right) + \frac{1}{\sin^2 \theta} \frac{\partial^2 Y}{\partial \varphi^2} + \lambda Y = 0,$$

and the function Y must be bounded for $0 \leq \varphi \leq 2\pi$, $0 \leq \theta \leq \pi$, and periodic in φ.

We shall also seek the solution of this problem for the function $Y(\theta, \varphi)$ by the method of separation of variables, setting $Y(\theta, \varphi) = \Theta(\theta)\Phi(\varphi)$. This leads to the equations

$$\Phi'' + \mu \Phi = 0,$$

$$\frac{1}{\sin \theta} \frac{d}{d\theta} \left(\sin \theta \frac{d\Theta}{d\theta} \right) + \left(\lambda - \frac{\mu}{\sin^2 \theta} \right) \Theta = 0.$$

It follows from the periodicity of the function $\Phi(\varphi)$ that $\mu = m^2$ and $\Phi(\varphi) = C_1 \cos m\varphi + C_2 \sin m\varphi$, where $m = 0, 1, \ldots$. Thus the function $\Theta(\theta)$ satisfies the equation

$$\frac{1}{\sin \theta} \frac{d}{d\theta} \left(\sin \theta \frac{d\Theta}{d\theta} \right) + \left(\lambda - \frac{m^2}{\sin^2 \theta} \right) \Theta = 0, \quad 0 \leq \theta \leq \pi,$$

and the function Θ must be bounded for $\theta = 0$ and $\theta = \pi$. Let $\cos \theta = t$ and $\Theta(\theta) = X(t)$. We then obtain the equation

$$\frac{d}{dt} \left[(1 - t^2) \frac{dX}{dt} \right] + \left(\lambda - \frac{m^2}{1 - t^2} \right) X = 0, \quad -1 \leq t \leq 1. \tag{9.1}$$

This equation has solutions that are bounded for $|t| \leq 1$ only for $\lambda = k(k+1)$, where k is an integer and $m = 0, \pm 1, \ldots, \pm k$. These solutions are called the *associated Legendre functions* and denoted $P_k^{(m)}(t)$. For each fixed $m = 0, 1, \ldots$ they form a complete orthogonal system on the closed interval $\{t : |t| \leq 1\}$.

The equation for the function $R(r)$ has a general solution of the form $C_1 r^k + C_2 r^{-k-1}$ for these values of λ. Since the problem under consideration is the interior Dirichlet problem, only the solution $C_1 r^k$ should be retained. Thus each solution of Laplace's equation in the ball of the form $R(r)Y(\theta, \varphi)$ coincides with one of the functions $u = r^k Y_k(\theta, \varphi)$, where $Y_k(\theta, \varphi) =$

$$\sum_{m=0}^{k} \left(A_{km} \cos m\varphi + B_{km} \sin m\varphi \right) P_k^{(m)}(\cos \theta).$$

The solution of the Dirichlet problem has the form

$$u(r, \theta, \varphi) = \sum_{k=0}^{\infty} r^k Y_k(\theta, \varphi),$$

where the coefficients A_{km} and B_{km} are chosen so that

$$\sum_{k=0}^{\infty} Y_k(\theta, \varphi) = f(\theta, \varphi).$$

Since the functions $Y_k(\theta, \varphi)$ form an orthogonal system on the sphere, it follows from this that

$$A_{km} = N_{km} \int_0^{2\pi} \int_0^{\pi} f(\theta, \varphi) P_k^{(m)}(\cos\theta) \cos m\varphi \, \sin\theta \, d\varphi \, d\theta,$$

$$B_{km} = N_{km} \int_0^{2\pi} \int_0^{\pi} f(\theta, \varphi) P_k^{(m)}(\cos\theta) \sin m\varphi \, \sin\theta \, d\varphi \, d\theta,$$

$$N_{k0} = \frac{2k+1}{4\pi}, \quad N_{km} = \frac{(2k+1)(k-m)!}{2\pi(k+m)!}, \quad m = 1, 2, \ldots$$

The spherical harmonics on $\mathbb{R}^n$ possess analogous properties. They are constructed most simply for $n = 2$. It is not difficult to verify that in this case $Y_k(\varphi) = a_k \cos k\varphi + b_k \sin k\varphi$, $k = 0, 1, \ldots$.

For an arbitrary $n \geq 2$ we introduce spherical coordinates by the formulas

$$\begin{aligned}
x_1 &= r\cos\theta_1, \\
x_2 &= r\sin\theta_1 \cos\theta_2, \\
&\cdots\cdots\cdots\cdots\cdots\cdots\cdots\cdots\cdots\cdots\cdots \\
x_{n-1} &= r\sin\theta_1 \sin\theta_2 \cdots \sin\theta_{n-2}\cos\theta_{n-1}, \\
x_n &= r\sin\theta_1 \sin\theta_2 \cdots \sin\theta_{n-2}\sin\theta_{n-1}.
\end{aligned}$$

The homogeneous harmonic polynomial $P_{k,n}(x)$ of degree k can be represented in the form

$$P_{k,n}(x) = r^k Y_{k,n}(\omega),$$

where ω is a point on the unit sphere with angular coordinates $\theta_1, \ldots, \theta_{n-1}$. A simple computation (cf. Mikhlin 1977) shows that the dimension of the space $\{Y_{k,n}(\omega)\}$ of spherical functions of order k is

$$m_{k,n} = \frac{(2k+n-2)(k+n-3)!}{k!(n-2)!}.$$

We note that $m_{k,n} = O(k^{n-2})$ as $k \to +\infty$.

If we write the Laplacian in spherical coordinates:

$$\Delta = \frac{\partial^2}{\partial r^2} + \frac{n-1}{r}\frac{\partial}{\partial r} + \frac{1}{r^2}\delta,$$

where δ is the Laplace-Beltrami operator on the sphere:

$$\delta v = \sum_{j=1}^{n-1} \frac{1}{q_j \sin^{n-j-1} \theta_j} \frac{\partial}{\partial \theta_j} \left(\sin^{n-j-1} \theta_j \frac{\partial v}{\partial \theta_j} \right),$$

$$q_1 = 1; \quad q_j = \left(\sin \theta_1 \sin \theta_2 \cdots \sin \theta_{j-1} \right)^2, \quad j \geq 2,$$

we find that the spherical functions $Y_{k,n}(\omega)$ satisfy the equation

$$\delta Y_{k,n}(\omega) - k(k+n-2) Y_{k,n}(\omega) = 0.$$

Thus $Y_{k,n}(\omega)$ are the eigenfunctions of the operator δ with eigenvalues $\lambda_k = k(k+n-2)$. The multiplicity of this eigenvalue is $m_{k,n}$. Being eigenfunctions of the symmetric operator δ, the spherical functions of different orders are orthogonal in $L^2(\mathbb{S}^{n-1})$. It can be shown that the system of spherical functions is complete in $L^p(\mathbb{S}^{n-1})$ for any p with $1 \leq p < \infty$ (cf. Mikhlin 1977).

Finally we note that if $Y_{k,n}(\omega)$ is normed so that $\int Y_{k,n}(\omega)^2 \, d\omega = 1$, then we have the estimate

$$|Y_{k,n}(\omega)| \leq C(n) k^{\frac{n}{2}-1}.$$

9.2. The Legendre Polynomials. The polynomials $P_k(t) = P_k^{(0)}(t)$ appeared in Sect. 9.1 in connection with the solution of the Dirichlet problem in a ball. The Legendre polynomials are closely connected with the Laplacian and can be defined as follows.

Let x and y be points of $\mathbb{R}^3$ and θ the angle between their radius vectors. Then $|x - y|^2 = |x|^2 + |y|^2 - 2|x||y| \cos \theta$. Set $|x| = r$, $|y| = r_0$, $\cos \theta = t$. A fundamental solution of the Laplacian in $\mathbb{R}^3$ has the form $-\dfrac{1}{4\pi |x - y|}$. We remark that

$$\frac{1}{|x-y|} = \frac{1}{\sqrt{r^2 + r_0^2 - 2 r r_0 t}} = \begin{cases} \dfrac{1}{r_0} \dfrac{1}{\sqrt{1 + \rho^2 - 2\rho t}} & \text{for } r < r_0, \ \rho = \frac{r}{r_0} < 1, \\[2ex] \dfrac{1}{r} \dfrac{1}{\sqrt{1 + \rho^2 - 2\rho t}} & \text{for } r_0 < r, \ \rho = \frac{r_0}{r} < 1. \end{cases}$$

The function $\psi(\rho, t) = \dfrac{1}{\sqrt{1 + \rho^2 - 2\rho t}}$, $(0 < \rho < 1, \ -1 \leq t \leq 1)$ is called the *generating function* for the Legendre polynomials. If we expand it in a power series in ρ:

$$\psi(\rho, t) = \sum_{k=0}^{\infty} P_k(t) \rho^k,$$

the coefficients $P_k(t)$ are the Legendre polynomials. An expansion for the fundamental solution of the Laplacian very similar to the expansion in spherical functions considered in Sect. 9.1 corresponds to this expansion:

$$\frac{1}{|x-y|} = \begin{cases} \dfrac{1}{r_0} \sum_{k=0}^{\infty} P_k(\cos\theta)\left(\dfrac{r}{r_0}\right)^k & \text{for } r < r_0, \\[3mm] \dfrac{1}{r} \sum_{k=0}^{\infty} P_k(\cos\theta)\left(\dfrac{r_0}{r}\right)^k & \text{for } r > r_0. \end{cases}$$

We remark that

$$(1 + \rho^2 - 2\rho t)^{-1/2} = 1 - \frac{1}{2}(\rho^2 - 2\rho t) + \frac{3}{8}(\rho^2 - 2\rho t)^2 + \cdots$$

and therefore the coefficient of ρ^k is indeed a polynomial of degree k in t. Moreover, for even k the polynomial $P_k(t)$ contains only even powers of t and for odd k only odd powers of t. In particular $P_k(-t) = (-1)^k P_k(t)$. We remark further that for $t = 1$

$$(1 + \rho^2 - 2\rho)^{-\frac{1}{2}} = \sum_{k=0}^{\infty} P_k(1)\rho^k,$$

so that $P_k(1) = 1$ for all $k = 0, 1, \ldots$.

If we differentiate $\psi(\rho, t)$ on ρ, we find that

$$(1 - 2\rho t + \rho^2)\frac{\partial \psi}{\partial \rho} = (t - \rho)\psi.$$

Since, on the other hand,

$$\frac{\partial \psi}{\partial \rho} = \sum_{k=0}^{\infty} k\rho^{k-1} P_k(t),$$

it follows from this that

$$(k + 1)P_{k+1}(t) - t(2k + 1)P_k(t) + kP_{k-1}(t) = 0. \qquad (9.2)$$

This recurrence formula makes it possible to find all $P_k(t)$ for $k \geq 3$ if we take account of the relation $P_0(t) = 1$ and $P_1(t) = t$.

If we differentiate $\psi(\rho, t)$ on t we find that

$$(1 - 2\rho t + \rho^2)\frac{\partial \psi}{\partial t} = \rho\psi.$$

Since, on the other hand,

$$\frac{\partial \psi}{\partial t} = \sum_{k=0}^{\infty} P_k'(t)\rho^k,$$

it follows from this that

$$P_{k+1}'(t) - 2tP_k'(t) + P_{k-1}'(t) = P_k(t). \qquad (9.3)$$

Combining the recurrence relations (9.2) and (9.3) just obtained, we arrive at a differential equation for $P_k(t)$:

$$(1 - t^2)P_k''(t) - 2tP_k'(t) + k(k+1)P_k(t) = 0. \tag{9.4}$$

Thus $P_k(t)$ are the eigenfunctions of the Sturm-Liouville problem for the operator L:

$$Ly = \frac{d}{dt}\left((1 - t^2)\frac{dy}{dt}\right), \quad -1 \le t \le 1,$$

with eigenvalues $-k(k+1)$. The role of the boundary conditions here is played by the condition that $y(1)$ and $y(-1)$ be finite, i.e., that the solution remain bounded as $t \to 1 - 0$ and $t \to -1 + 0$.

The degree of the polynomial $P_k(t)$ is k for $k = 0, 1, \dots$. Therefore the polynomials $P_k(t)$ form a complete system on $[-1, 1]$. In addition we have the equality $\int_{-1}^{1} P_k(t)P_l(t)\, dt = 0$ for $k \ne l$. It follows from this that the equation $Ly = \lambda y$ has no nontrivial solutions that are bounded on $[-1, 1]$ for $\lambda \ne -k(k+1)$.

Using the recurrence formula (9.2) one can easily verify that

$$\int_{-1}^{1} P_k^2(t)\, dt = \frac{2}{2k + 1}, \quad k = 0, 1, \dots.$$

By direct computation one can verify *Rodrigues' formula*:

$$P_k(t) = \frac{1}{2^k k!} \frac{d^k}{dt^k}\left[(t^2 - 1)^k\right].$$

To do this one must substitute the polynomial $Q_k(t) = \left[(t^2 - 1)^k\right]^{(k)}$ into Eq. (9.4) and then verify that $Q_k(1) = 2^k k!$.

It follows from Rodrigues' formula and Rolle's theorem that the polynomial $P_k(t)$ has exactly k zeros on the interval $(-1, +1)$, and its derivative of order i ($i \le k$) has exactly $k - i$ zeros on this interval.

It can be shown (cf. Tikhonov and Samarskij 1977) that

$$|P_k(t)| \le 1, \quad -1 \le t \le 1, \quad k = 0, 1, \dots.$$

The *associated Legendre functions* $P_k^{(m)}(t)$ were defined in Sect. 9.1 as the solutions of Eq. (9.1). If we make the substitution

$$X(t) = (1 - t^2)^{\frac{m}{2}} Y(t),$$

we obtain the equation

$$(1 - t^2)Y''(t) - 2(m + 1)tY'(t) + [\lambda - m(m + 1)]Y = 0.$$

This same equation can be obtained from Legendre's equation:

$$(1 - t^2)y'' - 2ty' + \lambda y = 0,$$

by differentiating m times on t:

$$(1 - t^2)y^{(m+2)} - 2(m + 1)ty^{(m+1)} + [\lambda - m(m + 1)]y^{(m)} = 0.$$

Thus the functions $(1 - t^2)^{\frac{m}{2}} \dfrac{d^m}{dt^m} P_k(t)$ satisfy Eq. (9.1) and are bounded for $|t| \leq 1$ if $\lambda = k(k+1)$.

It can be shown (cf. Tikhonov and Samarskij 1977) that Eq. (9.1) has no nontrivial solutions that are bounded on $[-1, 1]$ for $\lambda \neq k(k+1)$. If we set

$$P_k^{(m)}(t) = (1 - t^2)^{\frac{m}{2}} \frac{d^m}{dt^m} P_k(t), \quad m = 0, 1, \ldots, k,$$

then

$$\int_{-1}^{1} P_k^{(m)}(t) P_j^{(m)}(t)\, dt = \begin{cases} 0 & \text{for } k \neq j, \\ \dfrac{2}{2k+1} \dfrac{(k+m)!}{(k-m)!} & \text{for } k = j. \end{cases}$$

For $k = m, m+1, \ldots$ the functions $\{P_k^{(m)}(t)\}$ form a complete orthogonal system of functions on the interval $[-1, 1]$.

We note also *Laplace's formulas*:

$$P_k(t) = \frac{1}{\pi} \int_0^{\pi} \left(t \pm \sqrt{t^2 - 1} \cos \varphi \right)^k d\varphi,$$

$$P_k(t) = \frac{1}{\pi} \int_0^{\pi} \frac{d\varphi}{\left(t \pm \sqrt{t^2 - 1} \cos \varphi \right)^{k+1}}$$

and *Mehler's formulas*:

$$P_k(t) = \frac{2}{\pi} \int_0^{\arccos t} \frac{\cos \left(k + \frac{1}{2} \right) \varphi}{\sqrt{2(\cos \varphi - t)}} \, d\varphi,$$

$$P_k(t) = \frac{2}{\pi} \int_{\arccos t}^{\pi} \frac{\sin \left(k + \frac{1}{2} \right) \varphi}{\sqrt{2(t - \cos \varphi)}} \, d\varphi.$$

9.3. Cylindrical Functions. Many problems of mathematical physics lead to the ordinary differential equation

$$x^2 y'' + x y' + (x^2 - n^2) y = 0, \tag{9.5}$$

which is called *Bessel's equation*. The solutions of this equation are called *cylindrical functions* of order n.

Example 2.114. Consider the Dirichlet problem for Laplace's equation

$$\Delta u = 0$$

in the cylinder $Q = \{(x, y, z) \in \mathbf{R}^3 : x^2 + y^2 < 1, -1 < z < 1\}$, with the boundary conditions

$$u = 0 \quad \text{for } x^2 + y^2 = 1, \ -1 < z < 1,$$
$$u(x, y, -1) = f(x, y), \quad u(x, y, 1) = 0 \quad \text{for } x^2 + y^2 < 1.$$

Let (r, φ, z) be cylindrical coordinates. Consider solutions of Laplace's equation in Q having the form $u = v(r, \varphi)Z(z)$ and equal to zero for $r = 1$. Separating variables, we find that

$$\frac{\dfrac{\partial^2 v}{\partial r^2} + \dfrac{1}{r}\dfrac{\partial v}{\partial r} + \dfrac{1}{r^2}\dfrac{\partial^2 v}{\partial \varphi^2}}{v} = -\frac{Z''}{Z} = \lambda.$$

Thus the function v is an eigenfunction of the Dirichlet problem for Laplace's equation in the disk, i.e., it satisfies the equation

$$\frac{\partial^2 v}{\partial r^2} + \frac{1}{r}\frac{\partial v}{\partial r} + \frac{1}{r^2}\frac{\partial^2 v}{\partial \varphi^2} - \lambda v = 0 \tag{9.6}$$

in the disk $r < 1$ and the condition $v = 0$ for $r = 1$.

Consider the solutions of this problem of the special form $v = R(r)\Phi(\varphi)$. Separating variables, we find that

$$\frac{R'' + \dfrac{1}{r}R' - \lambda R}{\dfrac{1}{r^2}R} = -\frac{\Phi''}{\Phi} = \mu,$$

i.e.,

$$\Phi'' + \mu\Phi = 0, \quad 0 \le \varphi \le 2\pi,$$
$$R'' + \frac{1}{r}R' - \lambda R - \frac{\mu}{r^2}R = 0, \quad 0 \le r \le 1.$$

From the periodicity of the function Φ we find that $\mu = n^2$ and $\Phi(\varphi) = A_n \cos n\varphi + B_n \sin n\varphi$. The function R satisfies the equation

$$R'' + \frac{1}{r}R' - \left(\lambda + \frac{n^2}{r^2}\right)R = 0$$

or

$$\frac{1}{r}(rR')' - \frac{n^2}{r^2}R = \lambda R.$$

Multiplying both sides by rR and integrating over the closed interval $[0, 1]$, we find, taking account of the conditions $R(1) = 0$, $|R(0)| < \infty$, that

$$-\int_0^1 rR'^2 \, dr - n^2 \int_0^1 \frac{1}{r}R^2 \, dr = \lambda \int_0^1 rR^2 \, dr.$$

Therefore λ can assume only negative values, $\lambda = -\varkappa^2$. The substitution $\rho = \varkappa r$ leads to Bessel's equation:

$$\frac{1}{\rho}\frac{d}{d\rho}\left(\rho\frac{dR}{d\rho}\right) + \left(1 - \frac{n^2}{\rho^2}\right)R = 0,$$

with the boundary conditions

$$|R(0)| < \infty, \quad R(\varkappa) = 0.$$

We shall see below that $\varkappa$ can assume a countable number of values $\varkappa_1^{(n)}, \varkappa_2^{(n)}, \ldots, \varkappa_j^{(n)}, \ldots$, namely the zeros of the solution $J_n(t)$ of the equation

$$\frac{1}{t}\frac{d}{dt}\left(t\frac{dJ_n}{dt}\right) + \left(1 - \frac{n^2}{t^2}\right)J_n = 0,$$

defined for all $t \geq 0$ and satisfying the condition $|J_n(0)| < \infty$. In this situation the functions $\{J_n(\varkappa_j^{(n)}r)\}$ form a complete orthogonal system with weight r on the closed interval $[0, 1]$. Thus the solution V of the problem (9.6) has the form

$$V(r, \varphi) = \sum_{n=0}^{\infty}\sum_{j=1}^{\infty}\left(A_{nj}\cos n\varphi + B_{nj}\sin n\varphi\right)J_n(\varkappa_j^{(n)}r),$$

and the solution u of the original Dirichlet problem has the form

$$u(r, \varphi, z) = \sum_{n=0}^{\infty}\sum_{j=1}^{\infty}\left(A_{nj}\cos n\varphi + B_{nj}\sin n\varphi\right)J_n(\varkappa_j^{(n)}r)\sinh\left(\varkappa_j^{(n)}(z-1)\right),$$

where

$$A_{nj} = N_{nj}\int_0^{2\pi}\int_0^1 f(x,y)\cos n\varphi J_n(\varkappa_j^{(n)}r)r\,dr\,d\varphi,$$

$$B_{nj} = N_{nj}\int_0^{2\pi}\int_0^1 f(x,y)\sin n\varphi J_n(\varkappa_j^{(n)}r)r\,dr\,d\varphi,$$

with

$$N_{nj} = -\frac{1}{\pi\sinh(2\varkappa_j^{(n)})\int_0^1 rJ_n^2(\varkappa_j^{(n)}r)\,dr}\quad\text{for } n = 1, 2, \ldots$$

and

$$N_{0j} = -\frac{1}{2\pi\sinh(2\varkappa_j^{(0)})\int_0^1 rJ_0^2(\varkappa_j^{(0)}r)\,dr}.$$

The series just obtained converges in mean if $f \in L^2(K_1)$, where K_1 is the disk $\{r \leq 1\}$. This series converges in $C^2(\bar{Q})$ if, for example, $f \in C^3(\bar{K}_1)$ and f vanishes in a neighborhood of $r = 1$ (these conditions can be weakened).

9.4. Properties of the Cylindrical Functions. Consider Bessel's equation

$$x^2 y'' + xy' + (x^2 - \nu^2)y = 0 \tag{9.7}$$

for arbitrary real ν. A solution of it can be sought in the form

$$y = x^\sigma \sum_{j=0}^{\infty} a_j x^j.$$

Substituting this series in (9.7), we obtain the recurrence relations

$$a_0(\sigma^2 - \nu^2) = 0,$$
$$a_1[(\sigma+1)^2 - \nu^2] = 0,$$
$$a_j[(\sigma+j)^2 - \nu^2] + a_{j-2} = 0 \quad \text{for } j = 2, 3, \ldots.$$

If $a_0 \neq 0$, then $\sigma = \pm\nu$. When this happens, $a_1 = 0$ and $a_j = 0$ for all odd j. If $\sigma = \nu$, then

$$a_{2k} = -a_{2k-2}\frac{1}{4k(k+\nu)} \quad \text{for } k = 1, 2, \ldots,$$

i.e.,

$$a_{2k} = (-1)^k \frac{a_0}{4^k k!(\nu+1)(\nu+2)\cdots(\nu+k)}.$$

This formula can be simplified using the gamma function of Euler. We observe that

$$(\nu+1)(\nu+2)\cdots(\nu+k) = \Gamma(\nu+k+1)/\Gamma(\nu+1).$$

If $\nu \neq -m$, where m is a positive integer, then the coefficients a_{2k} are defined for all k. We set $a_0 = \dfrac{1}{2^\nu \Gamma(\nu+1)}$. Then

$$a_{2k} = (-1)^k \frac{1}{2^{2k+\nu}\Gamma(k+1)\Gamma(k+\nu+1)}.$$

For $\nu \geq 0$ the series

$$J_\nu(x) = \sum_{k=0}^{\infty}(-1)^k \frac{1}{\Gamma(k+1)\Gamma(k+\nu+1)}\left(\frac{x}{2}\right)^{2k+\nu} \tag{9.8}$$

converges on the entire line (and even in the entire complex plane). Its sum $J_\nu(x)$ is called the *Bessel function of first kind* of order ν.

If $\sigma = -\nu$, all the preceding calculations remain valid with ν replaced formally by $-\nu$ and lead to the function

$$J_{-\nu}(x) = \sum_{k=0}^{\infty}(-1)^k \frac{1}{\Gamma(k+1)\Gamma(k-\nu+1)}\left(\frac{x}{2}\right)^{2k-\nu} \tag{9.9}$$

This is a second solution of Eq. (9.7), linearly independent of $J_\nu(x)$.

The case $\nu = -m$, $m \in \mathbf{N}$, was excluded above. In this case one must set

$$a_0 = \frac{1}{2^m \Gamma(m+1)},$$

leading to the formulas

$$a_{2k} = (-1)^k \frac{1}{2^{m+2k}\Gamma(k+1)\Gamma(k+m+1)}.$$

Formula (9.9) defines $J_{-\nu}(x)$ only when $\nu \neq m$, $m \in \mathbb{N}$. If $\nu = m$, however, one can formally set $\Gamma(k - m + 1) = \infty$ for $k < m - 1$, so that in this case

$$J_{-m}(x) = \sum_{k=m}^{\infty} (-1)^k \frac{1}{\Gamma(k+1)\Gamma(k-m+1)} \left(\frac{x}{2}\right)^{2k-m} =$$

$$= (-1)^m \sum_{j=0}^{\infty} (-1)^j \frac{1}{\Gamma(j+1)\Gamma(j+m+1)} \left(\frac{x}{2}\right)^{2j+m} = (-1)^m J_m(x).$$

The functions that occur most commonly in applications are

$$J_0(x) = 1 - \left(\frac{x}{2}\right)^2 + \frac{1}{(2!)^2}\left(\frac{x}{2}\right)^4 - \frac{1}{(3!)^2}\left(\frac{x}{2}\right)^6 + \cdots,$$

$$J_1(x) = \frac{x}{2} - \frac{1}{2!}\left(\frac{x}{2}\right)^3 + \frac{1}{2!3!}\left(\frac{x}{2}\right)^5 - \cdots.$$

For $\nu > 0$ the function $J_\nu(x)$ has a zero of order ν at the point $x = 0$ and the function $J_{-\nu}(x)$ has a pole of order ν. For $\nu = 0$ the function $J_0(x)$ assumes a finite value at the point $x = 0$. Every solution of Eq. (9.7) with $\nu = 0$ that is linearly independent of $J_0(x)$ has a logarithmic singularity at the point $x = 0$.

By direct differentiation of the series (9.9) one can verify the following relations between the Bessel functions of different orders:

$$\frac{d}{dx}\left(\frac{J_\nu(x)}{x^\nu}\right) = -\frac{J_{\nu+1}(x)}{x^\nu},$$

$$\frac{d}{dx}\left(x^\nu J_\nu(x)\right) = x^\nu J_{\nu-1}(x).$$

If we carry out the differentiation and add these formulas termwise, we arrive at the recurrence relations

$$J_{\nu+1}(x) + J_{\nu-1}(x) = \frac{2\nu}{x} J_\nu(x), \qquad (9.10)$$

so that $J_{\nu+1}(x)$ is expressed in terms of $J_\nu(x)$ and $J_{\nu-1}(x)$.

We note that the Bessel functions of orders $n + \frac{1}{2}$ can be expressed in terms of elementary functions. In fact,

$$J_{\frac{1}{2}}(x) = \sum_{k=0}^{\infty} \frac{(-1)^k}{k!\Gamma(k+\frac{3}{2})}\left(\frac{x}{2}\right)^{2k+\frac{1}{2}} = \sqrt{\frac{2}{\pi x}} \sum_{k=0}^{\infty} \frac{(-1)^k}{(2k+1)!} x^{2k+1} = \sqrt{\frac{2}{\pi x}} \sin x,$$

since $\Gamma\left(k + \frac{3}{2}\right) = \left(k + \frac{1}{2}\right)\Gamma\left(k + \frac{1}{2}\right) = \cdots = \left(k + \frac{1}{2}\right)\left(k - \frac{1}{2}\right)\cdots\frac{1}{2}\Gamma\left(\frac{1}{2}\right) = \frac{(2k+1)(2k-1)\cdots 3 \cdot 1}{2^{k+1}}\Gamma\left(\frac{1}{2}\right)$ and $\Gamma\left(\frac{1}{2}\right) = \sqrt{\pi}$. Similarly

$$J_{-\frac{1}{2}}(x) = \sum_{k=0}^{\infty} \frac{(-1)^k}{k!\,\Gamma(k+\frac{1}{2})}\left(\frac{x}{2}\right)^{-\frac{1}{2}+2k} = \sqrt{\frac{2}{\pi x}}\sum_{k=0}^{\infty}\frac{(-1)^k}{(2k)!}x^{2k} = \sqrt{\frac{2}{\pi x}}\cos x.$$

From this and the recurrence formula (9.10) it follows that

$$J_{n+\frac{1}{2}}(x) = \sqrt{\frac{2}{\pi x}}\left\{ \sin\left(x - \frac{\pi n}{2}\right)P_n\left(\frac{1}{x}\right) + \cos\left(x - \frac{\pi n}{2}\right)Q_n\left(\frac{1}{x}\right)\right\},$$

where $P_n(y)$ is a polynomial of degree n and $Q_n(y)$ a polynomial of degree $n - 1$. Under the change of variable $y = v(x)x^{-1/2}$ Eq. (9.7) becomes the equation

$$v'' + \left(1 - \frac{\nu^2 - \frac{1}{4}}{x^2}\right)v = 0. \tag{9.11}$$

From this, in particular, we obtain the formulas written above for $J_{\pm\frac{1}{2}}(x)$. But one can draw conclusions about the behavior of $J_\nu(x)$ as $|x| \to \infty$ from Eq. (9.11) for other values of ν as well. As is known (cf., for example, Hartman 1964), the solutions of Eq. (9.11) for large $|x|$ have the form

$$v(x) = a\cos(x + \varphi) + O\left(\frac{1}{|x|}\right).$$

Therefore as $x \to +\infty$

$$J_\nu(x) = \frac{2\nu}{\sqrt{x}}\cos(x + \varphi_\nu) + O\left(\frac{1}{x\sqrt{x}}\right).$$

It can be shown (cf. Tikhonov and Samarskij 1977) that actually

$$J_\nu(x) = \sqrt{\frac{2}{\pi x}}\cos\left(x - \frac{\pi\nu}{2} - \frac{\pi}{4}\right) + O\left(\frac{1}{x\sqrt{x}}\right),$$

$$J_{-\nu}(x) = \sqrt{\frac{2}{\pi x}}\cos\left(x + \frac{\pi\nu}{2} - \frac{\pi}{4}\right) + O\left(\frac{1}{x\sqrt{x}}\right).$$

Using this asymptotic relation, we can distinguish other important classes of cylindrical functions. Thus, for example, a *Neumann function* or *cylindrical function of second kind* of order ν is a solution $N_\nu(x)$ of Bessel's equation for which

$$N_\nu(x) = \sqrt{\frac{2}{\pi x}}\sin\left(x - \frac{\pi\nu}{2} - \frac{\pi}{4}\right) + O\left(\frac{1}{x\sqrt{x}}\right)$$

as $x \to \infty$. The *Hankel functions* of first and second kinds $H_\nu^{(1)}(x)$ and $H_\nu^{(2)}(x)$ are the cylindrical functions for which

$$H_\nu^{(1)}(x) = \sqrt{\frac{2}{\pi x}}e^{i(x-\frac{\pi}{2}\nu-\frac{\pi}{4})} + O\left(\frac{1}{x\sqrt{x}}\right),$$

$$H_\nu^{(2)}(x) = \sqrt{\frac{2}{\pi x}}e^{-i(x-\frac{\pi}{2}\nu-\frac{\pi}{4})} + O\left(\frac{1}{x\sqrt{x}}\right)$$

as $x \to \infty$. It is clear that

$$H_\nu^{(1)}(x) = J_\nu(x) + iN_\nu(x),$$
$$H_\nu^{(2)}(x) = J_\nu(x) - iN_\nu(x).$$

Example 2.115. The solutions of the wave equation

$$u_{tt} = a^2 \left(u_{xx} + u_{yy} \right)$$

describing cylindrical waves can be expressed in terms of Hankel functions. Such solutions have the form

$$u(t, x, y) = v(x, y)e^{i\omega t},$$

where v is a radially-symmetric solution of the equation

$$\frac{\partial^2 v}{\partial x^2} + \frac{\partial^2 v}{\partial y^2} + k^2 v = 0, \quad k = \frac{\omega}{a}.$$

The solutions $v(r)$ of this Helmholtz equation depending only on $r = \sqrt{x^2 + y^2}$ satisfy the equation

$$\frac{1}{r}\frac{d}{dr}\left(r\frac{dv}{dr} \right) + k^2 v = 0,$$

i.e., the function $v(r)$ satisfies Bessel's equation of order 0. Thus the function

$$u(t, r) = H_0^{(1)}(kr)e^{i\omega t} = \sqrt{\frac{2}{\pi kr}}e^{i(\omega t + kr - \frac{\pi}{4})} + O\left(\frac{1}{r\sqrt{r}}\right)$$

is a solution describing divergent cylindrical waves, and the function

$$u_1(t, r) = H_0^{(2)}(kr)e^{i\omega t} = \sqrt{\frac{2}{\pi kr}}e^{i(\omega t - kr + \frac{\pi}{4})} + O\left(\frac{1}{r\sqrt{r}}\right)$$

corresponds to convergent waves.

The *Bessel functions of an imaginary argument* play an important role in mathematical physics. The function $I_\nu(x) = i^{-\nu}J_\nu(ix)$ can be defined as the sum of the series

$$I_\nu(x) = \sum_{k=0}^{\infty} \frac{1}{\Gamma(k+1)\Gamma(k+\nu+1)}\left(\frac{x}{2}\right)^{2k+\nu}$$

or as the solution of the equation

$$y'' + \frac{1}{x}y' - \left(1 + \frac{\nu^2}{x^2}\right)y = 0$$

that is bounded at $x = 0$ (for $\nu = 0$ the condition $I_\nu(0) = 1$ is imposed). As above, it can be shown that as $x \to \infty$

$$I_\nu(x) = e^x\left(\frac{1}{\sqrt{2\pi x}} + O\left(\frac{1}{x\sqrt{x}}\right)\right).$$

By studying the behavior at infinity, one can show easily that there exists a solution $K_\nu(x)$ of the same equation characterized by the relation

$$K_\nu(x) = e^{-x}\left(\sqrt{\frac{\pi}{2x}} + O\left(\frac{1}{x\sqrt{x}}\right)\right).$$

It is not difficult to verify that

$$K_\nu(x) = \frac{i\pi}{2} e^{\nu\frac{\pi}{2}i} H_\nu^{(1)}(ix).$$

The function $K_\nu(x)$ is called the *Macdonald function of order* ν.

Example 2.116. A diffusion process in the xy-plane for a gas with a steady-state source of power Q_0 located at the origin can be described by the equation

$$\Delta u - \varkappa^2 u = \frac{1}{r}\frac{\partial}{\partial r}\left(r\frac{\partial u}{\partial r}\right) + \frac{1}{r^2}\frac{\partial^2 u}{\partial \varphi^2} - \varkappa^2 u = 0,$$

where $\varkappa^2 = \dfrac{\beta}{D^2}$, D is the diffusion coefficient, and β is the dissipation coefficient. If a solution u of this equation depends only on r, then the function $y(r) = u(\varkappa r)$ satisfies the equation

$$\frac{1}{r}\frac{d}{dr}\left(r\frac{dy}{dr}\right) - y = 0.$$

We are interested in a solution of this equation that has a singularity at $r = 0$ and is bounded at infinity. Therefore

$$y(r) = aK_0(r),$$

where the constant a is determined by the condition

$$2\pi Da = Q_0.$$

Using the representation of the solution of this problem by Poisson's formula, we can obtain an integral representation for the function $K_0(x)$:

$$K_0(x) = \int_0^\infty e^{-x\cosh\eta}\, d\eta.$$

Example 2.117. Consider the solution $u(t, x, y)$ of the wave equation

$$u_{tt} = a^2(u_{xx} + u_{yy}),$$

which defines a planar wave moving along the y-axis. It is clear that $u(t, x, y) = v(x, y)e^{i\omega t}$, where v is a solution of the equation

$$v_{xx} + v_{yy} + k^2 v = 0, \quad ak = \omega,$$

and $v = e^{-iky} = e^{-ikr\sin\varphi}$. We expand this function in a Fourier series:

$$v(r, \varphi) = \sum_{n=-\infty}^{\infty} A_n(r)e^{-in\varphi}.$$

It turns out (cf. Tikhonov and Samarskij 1977) that $A_n(r) = J_n(r)$ and the integral representation

$$J_n(r) = \frac{1}{2\pi} \int_{-\pi}^{\pi} e^{-ir\sin\varphi + in\varphi} \, d\varphi \tag{9.12}$$

holds. If the index ν is not an integer, we have the formula

$$J_\nu(r) = \frac{1}{2\pi} \int_{-\pi}^{\pi} e^{-ir\sin\varphi + i\nu\varphi} \, d\varphi - \frac{\sin\nu\pi}{\pi} \int_0^\infty e^{-r\sinh\xi - \nu\xi} \, d\xi.$$

In some problems of mathematical physics it turns out to be convenient to apply the *Hankel transform* of order ν:

$$(H_\nu f)(\xi) = \int_0^\infty f(x) J_\nu(x\xi) x \, dx.$$

This transform is defined for functions $f(x)$ for which

$$\int_0^\infty x^{1/2} |f(x)| \, dx < \infty.$$

If in addition the integral $\int_0^a |f'(x)| \, dx$ converges for each $a > 0$, we have the inversion formula

$$(H_\nu^{-1} g)(x) = \int_0^\infty g(\xi) J_\nu(x\xi) \xi \, d\xi,$$

where $g = H_\nu f$. The most important property of the Hankel transform is the relation

$$H_\nu L_\nu f = -\xi^2 H_\nu f, \tag{9.13}$$

where $L_\nu = \dfrac{d^2}{dx^2} + \dfrac{1}{x}\dfrac{d}{dx} - \dfrac{\nu^2}{x^2}$ is the Bessel operator.

Plancherel's formula holds for the Hankel transform:

$$\int_0^\infty x f(x) g(x) \, dx = \int_0^\infty \xi (H_\nu f)(\xi)(H_\nu g)(\xi) \, d\xi.$$

Example 2.118. We shall find a fundamental solution of the Helmholtz operator in $\mathbb{R}^3$, i.e., a solution of the equation

$$(\Delta + k^2)G(R) = \delta(R), \quad R = \sqrt{x^2 + y^2 + z^2}.$$

To do this we apply the Fourier transform on x and y. Let

$$x = r\cos\varphi, \quad y = r\sin\varphi; \quad \xi = \rho\cos\theta, \quad \eta = \rho\sin\theta.$$

Consider the Fourier transform of a function f depending only on r:

$$g(\rho, \theta) = \iint f(r)e^{-i(x\xi + y\eta)}\,dx\,dy =$$

$$= \int_0^\infty \int_0^{2\pi} rf(r)e^{-ir\rho\cos(\varphi-\theta)}\,d\varphi\,dr = 2\pi(H_0 f)(\rho),$$

since by (9.12)

$$\frac{1}{2\pi}\int_0^{2\pi} e^{-i\rho\cos\varphi}\,d\varphi = J_0(\rho).$$

Using the commutation formula (9.13), we find that

$$\left(\frac{d^2}{dz^2} + k^2 - \rho^2\right)\tilde{G} = \delta(z),$$

where

$$\tilde{G}(\xi, \eta, z) = \iint G(x, y, z)e^{-i(x\xi + y\eta)}\,dx\,dy.$$

From this it is easy to deduce that

$$G(R) = -\frac{1}{4\pi}\int_0^\infty \frac{\rho}{\sqrt{\rho^2 - k^2}}e^{-\sqrt{\rho^2-k^2}|z|}J_0(\rho R)\,d\rho,$$

where the branch of the radical is taken so that $\sqrt{\alpha} > 0$ for $\alpha > 0$ and $\operatorname{Im}\sqrt{\alpha} < 0$ for $\alpha < 0$.

It can be shown that the following formula, known as *Sommerfeld's formula*, holds (cf. Tikhonov and Samarskij 1977):

$$\int_0^\infty J_0(\lambda\rho)\frac{e^{-\sqrt{\rho^2-k^2}|z|}}{\sqrt{\rho^2 - k^2}}\rho\,d\rho = \frac{e^{ik\sqrt{\lambda^2+z^2}}}{\sqrt{\lambda^2 + z^2}}.$$

Thus finally we conclude that

$$G(R) = -\frac{1}{4\pi}\frac{e^{ikR}}{R}.$$

Returning to Example 2.114, we discuss the question of the zeros of the function $J_\nu(x)$. Equation (9.11) and the asymptotic relation exhibited immediately after it show that $J_\nu(x)$ has infinitely many zeros as $x \to \infty$, and the distance between adjacent zeros is $\pi + o(1)$. All the zeros of the function $J_\nu(x)$ are simple and alternate with zeros of the function $J_{\nu+1}(x)$. Let $\alpha_{\nu,n}$

denote the nth root of the equation $J_\nu(x) = 0$. (It is assumed that these roots are arranged in increasing order.) Then for $\nu > -1$ we have

$$0 < \alpha_{\nu,1} < \alpha_{\nu+1,1} < \alpha_{\nu,2} < \alpha_{\nu+1,2} < \cdots.$$

We note further that

$$\sqrt{\nu(\nu+1)} < \alpha_{\nu,1} < \sqrt{\frac{4}{3}(\nu+1)(\nu+5)}.$$

For large roots of the equation

$$J_\nu(x)\cos\alpha - N_\nu(x)\sin\alpha = 0$$

we have the asymptotic expansion

$$\mu_n^{(\nu)} = \left(n+\frac{\nu}{2}-\frac{1}{4}\right)\pi-\alpha-\frac{4\nu^2-1}{8\left[(n+\frac{\nu}{2}+\frac{1}{4})\pi-\alpha\right]}-\frac{(4\nu^2-1)(28\nu^2-31)}{384\left[(n+\frac{\nu}{2}-\frac{1}{4}\pi)-3\right]^3}-\cdots$$

9.5. Airy's Equation. This equation has the form

$$y''(t) + \frac{1}{3}ty(t) = 0. \tag{9.14}$$

It is satisfied by the *Airy function*

$$\mathrm{Ai}(t) = \frac{1}{\pi}\int_0^\infty \cos\left(x^3 - tx\right)dx = \frac{1}{2\pi}\int_{-\infty}^\infty e^{i(x^3-tx)}dx.$$

Equation (9.14) can be reduced to Bessel's equation: setting

$$y = z\sqrt{t}, \quad t = 3\left(\frac{1}{2}s\right)^{2/3},$$

we obtain

$$\frac{d}{ds}\left(s\frac{dz}{ds}\right) + \left(s - \frac{1}{9s}\right)z = 0.$$

Therefore the general solution of Eq. (9.14) has the form

$$y = \sqrt{t}\left[C_1 J_{\frac{1}{3}}\left(2\left(\frac{t}{3}\right)^{3/2}\right) + C_2 J_{-\frac{1}{3}}\left(2\left(\frac{t}{3}\right)^{3/2}\right)\right].$$

The asymptotic behavior of the function $\mathrm{Ai}(t)$ as $t \to +\infty$ can be studied by the saddle-point method (cf. Hörmander 1983–1985, Sect. 7.6). After the change of variable $x \mapsto \sqrt{t}x$ we have

$$\mathrm{Ai}(t) = \frac{\sqrt{t}}{2\pi}\int_{-\infty}^\infty e^{it^{3/2}(x^{3/2}+x)}dx.$$

The function $x^{3/2} + x$ has two saddle points $x_{1,2} = \pm i$ of which the essential one is the point $x_1 = i$. Deforming the contour of integration so that it passes through the point x_1 in the direction of the line $\text{Im}\, x = 1$, one can show that

$$\mathrm{Ai}(t) = \frac{1}{2\pi} t^{-1/4} e^{-2t^{3/2}/3} \sum_{n=0}^{\infty} \frac{(-1)^n \Gamma\left(\frac{3n+1}{2}\right)}{3^{2n}(2n)!} t^{-3n/2} =$$

$$= \frac{1}{2\sqrt{\pi}} t^{-1/4} e^{-2t^{3/2}/3} \left[1 + O\left(t^{-3/2}\right)\right].$$

Similarly it can be verified that as $t \to -\infty$

$$\mathrm{Ai}(t) = \frac{1}{\sqrt{\pi}|t|^{1/4}} \left[\sin\left(\frac{2}{3}|t|^{3/2} + \frac{\pi}{4}\right) + O\left(|t|^{-3/2}\right)\right].$$

Every other solution of Eq. (9.14) that is not a multiple of $\mathrm{Ai}(t)$ grows exponentially as $t \to +\infty$:

$$y_2(t) = C t^{-1/4} e^{2t^{3/2}/3} \left[1 + O\left(t^{-3/2}\right)\right].$$

The following example plays an important role in the theory of boundary-value problems for second-order hyperbolic equations.

Example 2.119. Consider the equation

$$D_x^2 u - x D_{y_n}^2 u + D_{y_1} D_{y_n} u = 0, \quad x \geq 0, \, y \in \mathbf{R}^n,$$

with the condition

$$u = u_0(y) \quad \text{for } x = 0.$$

If

$$v(x, \eta) = \int e^{-i(y,\eta)} u(x, y)\, dy, \quad v_0(\eta) = \int e^{-i(y,\eta)} u_0(y)\, dy,$$

then

$$D_x^2 v = (x\eta_n^2 - \eta_1 \eta_n) v \quad \text{for } x \geq 0$$

and

$$v = v_0(\eta) \quad \text{for } x = 0.$$

After the change of variable $\zeta = 3^{1/3}(x\eta_n^2 - \eta_1 \eta_n)\eta_n^{-4/3}$ we obtain Airy's equation

$$\frac{d^2 v}{d\zeta^2} = \frac{1}{3}\zeta v(\zeta).$$

Setting $A_+(z) = \mathrm{Ai}(e^{2\pi i/3}z)$ and $A_-(z) = \mathrm{Ai}(e^{-2\pi i/3}z)$, we find that

$$v_\pm(x, \eta) = v_0(\eta)\rho(\eta)\frac{A_\pm(\zeta(x, \eta))}{A_\pm(\zeta(0, \eta))},$$

where ρ is a homogeneous cutoff function that justifies the application of the Fourier transform. Both solutions have a physical meaning: one of them corresponds to an incoming bicharacteristic, the other to an outgoing bicharacteristic.

This example shows that the complex plane is essential in the study of the behavior of the function Ai(t). It can be shown (cf. Hörmander 1983–1985, Sect. 7.6] that the Airy function decreases exponentially as $|z| \to \infty$ in the sector $|\arg z| < \frac{\pi}{3}$. It increases exponentially as $|z| \to \infty$ in the sector $\frac{\pi}{3} < |\arg z| < \pi$, and on the rays $\arg z = \pm \frac{\pi}{3}$ and $\arg z = \pi$ it oscillates. All the zeros of the Airy function are real, and it has infinitely many zeros on the half-line $(-\infty, 0)$.

9.6. Some Other Classes of Functions. At the present time hundreds of classes of special functions are known. We shall name just a few of them.

Example 2.120. The solutions of the equation

$$(1 - z^2)\frac{d^2 w}{dz^2} - z\frac{dw}{dz} + n^2 z = 0, \quad -1 < z < 1,$$

having the form

$$T_n(z) = \cos(n \arccos z), \quad U_n(z) = \sin(n \arccos z),$$

are called *Chebyshev polynomials* of first and second kinds.

The Chebyshev polynomials of first kind can be defined using the following generating function:

$$\frac{1 - t^2}{1 - 2tz + t^2} = T_0(z) + 2 \sum_{n=1}^{\infty} T_n(z) t^n.$$

They satisfy the recurrence relations

$$T_{n+1}(z) - 2z T_n(z) + T_{n-1}(z) = 0$$

and the orthogonality relations

$$\int_{-1}^{1} \frac{T_n(z) T_m(z)}{\sqrt{1 - z^2}} \, dz = \begin{cases} 0, & \text{if } m \neq n, \\ \frac{\pi}{2}, & \text{if } m = n \neq 0, \\ \pi, & \text{if } m = n = 0. \end{cases}$$

The polynomial $2^{1-n} T_n(z)$ is distinguished in the set of polynomials of degree n with leading coefficient 1 by being the best approximation to zero on the closed interval $[-1, 1]$.

Here are the first few Chebyshev polynomials:

$$T_0(z) = 1, \quad T_1(z) = z, \quad T_2(z) = 2z^2 - 1,$$
$$T_3(z) = 4z^3 - 3z, \quad T_4(z) = 8z^4 - 8z^2 + 1.$$

Since the degree of the polynomial $T_n(z)$ is n, these polynomials form a complete system on $[-1, 1]$ (and on any finite closed interval).

Example 2.121. The polynomial solutions of the differential equation

$$z\frac{d^2w}{dz^2} + (\alpha + 1 - z)\frac{dw}{dz} + nw = 0,$$

where $n = 0, 1, 2, \ldots$ and $\alpha \in \mathbb{C}$, are called *Laguerre polynomials*. In particular this equation is satisfied by the function

$$L_n^{(\alpha)}(z) = e^z z^{-\alpha}\frac{d^n}{dz^n}(e^{-z}z^{n+\alpha}).$$

For example, with $\alpha = 0$ we obtain the solution

$$L_n(z) = 1 - C_n^1\frac{z}{1!} + C_n^2\frac{z^2}{2!} - \cdots + (-1)^n\frac{z^n}{n!}.$$

The Laguerre polynomials can be found using the generating function

$$e^{-zt}(1+t)^\alpha = \sum_{n=0}^{\infty} L_n^{(\alpha-n)}(z)\frac{t^n}{n!}.$$

We have the recurrence relations

$$L_n^{(\alpha)}(z) = (2n + \alpha - 1 - z)L_{n-1}^{(\alpha)}(z) - (n-1)(n+\alpha-1)L_{n-2}^{(\alpha)}(z)$$

and (for $\alpha > -1$) the orthogonality relations:

$$\int_0^\infty e^{-x}x^\alpha L_n^{(\alpha)}(x)L_m^{(\alpha)}(x)\,dx = \begin{cases} 0, & \text{if } m \neq n, \\ \Gamma(1+\alpha)(n!)^2 C_n^{n+\alpha} & \text{if } m = n. \end{cases}$$

Example 2.122. The differential equation

$$\frac{d^2w}{dz^2} - 2z\frac{dw}{dz} + 2nw = 0$$

for $n = 0, 1, 2, \ldots$ defines the *Hermite polynomials*

$$H_n(z) = (-1)^n e^{z^2}\frac{d^n}{dz^n}(e^{-z^2}).$$

The Hermite polynomials are connected with the Laguerre polynomials by the relations

$$H_{2m}(z) = (-1)^m 2^{2m} L_m^{(-1/2)}(z^2),$$
$$H_{2m+1}(z) = (-1)^m 2^{2m+1} z L_m^{(1/2)}(z^2).$$

The generating function for the Hermite polynomials is

$$e^{2zt-t^2} = \sum_{n=0}^{\infty} H_n(z)\frac{t^n}{n!}.$$

The following recurrence relations hold:

$$H_{n+1}(z) = 2zH_n(z) - 2nH_{n-1}(z),$$

$$\frac{dH_n(z)}{dz} = 2nH_{n-1}(z).$$

All the zeros of the Hermite polynomials are real and simple. The following orthogonality relations hold:

$$\int_{-\infty}^{\infty} e^{-z^2} H_n(z)H_m(z)\, dz = \begin{cases} 0 & \text{for } m \neq n, \\ 2^n \sqrt{\pi} n! & \text{for } m = n, \end{cases}$$

The functions

$$D_n(z) = 2^{-\frac{n}{2}} e^{-\frac{z^2}{4}} H_n\left(\frac{z}{\sqrt{2}}\right)$$

are called the *parabolic cylinder functions*. They satisfy the equation

$$\frac{d^2w}{dz^2} + \left(n + \frac{1}{2} - \frac{z^2}{4}\right)w = 0.$$

Like the Hermite polynomials, these functions form a complete orthogonal system of functions on the line.

Example 2.123. The *Mathieu functions* or *elliptic cylinder functions* are solutions of *Mathieu's equation*

$$\frac{1}{4}\frac{d^2w}{dz^2} + (\alpha - 4q\cos 2z)w = 0.$$

The solutions of this equation having period 2π are called the periodic Mathieu functions. In this case α can be regarded as an eigenvalue of the operator $\frac{1}{4}\frac{d^2}{dz^2} - 4q\cos 2z$ with periodic boundary conditions. For every real q there is an infinite sequence of eigenvalues corresponding to eigenfunctions $\varphi_n(z,q)$. These functions are entire functions of z and form a complete orthogonal system on $[0, 2\pi]$.

Example 2.124. The *confluent hypergeometric functions* are the solutions of the *degenerate hypergeometric equation*

$$z\frac{d^2w}{dz^2} + (c - z)\frac{dw}{dz} - aw = 0.$$

For $c = 2a$ these functions are the Bessel functions; for $c = \frac{1}{2}$ they are the parabolic cylinder functions; and for $a = -n$ they are the Laguerre polynomials.

If $c \neq 0, -1, -2, \ldots$, this equation is satisfied by the *Kummer function*

$$\varphi(a, c; z) = 1 + \frac{a}{c}\frac{z}{1!} + \frac{a(a+1)}{c(c+1)}\frac{z^2}{2!} + \frac{a(a+1)(a+2)}{c(c+1)(c+2)}\frac{z^3}{3!} + \cdots.$$

Example 2.125. The *hypergeometric functions* are the solutions of the *hypergeometric equation*

$$z(1-z)\frac{d^2w}{dz^2} + [c - (a+b+1)z]\frac{dw}{dz} - abw = 0,$$

where a, b, c are complex parameters.

This equation was studied by Euler, Gauss, Riemann, Klein, and many others. It is satisfied by *Gauss' hypergeometric series*

$$F(a,b,c,z) = 1 + \sum_{n=1}^{\infty} \frac{(a)_n(b)_n}{(c)_n(1)_n} z^n,$$

which converges for $|z| < 1$. Here

$$(a)_0 = 1, \ (a)_n = \frac{\Gamma(a+n)}{\Gamma(a)} = a(a+1)\cdots(a+n-1), \quad n = 1, 2, \ldots.$$

The Kummer function is obtained from F by passing to the limit:

$$\varphi(a,c,z) = \lim_{b \to \infty} F\left(a, b, c, \frac{z}{b}\right).$$

We note that for positive integers n the function $F\left(n+1, -n, 1, \frac{1}{2} - \frac{z}{2}\right)$ coincides with the Legendre polynomial. The associated Legendre functions are obtained from F if $2c = a + b + 1$.

References *

We note at the outset that the list of literature cited here is of necessity brief and subjective. Besides the works cited in the text it includes only a few textbooks and monographs where one can find material that clarifies and develops the contents of the present work. A more complete bibliography on particular topics of the present work will be given in subsequent volumes of this series.

The books Petrovskij 1961, Bers, John, and Schechter 1964, Vladimirov 1967, Trèves 1975, Mikhlin 1977, Tikhonov and Samarskij 1977, Smirnov 1981, and Mikhailov 1983 are textbooks suitable for a first acquaintance with linear partial differential equations (naturally there exist many other textbooks of similar type). The more advanced textbooks are the classical monographs Courant and Hilbert 1931, Courant and Hilbert 1962, and Mizohata 1973.

Through the works of I. G. Petrovskij and commentaries on them (cf. Petrovskij 1986) the reader can follow the rise of the general theory of linear partial differential equations.

In the books Landau and Lifshits 1973, Landau and Lifshits 1974, Berestetskij, Lifshits, and Pitaevskij 1980, Landau and Lifshits 1982, Landau and Lifshits 1987, and in other textbooks of theoretical physics one can find a large collection of physical problems that lead to partial differential equations. See also Reed and Simon 1972–1979 and Berezin and Shubin 1983 for information on the physical and mathematical aspects of quantum mechanics and the theory of the Schrödinger equation.

The classical and modern aspects of the theory of linear partial differential equations are discussed quite widely and variously in the multivolume monograph Hörmander 1983–1985. See also Hörmander 1958.

Various questions of the theory of elliptic equations and boundary-value problems are discussed in Agmon, Douglis, and Nirenberg 1959, Palais 1965, Ladyzhenskaya and Ural'tseva 1967, Lions and Magenes 1968, Miranda 1970, Eskin 1973, Ladyzhenskaya 1973, and Rempel and Schulze 1982. Potential theory is discussed in Berlot 1959 and Landkof 1966.

In Il'in, Kalashnikov, and Olejnik 1962, Eidelman 1964, Friedman 1964, Solonnikov 1965, Ladyzhenskaya, and Solonnikov, and Ural'tseva 1967 one can find an exposition of a wide circle of questions of the theory of parabolic equations.

The monograph Landis 1971 contains an exposition of the qualitative theory of elliptic and parabolic equations.

The classical aspects of the theory of hyperbolic equations is discussed in Gårding 1951 and Ladyzhenskaya 1953.

The abstract approach to evolution equations is discussed in detail in Krejn 1967.

The theory of distributions (including the most advanced questions of the theory) is discussed in Schwartz 1950, Gel'fand and Shilov 1958–1959, Edwards 1965, Shilov 1965, Vladimirov 1967, Reed and Simon 1972–1979, Rudin 1973, Vladimirov 1979, and Hörmander 1983–1985.

For more details on equations and systems with constant coefficients see Pala modov 1967.

Various questions of the theory of function spaces and the theory of generalized solutions of equations and boundary-value problems connected with them are discussed in Sobolev 1950, Lions and Magenes 1968, Nikol'skij 1969, Stein 1970, Eskin 1973, Ladyzhenskaya 1973, Birman and Solomyak 1974, Sobolev 1974, Besov, Il'in, and Nikol'skij 1975, Mikhlin 1977, Triebel 1978, Hörmander 1983–1985, Triebel 1983, and Maz'ya 1985.

---

* For the convenience of the reader, references to reviews in Zentralblatt für Mathematik (Zbl.), compiled using the MATH database, have been included as far as possible.

For the theory of exterior problems and scattering theory see the monographs Lax and Phillips 1960, Newton 1966, Taylor 1972, Vajnberg 1982, Berezin and Shubin 1983, and Hörmander 1983–1985.

The spectral theory of one-dimensional differential operators and certain general aspects of spectral theory connected with it are discussed in the books Titchmarsh 1946, Coddington and Levinson 1955, Dunford and Schwartz 1958–1971, Glazman 1963, Berezanskij 1965, Najmark 1969, Levitan and Sargsyan 1970, Marchenko 1977, Kostyuchenko and Sargsyan 1979, Birman and Solomyak 1980, Smirnov 1981, and Berezin and Shubin 1983.

For information on special functions we refer the reader to the monographs and handbooks Erdélyi et al. 1953–1955, Jahnke, Emde and Lösch 1960, Abramowitz and Stegun 1964, Vilenkin 1965, and Miller 1977.

Abramowitz, M. and Stegun, I. A., eds. (1964): Handbook of Mathematical Functions with Formulas, Graphs, and Mathematical Tables. National Bureau of Standards, Applied Mathematics Series 55, Zbl. 171,385.

Agmon, S., Douglis, A., and Nirenberg, L. (1959): Estimates near the boundary for solutions of elliptic partial differential equations satisfying general boundary conditions. I. Commun. Pure Appl. Math. 12, 623–727, Zbl. 93,104.

Agranovich, M. S. (1969): Boundary-value problems for systems of first-order pseudodifferential operators . Usp. Mat. Nauk 24, No. 1(145), 61–125. English translation: Russ. Math. Surv. 24, No. 1, 59–126 (1969), Zbl. 175,108.

Agranovich, M. S. (1970): Boundary-value problems for systems with a parameter. Mat. Sb., Nov. Ser. 84(126), 27–65. English translation: Math. USSR, Sb. 13, 25–64 (1971), Zbl. 207,108.

Agranovich, M. S. and Vishik, M. I. (1964): Elliptic problems with a parameter and parabolic problems of general form. Usp. Mat. Nauk 19, No. 3(117), 53–161. English translation: Russ. Math. Surv. 19, No. 3, 53–157 (1964), Zbl. 137,296.

Arnol'd, V. I. (1974): Mathematical Methods of Classical Mechanics. Nauka, Moscow. English translation: Springer-Verlag, Berlin - Heidelberg - New York (1978), Zbl. 386.70001.

Atiyah, M., Bott, R., and Gårding, L. (1970): Lacunas for hyperbolic differential operators with constant coefficients. I. Acta Math. 124, 109–189, Zbl. 191,112.

Berestetskij, V. B., Lifshits, E. M., and Pitaevskij, L. P. (1980): Quantum Electrodynamics. Nauka, Moscow. English translation: Pergamon, Oxford (1982).

Berezanskij, Yu. M. (1965): Eigenfunction Expansions of Self-Adjoint Operators. Naukova Dumka, Kiev. English translation: Am. Math. Soc., Providence (1968), Zbl. 142,372.

Berezin, F. A. and Shubin, M. A. (1983): The Schrödinger Equation. Moscow University Press. English translation: Kluwer Acad. Publishers, Dordrecht (1991), Zbl. 546.35002.

Bers, L., John, F., and Schechter, M. (1964): Partial Differential Equations. Interscience, New York, Zbl. 128,93 and Zbl. 128,94.

Besov, O. V., Il'in, V. P., and Nikol'skij, S. M. (1975): Integral Representations of Functions and Imbedding Theorems. Nauka, Moscow. English translation: J. Wiley, New York (1978, Part I; 1979, Part II), Zbl. 352.46023.

Birman, M. Sh. and Solomyak, M. Z. (1974): Quantitative analysis in the Sobolev imbedding theorems and an application to spectral theory. In: The Tenth Mathematical School, 5-189, Kiev. English translation: Am. Math. Soc., Transl., Ser. 2, Vol. 114, 132 pp. (1980), Zbl. 426.46019.

Birman, M. Sh. and Solomyak, M. Z. (1980): Spectral Theory of Self-Adjoint Operators in Hilbert Space. Leningrad University Press [Russian]. English translation: D. Reidel, Dordrecht (1987).

Brelot, M. (1959): Éléments de la Théorie Classique du Potentiel. Centre de Documentation Universitaire, Paris, Zbl. 84,309.

Coddington, E. A. and Levinson, N. (1955): Theory of Ordinary Differential Equations. McGraw-Hill, New York, Zbl. 64,330.

Courant, R. and Hilbert, D. (1931): Methoden der Mathematischen Physik, 1, 2. Springer-Verlag, Berlin - Heidelberg - New York. English translation: Methods of Mathematical Physics. Interscience, New York, Zbl. 1,5.

Courant, R. and Hilbert, D. (1962): Partial Differential Equations. Interscience, New York, Zbl. 99,295.

Dunford, N. and Schwartz, J. T. (1958–1971): Linear Operators, 1, 2, 3. Interscience, New York, Zbl. 84,104; Zbl. 128,348; Zbl. 243.47001.

Edwards, R. E. (1965): Functional Analysis. Theory and Applications. Holt, Rinehart, and Winston, London, Zbl. 182,161.

Egorov, Yu. V. (1984): Linear Differential Equations of Principal Type. Nauka, Moscow. English translation: Contemp. Sov. Math., New York (1986), Zbl. 574.35001.

Egorov, Yu. V. (1985): Lectures on Partial Differential Equations: Supplementary Chapters. Moscow University Press [Russian], Zbl. 615.35001.

Eidelman, S. D. (1964): Parabolic Systems. Nauka, Moscow. English translation: North Holland Publ., London etc. (1969), Zbl. 121,319.

Eidus, D. M. (1969): The principle of limiting amplitude. Usp. Mat. Nauk 24, No. 3 (147), 91–154. English translation: Russ. Math. Surv. 24, No. 3, 97–167 (1969), Zbl. 177,142.

Eskin, G. I. (1973): Boundary-Value Problems for Elliptic Pseudodifferential Equations. Nauka, Moscow. English translation: Transl. Math. Monogr., Vol. 52, Providence (1981), Zbl. 262.35001.

Erdélyi, A., Magnus, W., Oberhettinger, F., and Tricomi, F. G., eds. (1953–1955): Higher Transcendental Functions, 1,2,3 (Bateman Manuscript Project). McGraw-Hill, New York, Zbl. 51,303; Zbl. 52,295; Zbl. 64,63.

Faddeev, L. D. (1959): The quantum inverse scattering problem I, II. Usp. Mat. Nauk 14, No. 4(88), 57–119; English translation: J. Math. Phys. 4, 72–104 (1963), Zbl. 112,451; Itogi Nauki Tekh., Ser. Sovrem. Probl. Mat. 3 (1974), 93–180. English translation: J. Sov. Math. 5, 334–396 (1976), Zbl. 373.35014.

Fedosov, B. V. (1974–1976): An analytic formula for the index of an elliptic boundary-value problem. I, II, III. Mat. Sb., Nov. Ser. 93, 62–89; 95 (1975), 525–550; 101 (1976), 380–401. English translation: Math. USSR, Sb. 22, 61–90 (1975), Zbl. 306.58016; 24, 511–535 (1976), Zbl. 312.58010; 30, 341–359 (1978), Zbl. 349.58007.

Flaschka, H. and Strang, G. (1971): The correctness of the Cauchy problem. Adv. Math. 6, 347–379, Zbl. 144,349.

Friedman, A. (1964): Partial Differential Equations of Parabolic Type. Prentice-Hall, Englewood Cliffs, New Jersey, Zbl. 144,349.

Gårding, L. (1951): Linear hyperbolic partial differential equations with constant coefficients. Acta Math. 85, 1–62, Zbl. 45,202.

Gel'fand, I. M. and Shilov, G. E. (1958–1959): Generalized Functions, Vols. 1,2,3. Fizmatgiz, Moscow. English translation: Acad. Press, London (1964, 1968, 1967), Zbl. 91,111; Zbl. 103,92.

Glazman, I. M. (1963): Direct Methods of Qualitative Spectral Analysis of Singular Differential Operators. Fizmatgiz, Moscow. English translation: Oldbourne Press, London (1965), Zbl. 143,365.

Gokhberg, I. Ts. and Krejn, M. G. (1965): Introduction to the Theory of Linear Nonself-Adjoint Operators. Nauka, Moscow. English translation: Am. Math. Soc., Providence (1969), Zbl. 138,78.

Hartman, P. (1964): Ordinary Differential Equations. Wiley, New York, Zbl. 125,321.

Hörmander, L. (1958): On the division of distributions by polynomials. Ark. Mat. 3, 555–568, Zbl. 131,119.

Hörmander, L. (1963): Linear Partial Differential Operators. Springer-Verlag, Berlin - Heidelberg - New York, Zbl. 108,93.

Hörmander, L. (1973): An Introduction to Complex Analysis in Several Variables, 2nd ed., North Holland Publishing Company, Amsterdam, Zbl. 271.32001.

Hörmander, L. (1983–1985): The Analysis of Linear Partial Differential Operators I, II, III, IV. Springer-Verlag, Berlin - Heidelberg - New York, Zbl. 521.35001; Zbl. 521.35002; Zbl. 601.35001; Zbl. 612.35001.

Il'in, A. M., Kalashnikov, A. S., and Olejnik, O. A. (1962): Second-order linear equations of parabolic type. Usp. Mat. Nauk 17, No. 3(105), 3–141. English translation: Russ. Math. Surv. 17, No. 3, 1–146 (1962), Zbl. 108,284.

Jahnke, E., Emde, F., and Lösch, F. (1960): Tafeln Höherer Funktionen. Teubner, Stuttgart, Zbl. 87,128.

John, F. (1955): Plane Waves and Spherical Means Applied to Partial Differential Equations. Interscience, New York, Zbl. 67,321.

Komatsu, H. (1977): Ultradistributions. I. Structure theorems and a characterization. II. The kernel theorem and ultradistributions with support in a submanifold. J. Fac. Sci., Univ. Tokyo, Sect. I A. 20, 25–105, Zbl. 258.46039; 24 ((1977), 607–628, Zbl. 385.46027.

Kostyuchenko, A. G. and Sargsyan, I. S. (1979): The Distribution of Eigenvalues. Self-Adjoint Ordinary Differential Operators. Nauka, Moscow [Russian], Zbl. 478.34022.

Krejn, S. G. (1967): Linear Differential Equations in a Banach Space. Nauka, Moscow. English translation: Trans. Math. Monogr., Vol. 29, Providence (1972), Zbl. 172,419.

Kreiss, H.-O. (1958): Über sachgemäße Cauchyprobleme für Systeme von linearen partiellen Differentialgleichungen. Tekn. Högskol. Handl. 127, Zbl. 84,298.

Kreiss, H.-O. (1970): Initial boundary value problem for hyperbolic systems. Commun. Pure Appl. Math. 23, 277–298, Zbl. 188,411.

Ladyzhenskaya, O. A. (1953): The Mixed Problem for a Hyperbolic Equation. Gosteorizdat, Moscow [Russian], Zbl. 52,325.

Ladyzhenskaya, O. A. (1973): Boundary-Value Problems of Mathematical Physics. Nauka, Moscow. English translation: Appl. Math. Sci. 49, Springer-Verlag, Berlin - Heidelberg - New York (1985), Zbl. 284.35001.

Ladyzhenskaya, O. A., Solonnikov, V. A., and Ural'tseva, N. N. (1967): Linear and Quasilinear Equations of Parabolic Type. Nauka, Moscow. English translation: Am. Math. Soc., Providence (1968), Zbl. 164,123.

Ladyzhenskaya, O. A. and Ural'tseva, N. N. (1973): Linear and Quasilinear Equations of Elliptic Type. Nauka, Moscow [Russian]. English translation: Academic Press, New York (1968), Zbl. 269.35029.

Landau, L. D. and Lifshits, E. M. (1973): Field Theory. Nauka, Moscow. German translation: Akademie-Verlag, Berlin (1989). English translation: The classical Theory of Fields, Pergamon Press, London (1961), Zbl. 652.70001.

Landau, L. D. and Lifshits, E. M. (1974): Quantum Mechanics. Nonrelativistic Theory. Nauka, Moscow [Russian]. English translation: Pergamon Press, London (1959).

Landau, L. D. and Lifshits, E. M. (1982): Electrodynamics of continuous media. Nauka, Moscow [Russian]. English translation: Pergamon Press, London (1960).

Landau, L. D. and Lifshits, E. M. (1987): Theory of Elasticity. 4th ed., Nauka, Moscow. German translation: Akademie-Verlag, Berlin (1989). English translation: Pergamon Press, London (1959), Zbl. 621.73001.

Landis, E. M. (1971): Second-Order Equations of Elliptic and Parabolic Types. Nauka, Moscow [Russian], Zbl. 226.35001.

Landkof, N. S. (1966): Foundations of Modern Potential Theory. Nauka, Moscow [Russian]. English translation: Springer-Verlag, Berlin (1972), Zbl. 148,103.

Lax, P. N. and Phillips, R. S. (1960): Local boundary conditions for dissipative symmetric linear differential operators. Commun. Pure Appl. Math. 13, 427–455, Zbl. 94,75.

Lax, P. N. and Phillips, R. S. (1967): Scattering Theory. Academic Press, New York, Zbl. 186,163.

Levitan, B. M. and Sargsyan, I. S. (1970): Introduction to Spectral Theory. Nauka, Moscow. English translation: Trans. Math. Monogr., Providence (1975), Zbl. 225.47019.

Lions, J. L. and Magenes, E. (1968): Problèmes aux Limites Non-Homogènes et Applications, 1. Dunod, Paris, Zbl. 165,108.

Marchenko, V. A. (1977): Sturm-Liouville Operators and Their Applications. Naukova Dumka, Kiev. English translation: Birkhäuser (1986), Zbl. 399.34022.

Maz'ya, V. G. (1985): Sobolev Spaces. Leningrad University Press. English translation: Springer-Verlag, Berlin - Heidelberg - New York (1985), Zbl. 692.46023.

Mikhailov, V. P. (1983): Partial Differential Equations. Nauka, Moscow. 1st ed. 1976. English translation: Moscow: Mir (1978), Zbl. 388.35002.

Mikhlin, S. G. (1977): Linear Partial Differential Equations. Vyshaya Shkola, Moscow [Russian].

Miller, W., Jr. (1977): Symmetry and separation of variables. Addison-Wesley, Reading, Massachusetts, Zbl. 368.35002.

Miranda, C. (1970): Partial Differential Equations of Elliptic Type (translated from the Italian). Springer-Verlag, Berlin - Heidelberg - New York, Zbl. 198,141.

Mizohata, S. (1973): Theory of Partial Differential Equations. Cambridge University Press, Zbl. 263.35001.

Najmark, M. A. (1969): Linear Differential Operators, 2nd ed., Nauka, Moscow. English translation: Frederick Ungar Publ. Co., New York (1967, Part I; 1968, Part II), Zbl. 193,41.

Newton, R. G. (1966): Scattering Theory of Waves and Particles. New York.

Nikol'skij, S. M. (1969): Approximation of Functions of Several Variables and Imbedding Theorems. Nauka, Moscow. English translation: Springer-Verlag, Berlin - Heidelberg - New York (1975), Zbl. 185,379 and Zbl. 496.46020.

Nirenberg, L. (1973): Lectures on Linear Partial Differential Equations. Regional Conf. Ser. math. 17, Am. Math. Soc., Providence, Zbl. 267.35001.

Palais, R. S. (1965): Seminar on the Atiyah-Singer Index Theorem. Princeton University Press, Zbl. 137,170.

Palamodov, V. P. (1967): Linear Differential Operators with Constant Coefficients. Nauka, Moscow. English translation: Springer-Verlag, Berlin - Heidelberg - New York (1970), Zbl. 191,434.

Peetre, J. (1960): Rectification à l'article "Une caractérisation abstraite des opérateurs différentiels" Math. Scand. 8, 116–120, Zbl. 67,104.

Petrovskij, I. G. (1961): Lectures on Partial Differential Equations, 3rd ed., Fizmatgiz, Moscow. English translation: London (1967), Zbl. 115,81.

Petrovskij, I. G. (1986): Selected Works. Systems of Partial Differential Equations. Algebraic Geometry. Nauka, Moscow [Russian], Zbl. 603.01018.

Reed, M. and Simon, B. (1972–1979): Methods of Modern Mathematical Physics 1,2,3,4. Academic Press, New York, Zbl. 242.46001; Zbl. 308.47002; Zbl. 401.47007; Zbl. 401.47001.

Rempel, S. and Schulze, B.-W. (1982): Index Theory of Elliptic Boundary Problems. Akademie-Verlag, Berlin, Zbl. 504.35002.

Rudin, W. (1973): Functional Analysis. McGraw-Hill, New York, Zbl. 253.46001.

Sakamoto, R. (1970): Mixed problems for hyperbolic equations. J. Math. Kyoto Univ. 10, 349–373, 403–427, Zbl. 203,100; Zbl. 206,401.

Schwartz, L. (1950–1951): Théorie des Distributions 1,2. Hermann, Paris, Zbl. 37,73; Zbl. 42,114.

Shilov, G. E. (1965): Mathematical Analysis. A Second Special Course. Nauka, Moscow. English translation: Pergamon Press, New York (1968), Zbl. 177,363.

Shubin, M. A. (1978): Pseudodifferential Operators and Spectral Theory. Nauka, Moscow. English translation: Springer-Verlag, Berlin - Heidelberg - New York (1987), Zbl. 451.47064.

Smirnov, V. I. (1981): A Course of Higher Mathematics, Vol. 4, Part 2, 6th ed., Nauka, Moscow. German translation: VEB, Berlin (1988) English translation of 3rd ed.: Pergamon Press, New York (1964), Zbl. 301.45001 and Zbl. 44,320.

Sobolev, S. L. (1950): Some Applications of Functional Analysis in Mathematical Physics. Leningrad University Press [Russian]. German translation: Einige Anwendungen der

Funktionalanalysis auf Gleichungen der mathematischen Physik. Akademie-Verlag, Berlin (1964).

Sobolev, S. L. (1974): Introduction to Cubature Formulas. Nauka, Moscow [Russian], Zbl. 294.65013.

Solonnikov, V. A. (1965): Boundary-value problems for linear parabolic systems of differential equations of general form. Tr. Mat. Inst. Akad. Nauk SSSR, No. 83 [Russian], Zbl. 161,84.

Stein, E. M. (1970): Singular Integrals and Differentiability Properties of Functions. Princeton University Press, Zbl. 207,135.

Taylor, J. R. (1972): Scattering Theory. The Quantum Theory of Nonrelativistic Collisions. Wiley, New York.

Tikhonov, A. N. and Arsenin, V. Ya. (1979): Methods of Solving Ill-Posed Problems. Nauka, Moscow. English translation: J. Wiley, New York (1977), Zbl. 309.65002 (1st ed.); Zbl. 499.65030.

Tikhonov, A. N. and Samarskij, A. A. (1977): The Equations of Mathematical Physics. Nauka, Moscow. German translation: VEB, Berlin (1959). English translation: Macmillan, New York (1963), Zbl. 44,93.

Tikhonov, A. N. and Samarskij, A. A. (1948): On the principle of radiation. Zh. Ehksper. Teor. Fiziki 18, 243–248 [Russian].

Titchmarsh, E. C. (1946): Eigenfunction Expansions Associated with Second-Order Differential Equations, 1,2. Clarendon Press, Oxford, Zbl. 61,135.

Trèves, F. (1975): Basic Linear Partial Differential Equations. Academic Press, New York, Zbl. 305.35001.

Triebel, H. (1978): Interpolation Theory, Function Spaces, Differential Operators. Deutscher Verlag der Wissenschaften, Berlin, Zbl. 387.46033.

Triebel, H. (1983): Theory of Function Spaces. Birkhäuser, Leipzig, Zbl. 546.46027.

Vajnberg, B. R. (1966): The principle of radiation, limiting absorption, and limiting amplitude in the general theory of partial differential equations. Usp. Mat. Nauk 21, 115–194. English translation: Russ. Math. Surv. 21, No. 3, 115–193 (1966), Zbl. 172,137.

Vajnberg. B. R. (1975): The short-wave asymptotics of solutions of steady-state problems and the asymptotics as $t \to \infty$ of solutions of nonsteady-state problems. Usp. Mat. Nauk 30, No. 2(182), 3–55. English translation: Russ. Math. Surv. 30, No. 2, 1–58 (1975), Zbl. 308.35011.

Vajnberg, B. R. (1982): Asymptotic Methods in the Equations of Mathematical Physics. Moscow University Press [Russian]. English translation: Gordon and Breach (1989), Zbl. 518.35002.

Vilenkin, N. Ya. (1965): Special Functions and the Theory of Group Representations. Nauka, Moscow. English translation: Am. Math. Soc., Providence (1968), Zbl. 144,380.

Vladimirov, V. S. (1967): The Equations of Mathematical Physics. Nauka, Moscow. English translation: M. Dekker, New York (1971), Zbl. 207,91.

Vladimirov, V. S. (1979): Distributions in Mathematical Physics. Nauka, Moscow. English translation: Moscow: Mir (1979), Zbl. 403.46036; Zbl. 515.46033.

Warner, F. W. (1983): Foundations of Differentiable Manifolds and Lie Groups. Springer-Verlag, Berlin - Heidelberg - New York, Zbl. 516.58001.

Yosida, K. (1965): Functional Analysis. Springer-Verlag, Berlin - Heidelberg - New York (6th ed. 1980), Zbl. 126.115; Zbl. 435.46002.

Zakharov, V. E., Manakov, S. V., Novikov, S. P., and Pitaevskij, L. P. (1980): Theory of Solitons. The Inverse Method. Nauka, Moscow. English translation: New York - London (1984), Zbl. 598.35003.

Author Index

Subject Index

Druck- und Bindearbeiten: Legoprint, Italien